AI for Robotics

Toward Embodied and General Intelligence in the Physical World

Alishba Imran
Keerthana Gopalakrishnan

Apress®

AI for Robotics: Toward Embodied and General Intelligence in the Physical World

Alishba Imran
San Francisco, CA, USA

Keerthana Gopalakrishnan
San Francisco, CA, USA

ISBN-13 (pbk): 979-8-8688-0988-0
https://doi.org/10.1007/979-8-8688-0989-7

ISBN-13 (electronic): 979-8-8688-0989-7

Copyright © 2025 by Alishba Imran, Keerthana Gopalakrishnan

This work is subject to copyright. All rights are reserved by the Publisher, whether the whole or part of the material is concerned, specifically the rights of translation, reprinting, reuse of illustrations, recitation, broadcasting, reproduction on microfilms or in any other physical way, and transmission or information storage and retrieval, electronic adaptation, computer software, or by similar or dissimilar methodology now known or hereafter developed.

Trademarked names, logos, and images may appear in this book. Rather than use a trademark symbol with every occurrence of a trademarked name, logo, or image we use the names, logos, and images only in an editorial fashion and to the benefit of the trademark owner, with no intention of infringement of the trademark.

The use in this publication of trade names, trademarks, service marks, and similar terms, even if they are not identified as such, is not to be taken as an expression of opinion as to whether or not they are subject to proprietary rights.

While the advice and information in this book are believed to be true and accurate at the date of publication, neither the authors nor the editors nor the publisher can accept any legal responsibility for any errors or omissions that may be made. The publisher makes no warranty, express or implied, with respect to the material contained herein.

Managing Director, Apress Media LLC: Welmoed Spahr
Acquisitions Editor: Celestin Suresh John
Development Editor: Laura Berendson
Coordinating Editor: Kripa Joseph
Copy Editor: Kezia Endsley

Cover designed by eStudioCalamar

Distributed to the book trade worldwide by Springer Science+Business Media New York, 1 New York Plaza, New York, NY 10004. Phone 1-800-SPRINGER, fax (201) 348-4505, e-mail orders-ny@springer-sbm.com, or visit www.springeronline.com. Apress Media, LLC is a Delaware LLC and the sole member (owner) is Springer Science + Business Media Finance Inc (SSBM Finance Inc). SSBM Finance Inc is a **Delaware** corporation.

For information on translations, please e-mail booktranslations@springernature.com; for reprint, paperback, or audio rights, please e-mail bookpermissions@springernature.com.

Apress titles may be purchased in bulk for academic, corporate, or promotional use. eBook versions and licenses are also available for most titles. For more information, reference our Print and eBook Bulk Sales web page at http://www.apress.com/bulk-sales.

Any source code or other supplementary material referenced by the author in this book can be found here: https://www.apress.com/gp/services/source-code.

If disposing of this product, please recycle the paper

*Dedicated to my father,
for his unconditional love and unwavering support*

—Keerthana

*Dedicated to my family and friends who have
supported me on this journey*

—Alishba

Table of Contents

About the Authors xiii

About the Technical Reviewers xv

Acknowledgments xvii

Introduction xix

Chapter 1: Introduction to General Purpose Robotics 1

    A Robot System 5

    Common Types of Robots 6

    Common Concepts in Robot Design 10

        Robotic Manipulators 11

        Degrees of Freedom 12

        End Effectors and Workspaces 13

        Kinematics 13

    Deep Learning for Robotics 15

    Deep Learning Frameworks 19

    Robot Learning Frameworks and Objectives 22

    Toward Embodied General Intelligence 23

        Environment Is Deeply Tied to the Definition of Intelligence 24

    Summary 26

    References 28

TABLE OF CONTENTS

Chapter 2: Robot Perception: Sensors and Image Processing 35
    Sensors 36
        Vision Sensors (Cameras) 37
        Depth Sensors 42
        Range Sensors 45
        Inertial Measurement Units (IMUs) 51
    Problems in Perception 53
        Classification 54
        Segmentation 54
        Object Detection 60
    Convolutional Neural Nets Overview 61
        Convolutional Layers 62
        Pooling Layers 65
        Fully Connected Layers 65
    CNNs for Perception 65
        R-CNN 67
        Fast R-CNN 68
        Faster R-CNN 69
        Mask R-CNN 71
    ResNet 72
        Skip Connection: The Strength of ResNet 73
        U-Net 74
    EfficientNet 76
    One-Stage Detectors 77
        YOLO 77
        SSD 78
    Model Comparison 79

TABLE OF CONTENTS

Transformers for Perception 81
    Transformer Introduction 82
    The Transformer 84
    Transformers for Vision 87
Summary 97
References 98

Chapter 3: Robot Perception: 3D Data and Sensor Fusion 107

3D Data Processing 108
    Data Representation 108
    Processing Point Clouds 110
    Research Opportunities 115
Multimodal Perception and Sensor Fusion 116
    Fusion Strategies 116
    LiDAR-Camera Fusion 123
Summary 133
References 134

Chapter 4: Foundation Models in Robotics 139

Large Foundation Models 139
    Scaling Laws for Language Models 144
    Evaluating Language Models 148
Language as a Connective Tissue in Robotics 151
    Language for Planning 151
    Language for Mapping 160
    Language for Reward 162
    Language for Robot Code 164

vii

TABLE OF CONTENTS

End-to-End Robot Control 167
    End-to-End Robot Control with Autoregressive Transformers 169
    End-to-End Robot Control with Diffusion Models 176
    DDPMs (Denoising Diffusion Probabilistic Models) 180
    DDIM (Denoising Diffusion Implicit Models) 183
    Conditioned Generation 186
    Action Diffusion for Robot Control 188
    Combining VLMs and Diffusion Models 190
Learning from Video Demonstrations 191
    World Modeling Using Video Data 192
AI Safety for Robotics 194
Summary 198
References 199

Chapter 5: Simulation 211

Simulation for Robots 211
    Considerations for Simulation in Robotics 212
Components of a Robot Simulator 214
    The PyBullet Module 219
    MuJoCo 221
    Gazebo 221
Concepts in Sim2Real 222
    Domain Adaptation 223
    Domain Randomization 224
    Guided Domain Randomization 227
Closing the Sim2Real Gap for RL 230
    CycleGAN 231
    RL-CycleGAN 233

TABLE OF CONTENTS

Learning from Simulation 235
    Simulation for Bootstrapping RL 236
    Foundation Agents in Simulation 238
    Simulation for Reward Design 244
    Simulation for World Modelling 246
    Simulation for Imitation Learning 249
Summary 254
    Tutorials 255
References 258

Chapter 6: Mapping, Localization, and Navigation 265

Why Use Deep Learning? 266
    Traditional Methods 266
    Deep Learning Methods 267
    A Hybrid Approach 267
Typical Mobile Robot Setup 269
Mapping 270
    Geometric Mapping 270
    Semantic Mapping 277
Localization 280
    2D-to-2D Localization 281
    2D-to-3D Localization 284
    3D-to-3D Localization 286
Navigation 287
    Navigation and Exploration 291
    Locomotion for Legged Robots 294
Summary 303
References 305

TABLE OF CONTENTS

Chapter 7: Reinforcement Learning and Control 311

Reinforcement Learning Basics 314

    Solving a Markov Decision Process 316

    Considerations 316

Model-Free vs Model-Based RL 318

    Model-Free Reinforcement Learning 320

    Model-Based Reinforcement Learning 327

Offline Reinforcement Learning 331

Applications and Challenges 333

    Scaling Up RL in the Real World 334

    Reinforcement Learning and Large Language Models 340

Challenges in RL for Robotics 343

Emerging Trends in RL for Robotics 344

Conclusions 345

Summary 346

References 347

Chapter 8: Self-Driving Vehicles 353

Economic Opportunity 353

System Design 355

    End-to-End Self-Driving (E2E) 357

Perception 358

Prediction 362

Planning 363

Safety 366

    AI Safety and Systems 366

    Safety Considerations 367

Summary 370

References 371

TABLE OF CONTENTS

Chapter 9: Industrial Robotics 375

    Common Tasks 376

        Pick-and-Place 376

        Peg-in-Hole 377

        Welding 378

        Warehouse Tasks 379

    Common Robots 380

        Standalone Industrial Robots 380

        Collaborative Robots (Cobots) 381

        Mobile Robots 382

        Humanoids 384

    Market Opportunity 384

    System Design for Pick-and-Place Robots 387

        Hardware Components 387

        Software Components 390

        Scaling RL for Robotic Grasping 393

    Safety Considerations 398

    Summary 398

    References 400

Chapter 10: Humanoid Robotics 405

    The Case for Humanoids 405

    Alternative Approaches 406

    Humanoid Markets 407

    How to Build a Humanoid 410

        Hardware 410

        Software 415

TABLE OF CONTENTS

Conclusion 421

Summary 422

References 423

Chapter 11: Data-Driven Robotics in Practice 427

Robot Operations 427

Data Infrastructure 429

The Training and Deployment Infrastructure 431

Robot Data Flywheels 433

Large-Scale Robotic Data Collection 434

Recipes for the Future 438

References 439

Index 443

About the Authors

Alishba Imran is a machine learning and robotics developer specializing in robot learning for manipulation and perception. She is currently conducting research in reinforcement learning and unsupervised learning with Pieter Abbeel at the Berkeley AI Research Lab. Her past experience spans across top robotics companies—she's worked on advanced perception systems at Cruise, developed simulation-based manipulation methods at NVIDIA, and led impactful initiatives to reduce the cost of prosthetics. At Hanson Robotics—the creators of Sophia the Robot, the world's most advanced human-like robot—Alishba co-led neuro-symbolic AI research and developed low-cost hardware system for humanoids.

Keerthana Gopalakrishnan is a Senior Research Scientist at Google DeepMind working on robot manipulation and the Gemini project. She was educated at Carnegie Mellon University and the Indian Institute of Technology. Her research concerns large language models for robotic planning, scaling visual language models for low-level control, and cross-embodiment robot learning.

About the Technical Reviewers

Lentin Joseph is the author of ten books in ROS and the cofounder/CTO of RUNTIME Robotics from India/Kerala. He is an accomplished robotics engineer with 14 years of experience in the field. He specializes in the Robot Operating System (ROS), contributing significantly to the development of various robotics applications and solutions. He holds a degree in electronics and communication engineering and an M.Tech in robotics and automation, from which he graduated with honors. Lentin has authored and reviewed more than ten books on ROS, including *Mastering ROS for Robotics Programming, ROS Robotics Projects,* and *Robot Operating System for Absolute Beginners,* making significant contributions to the robotics community.

LinkedIn profile: `https://www.linkedin.com/in/lentinjoseph/`
Company website: `https://runtimerobotics.com/`

Sean Kirmani is a research scientist at Google DeepMind working on problems in vision, language, and action. He was also the technical lead for semantic perception at The Everyday Robot Project within Google[x]. He holds degrees in electrical engineering and computer science from The University of Texas at Austin, where he worked in several robotics labs. Sean has co-authored over 20 publications in robotics and AI.

Acknowledgments

We'd like to acknowledge Chris Paxton for his valuable contribution to the Reinforcement Learning chapter of this book. We'd like to thank the robotics research community for allowing us to tell the story of this field, which was collectively built, and for graciously permitting us to feature their work in this book. We are also grateful for Marco Mascorro and Vivek Aithal for reviews on earlier versions of this book.

Introduction

What This Book Is About

AI for Robotics is the reimagination of robotics as an artificial intelligence problem. Modern robotics is steadily transformed by breakthroughs in AI. This book is your comprehensive guide to framing traditional robotics problems as AI problems and approaching them with deep learning techniques. Whether you're a coder, an enthusiast, or an investor, *AI for Robotics* gives you the blueprint to create generalizable and data-driven robotic intelligence that learns, evolves, and tackles challenges we once thought impossible in dynamic, real-world environments.

Who This Book Is For

This book empowers:

- **Software and AI engineers:** If you have a background in machine learning but are new to robotics, this book bridges the gap, showing you how to build robots that learn and adapt.

- **Robotics and mechanical engineers:** Stay ahead of the curve by learning how to integrate AI and data-driven approaches into your designs, ensuring your robots are at the forefront of innovation.

- **Investors, executives, and decision-makers:** Gain a clear understanding of the AI-robotics landscape. Make informed choices about which technologies to bet on.

INTRODUCTION

No matter your background, if you're ready to shape the future of robotics, this book is your guide.

The Structure of the Book

This book is structured to gradually build your understanding of the use of artificial intelligence for robotics, starting with fundamental concepts and progressing to advanced applications.

- **Chapter 1: Introduction to General Purpose Robotics**

  Provides an overview of the current state and future directions of robotics, emphasizing the role of machine learning in enabling more versatile and intelligent systems.

- **Chapter 2: Robot Perception: Sensors and Image Processing**

  Covers the basics of how robots perceive their environment through sensors and image-processing techniques, focusing on learning representations for vision tasks.

- **Chapter 3: Robot Perception: 3D Data and Sensor Fusion**

  Explores how to process and integrate 3D data from various sensors to create a coherent understanding of the robot's surroundings.

- **Chapter 4: Foundation Models in Robotics**

  Discusses the application of large, pretrained models in robotics, including language models and visual language models, and how they can be adapted for robotics.

- **Chapter 5: Simulation**

  Details the use of synthetic data and simulation environments for training and testing robots, including simulated-to-real transfer techniques.

- **Chapter 6: Mapping, Localization, and Navigation**

  Focuses on the techniques robots use to map the surroundings, understand where they are, and navigate their environments.

- **Chapter 7: Reinforcement Learning and Control**

  Introduces reinforcement learning and control strategies for teaching robots to self-improve and learn from trial and error.

- **Chapter 8: Self-Driving Vehicles**

  Explains the design, safety considerations, and technical challenges involved in building autonomous vehicles.

- **Chapter 9: Industrial Robotics**

  Covers the application of robotics in industrial settings, including manufacturing and warehouse automation, and the integration of machine learning to enhance these processes.

- **Chapter 10: Humanoid Robotics**

  Delves into the unique challenges and opportunities in developing humanoid robots, including perception, hardware, and software design.

- **Chapter 11: Data-Driven Robotics in Practice**

    Discusses the infrastructure required to support data-driven robotics, including important considerations, safety issues, and future directions.

What You Will Learn

By the end of this book, you'll gain expertise in the following:

- Applying machine learning to key robotics areas, including perception, mapping, control, and decision-making.

- Designing and implementing robotic systems for diverse industries, including self-driving cars, manufacturing, and humanoid robots.

- Overcoming the specific hurdles of integrating machine learning with robotics, understanding the future trends of robotics, and learning about the ongoing impact of machine learning.

What You Need to Know Before You Start

This book requires some Python programming knowledge and familiarity with libraries like NumPy, PyTorch/Jax, or ROS. A basic understanding of neural networks and machine learning is also necessary, either through an introductory course or self-study. If you lack this background, consider taking Andrew Ng's ML course or the Deep Learning Specialization on Coursera.

INTRODUCTION

Why We Wrote This Book

The last half decade has shown robotics being disrupted by machine learning methods and evidence is stronger than ever that the path to building generally intelligent robots is paved heavily with AI. We believe that the field of robotics is at a special moment today: one that is ready to be disrupted by breakthroughs in AI research. Machine learning has fundamentally transformed how we design and build robots, opening up a world of possibilities to create intelligent machines that effortlessly navigate and interact with our complex world.

Innovation at a rapid pace has created a gap in literature, where most textbooks on robotics taught at schools tread in classical methods and most ML textbooks rarely address embodied AI, therefore restricting the knowledge of designing data-driven robotics to privileged conferences, research labs, and academic papers. We are writing this book to democratize access to the practice and know-how of modern robotics.

We aim to break down barriers, making the fusion of AI and robotics comprehensible to a broader audience and inspiring a new generation of roboticists. We write this book to spark innovation, ignite new ideas, and invite more people to contribute to this thrilling field.

It's time to learn the new robotics, where AI leads the way.

CHAPTER 1

Introduction to General Purpose Robotics

People have dreamt of making intelligent machines that behave and think like humans for centuries. From the industrial revolution to Asimov's "I, Robot" and the world's first humanoids built a century ago, robots have occupied our collective imagination for a long time. Robots have transcended from being a figment of science fiction to being realized in the present, with accelerating capabilities.

What led to this transformation? Advances in artificial intelligence have disrupted various industries in the last decade by unlocking new capabilities with machine learning. Robotics has escaped its constrained and narrow applications in well-structured industrial and research environments and is now integrated into our daily lives. Robots drive competition through automation in large-scale manufacturing[1], space and underwater exploration[2], agriculture[3], and healthcare[4], among other industries. In the future, we expect to see robots handling fine-manipulation tasks in industries, performing household chores in homes, and autonomously operating on public roads and in hospitals.

As the capability has increased in the last few decades, the cost has decreased. Over the past 30 years, the average cost of robotics has fallen by half in Consumer Price Index (CPI)-adjusted terms (after accounting

CHAPTER 1   INTRODUCTION TO GENERAL PURPOSE ROBOTICS

for inflation), according to a recent McKinsey & Company report[5]. Costs have fallen even further in relation to their capabilities due to Moore's law, the ubiquity of GPUs, and the falling cost yet rising capacity of batteries and onboard computers. The widespread adoption of robots is motivated by increased economic expansion, the rising cost of human labor, the falling cost of robots, and the increase in their capabilities.

However, as robots move from research labs and constrained industrial settings to the real world, they face new challenges. Consider, for example, a household cleaning robot. This robot would have to engage in many tasks, including cleaning the floors, dusting counters, and washing dishes. To accomplish this, it must know how to:

- Traverse indoor environments while perceiving and avoiding obstacles.
- Handle fragile, soft, and sometimes heavy objects with irregular shapes, including objects it may never have encountered before.
- Manipulate scenes it may have never experienced before, since each home looks different, has different lighting, layouts, and so on.
- Reason about interactions with household objects, humans, and pets and past configurations of the space.

The challenge is designing an approach that can adapt to changes in the real world and the variety of situations it will encounter. Before the advent of deep learning, a software stack to solve any of these tasks would be written as a state machine with "hard-wired" motion primitives resembling traditional controls for that particular task. This approach cannot handle unseen situations very well, doesn't scale, and limits the utility of hard-programmed robots. Additionally, even for simple pick-and-place, translating the wide repository of human intuitions into transitional controls is challenging, if not close to impossible.

CHAPTER 1   INTRODUCTION TO GENERAL PURPOSE ROBOTICS

The breadth and universality of perception, reasoning, and controls required for general-purpose robotics is best handled by universal function approximators: *neural networks.* Instead of hand-coding a control system, we use machine learning to allow a robot to learn the relevant features and their relationships from training data.

This approach has yielded results in many other areas. The recent success of ChatGPT and language models in general has minted multibillion dollar AI companies. The research behind these products shows that scaling data, compute, and models programmatically leads to general capabilities in the language/vision/audio spaces. By converting data into tokens, similarities between them can be identified, and those similarities can be transferred to other domains. A wide array of capabilities have been unlocked as a result:

1. Creation of custom and realistic images[37] and videos[38] on demand, which promises to transform the film, marketing, and advertising industries.

2. Language generation, including translations, creative writing, copywriting, code generation[41], and transcription[40].

3. Audio generation. Creating on-demand podcasts[42] and music[43].

4. Multimodal reasoning. Solving mathematics[45], graduate-level science problems, and law and medical problems[44].

These capabilities have led to an AI spring, with generative AI companies raising 25.9 billion dollars in funding in 2023 alone[39]. These trends, as well as recent breakthroughs in spatial intelligence and robot foundation models, show that robotics can also be framed and solved as an AI problem. In an era in which we have generic intelligence, the ingredients needed to build generally useful robots are mostly present.

3

CHAPTER 1   INTRODUCTION TO GENERAL PURPOSE ROBOTICS

The success of AI and the promise of emergent capabilities has led to a boon in machine learning powered robotics and a rising demand for talent in the labor market. According to research from *Mordor Intelligence*[6], the Global Robotics Market was valued at USD 27.73 billion in 2020 and is expected to reach USD 74.1 billion by 2026, registering a Compound Annual Growth Rate (CAGR) of 17.45 percent. While this increase mostly accounts for the boon in industrial robots, the AI robotics market is expected to grow at a CAGR of 38.6 percent from 6.9 billion USD in 2021 to 35.3 billion USD in 2026 according to this[7] report. To capitalize on this opportunity, large tech companies, startups, and research labs are increasingly seek qualified AI and robotics engineers for their robotics R&D, autonomous cars R&D, and manufacturing divisions. To start contributing to these companies' machine learning efforts, you'll need to understand:

- How to formulate a robotics problem in the context of machine learning
- Which machine learning methods can be used to solve different problems in robotics and the tradeoffs between them
- At what point in the robotics stack you should use machine learning

Moravec's paradox is one of the main challenges of machine learning for robotics. As Steven Pinker described in 1994[8], "The main lesson of 35 years of AI research is that the hard problems are easy and the easy problems are hard." Artificial intelligence, especially neural nets, is a fairly different form of intelligence than the human brain and, as such, has different strengths. Things that may seem very difficult for humans—such as generative imaging, language compression, and sequential projection like stock analytics—are quite easy for AI. However, tasks that even a four year old child can do easily via sensorimotor and perceptual reasoning—such as taking a walk and lifting a pencil—are much harder. In the history of scientific innovation, all problems seem hard before they are solved, and the authors of this book are optimistic that mapping and fixing the real challenges in robotic learning can put a dent in advancing physical intelligence.

This chapter starts with defining the two premises of this book: robots and AI. We present general motivations for why one needs to use AI for robotics and the challenges in doing so. Subsequent chapters map out key areas in the development of AI for robotics, such as machine learning perception, language in robotics, training robots in simulations, and building infrastructures for scalable robot learning. Then we explain how to practically design and implement these principles in a few select applications—self-driving cars, industrial robots, and humanoids.

Let's get into it!

A Robot System

A robot is defined as an interactive machine that takes in a world model and outputs actions. Unlike many machine learning applications, a robot is characterized by agency and a closed loop feedback in a real or simulated world.

A robot typically senses the world through its suite of sensors, including cameras, LiDAR, inertial measurement units (IMU), voice detectors, and/or radars, as a few examples. A robot brain, typically executing on an onboard microcontroller, processes the inputs from sensors and calculates actions, which are sent as signals to the robot's actuators. These actuators can be direct current motors that cause its joints to move or compliant materials in the case of soft robotics.

The action space of a robot is determined by its application. For example, the action space for self-driving cars and navigating robots is the acceleration and steering angle. A robot arm could be designed as

CHAPTER 1   INTRODUCTION TO GENERAL PURPOSE ROBOTICS

positions or velocities of the joints on the arm. Additional action spaces for robots that interact with humans could include natural language via a chat interface, gestures, and facial expressions.

Figure 1-1 shows a high-level diagram of a robot system.

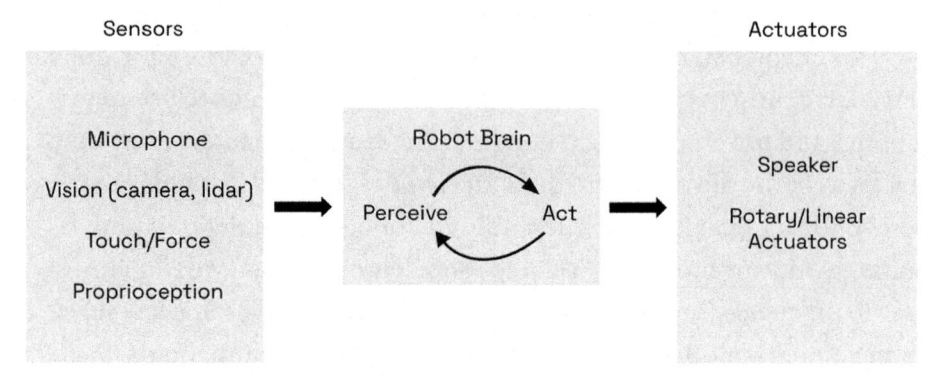

Figure 1-1. *The main components of a robot: sensors (microphones, vision systems like cameras and LiDAR, touch/force sensors, and proprioception), which are used to perceive the environment; the robot brain where perception data is processed in a continuous perception-action loop; and actuators (speakers and rotary/linear actuators), which carry out the robot's actions. This perception-action loop is critical for robotic learning*

Common Types of Robots

Robots come in many sizes and shapes. We can segregate them by vertical (or the sector in which they're deployed), as shown in Table 1-1 (curated with assistance from AI).

CHAPTER 1 INTRODUCTION TO GENERAL PURPOSE ROBOTICS

Table 1-1. *Types of Robots by Vertical*

	Type	Definition	Examples Industrial robots
1.	Industrial robots	Robots used in manufacturing processes such as assembly, painting, welding, and packaging	Robotic arms, gantry robots Service robots
2.	Service robots	Robots that perform tasks to assist humans in various environments such as hospitals, hotels, and restaurants	Delivery robots, cleaning robots, telepresence robots
3.	Medical robots	Robots used in healthcare settings to assist with surgeries, diagnostics, and patient care	da Vinci surgical system, rehabilitation robots, pharmacy automation systems
4.	Military and defense robots	Robots designed for use in military applications, such as reconnaissance, surveillance, and combat support	Unmanned aerial vehicles (UAVs), unmanned ground vehicles (UGVs), bomb disposal robots
5.	Agricultural robots	Robots used in farming to automate tasks like planting, harvesting, and monitoring crop health	Autonomous tractors, drones for crop monitoring, fruit-picking robots
6.	Domestic robots	Robots designed for use in homes to help with chores and other tasks	Roomba vacuuming robots, lawn-mowing robots, personal assistant robots like Jibo

(*continued*)

7

CHAPTER 1   INTRODUCTION TO GENERAL PURPOSE ROBOTICS

Table 1-1. (*continued*)

Type	Definition	Examples Industrial robots
7. Educational robots	Robots used in educational settings to help teach various subjects or skills	LEGO Mindstorms, social robots like Pepper, Sphero
8. Research robots	Robots used in scientific research, including exploring remote or hazardous environments and developing new robotic technologies	Underwater robots, Mars Rovers, humanoid robots like ASIMO
9. Entertainment robots	Robots designed for amusement or companionship	Robotic pets like Aibo, interactive toys like Furby, robots used in theme parks or movies
10. Swarm robots	Robots that work together in large groups, coordinating their actions to complete tasks more efficiently	Swarm robots used in research, agriculture, search and rescue, and environmental monitoring

A second way to split robots is by the nature of their embodiment, as shown in Table 1-2 (curated with assistance from AI).

8

CHAPTER 1　INTRODUCTION TO GENERAL PURPOSE ROBOTICS

Table 1-2. Types of Robots by Embodiment

Serial Num	Embodiment	Explanation	Examples
1	Wheeled robots	Robots using wheels for locomotion, often used on flat surfaces	Roomba, TurtleBot, self-driving cars like Waymo and Cruise
2	Tracked robots	Robots utilizing tracks for movement, providing greater traction and stability on rough or uneven terrain	Mars rovers, bomb disposal robots
3	Legged robots	Robots using legs for locomotion, navigating complex environments like stairs and uneven terrain	Boston Dynamics' Spot, ASIMO
4	Flying robots	Robots capable of flight, typically using rotors or wings, for aerial surveillance, inspection, and photography.	Quadcopter drones, fixed-wing UAVs
5	Underwater robots	Robots designed for underwater operation, used for exploration, inspection, and monitoring tasks	Bluefin Robotics AUV, SeaBED
6	Snake robots	Robots with long, flexible bodies, for moving through tight spaces and navigating around obstacles	CMU's Biorobotics Lab's snake robots, OC Robotics' snake-arm robots

(*continued*)

Table 1-2. (*continued*)

Serial Num	Embodiment	Explanation	Examples
7	Robotic arms	Robots consisting of a series of joints and links, resembling a human arm, used in industrial settings	KUKA, Fanuc, and ABB robotic arms
8	Humanoid robots	Robots with human-like forms, used in research, entertainment, and service applications	SoftBank Robotics' Pepper, Hanson Robotics' Sophia
9	Soft robots	Robots that mimic locomotion mechanisms of deformable matter such as fluids, gels, and elastomers for greater flexibility. Commonly used in biomedical applications such as soft tools for surgery, rehabilitation devices, and drug delivery	Harvard's Wyss Institute's soft robots, Octobot

Despite the variety in robots, they share many similarities that can be used to build a common framework and science for robotics, which is extensible with modifications to fit the deployment conditions of a robot.

Common Concepts in Robot Design

This section explains a few ubiquitous concepts that are used in robot design.

CHAPTER 1 INTRODUCTION TO GENERAL PURPOSE ROBOTICS

Robotic Manipulators

A common form of robots are robotic arms/manipulators. These robots can be found in the form of industrial robotics arms, assistive robots, and medical robots, which are used to complete various tasks in their environment. A robotic arm is a series of joints and links, such as the one depicted in Figure 1-2. Here, a link connects the joints and is a rigid body. A joint connects two or more links and allows for relative motion between the links.

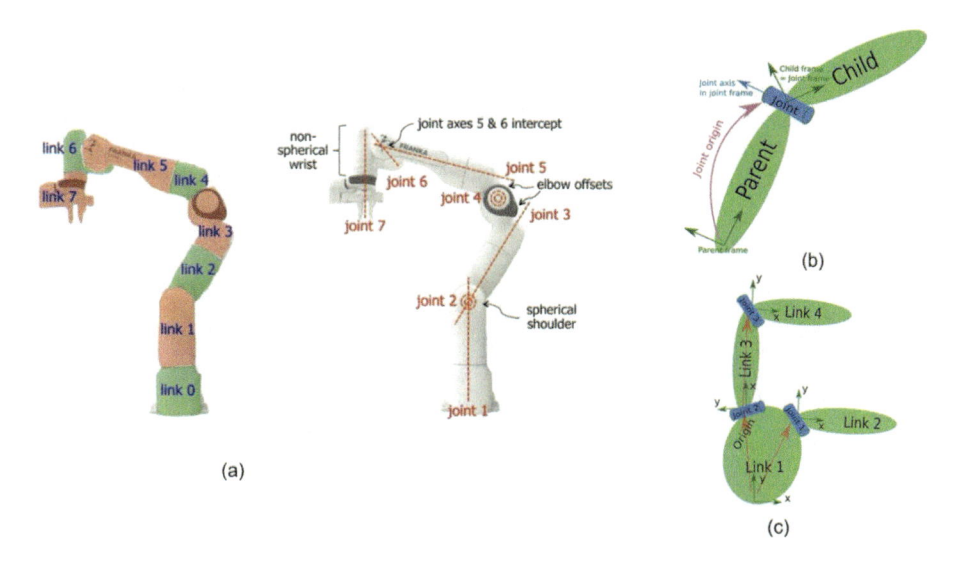

Figure 1-2. *Links and joints in a robot arm/manipulator. (a) A seven-link robot arm labeled with its joints. (b) Representation of the parent-child relationship between links through a joint. (c) Visualization of links connected through joints with rotational axes. Used with permission, source:* `https://robocademy.com/2020/04/21/robot-kinematics-in-a-nutshell/`*[49]*

11

CHAPTER 1   INTRODUCTION TO GENERAL PURPOSE ROBOTICS

Degrees of Freedom

Degrees of freedom (DoF) is a measure of the different components of motion a robot can undertake. For example, a rigid body in space has six DoF: translatory motion along the X, Y, and Z axes and rotary motion about X, Y, and Z axes, as shown in Figure 1-3. Every additional link adds to the degrees of freedom of a robot and every joint takes away from it by constraining motion in one or more directions. Each degree of freedom can be modeled as an independent, bounded variable that a robot brain needs to predict as an action target to control the robot.

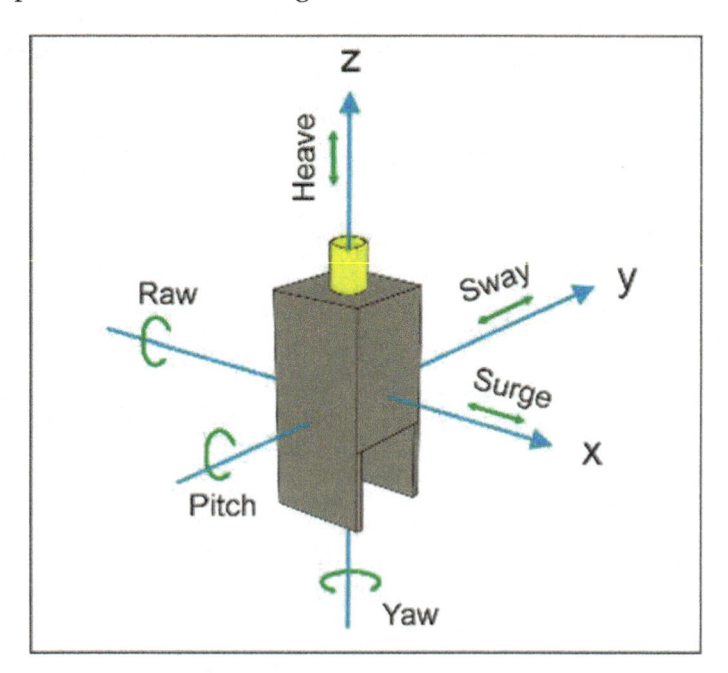

Figure 1-3. Any rigid body has six degrees of freedom (DoF): three translational (surge, sway, heave) and three rotational (roll, pitch, yaw). Used with permission, source: `https://www.researchgate. net/publication/340403456_Efficiency_and_Survivability_ of_a_Floating_Oscillating_Water_Column_Wave_Energy_ Converter_Moored_to_the_Seabed_An_Overview_of_the_EsflOWC_ MaRINET2_Database[50]`

12

End Effectors and Workspaces

An end effector is a device attached to the end of the arm. A gripper, or dexterous hand, is the most common type of end effector and is a form factor that generalizes to a wide variety of tasks. Sometimes robots also have task specific end effectors like wipers, squeezers, and so on.

The union of the three-dimensional space occupied by a robot is defined as its *workspace*. A subset of this space, that which can be reached by the end effector, is defined as the reachable space of a robot. Reachable space and workspace bounds drive robot design considerations given that they drive the utility of a robot and its ability to manipulate objects within their placement in space.

Kinematics

A key concept in robot control is robot kinematics. Etymologically, the term refers to the study of the motion of a body or a system of bodies. In this case, it is the joint motion of a robot's joints and links. With respect to robot control, two types of kinematics are relevant:

1. **Forward kinematics**: Here, given input joint angles, we determine the position and orientation of the end effector when all other joint parameters are known in a constellation of links and joints.

2. **Inverse kinematics**: Here, given a specific position or velocity that an end effector intends to achieve, we calculate the required motions/orientations of the previous joints to achieve that motion.

The sequence of links in the robot's physical body, their properties (e.g., mass, moment of inertia, length), and the properties of joints (e.g., constraints imposed, torque of the joint) determine the kinematic

CHAPTER 1 INTRODUCTION TO GENERAL PURPOSE ROBOTICS

system/chain representing the robot. Figure 1-4 shows a depiction of the kinematics chain of a typical robot arm. Figure 1-5 represents the transformation between link i and link i+1.

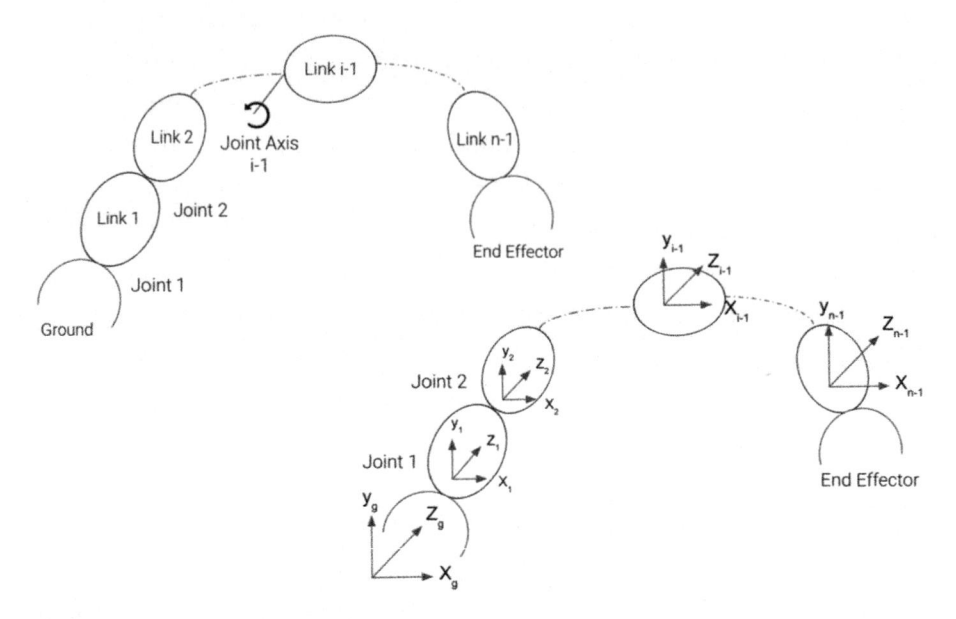

Figure 1-4. Kinematic chain of a robot arm showing multiple links connected by joints, with joint axes and coordinate frames

$$\begin{bmatrix} x_i \\ y_i \\ z_i \end{bmatrix} = \begin{bmatrix} \cos\gamma & -\sin\gamma & 0 \\ \sin\gamma & \cos\gamma & 0 \\ 0 & 0 & 1 \end{bmatrix} \begin{bmatrix} \cos\beta & 0 & \sin\beta \\ 0 & 1 & 0 \\ -\sin\beta & 0 & \cos\beta \end{bmatrix} \begin{bmatrix} 1 & 0 & 0 \\ 0 & \cos\alpha & -\sin\alpha \\ 0 & \sin\alpha & \cos\alpha \end{bmatrix} \begin{bmatrix} x_{i+1} \\ y_{i+1} \\ z_{i+1} \end{bmatrix}$$

α, β, γ are yaw, pitch, and roll of link i+1 with respect to link i.

$$x_i = R\ x_{i+1}$$
$$x_{i+1} = R^{-1}\ x_i$$

Figure 1-5. The transformation matrix between link i and link i+1 using yaw, pitch, and roll angles α, β, and γ, respectively

CHAPTER 1   INTRODUCTION TO GENERAL PURPOSE ROBOTICS

Deep Learning for Robotics

In the last decade, several robotics companies have bloomed and perished in the market for a variety of reasons[9]. These reasons range from product market fit to revenue/financing and technological issues.

Despite these failures, the robotics outlook for 2030 remains positive, partly due to the fact that most breakthroughs in deep learning research are happening in the current decade and are yet to be productized/deployed on robots. Improving upon the failures of the last-gen classical robots and a fundamental rethinking of robot learning paradigms are both key to bringing to market more powerful and generalizable robots. Deep learning is a promising prospect[10] toward that goal, as shown in Figure 1-6.

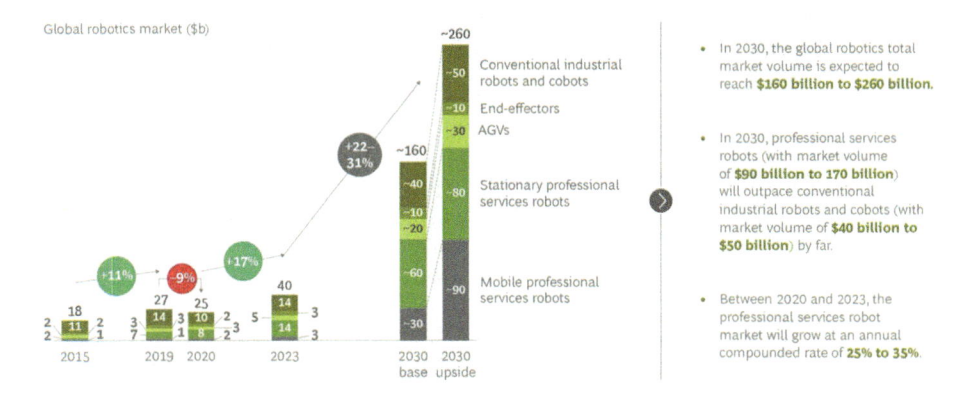

Figure 1-6.   Robotics market outlook for 2030, projecting a total market volume of $160 billion to $260 billion, from BCG. Used with permission, source: https://www.bcg.com/publications/2021/how-intelligence-and-mobility-will-shape-the-future-of-the-robotics-industry[51]

15

Some of the benefits that deep learning provides include:

- **Generalization:** Deep networks can learn nonlinear functions with enough parameters in the model and robust training methods to avoid overfitting. These functions are otherwise impossible to model with hand engineering. This is useful for dealing with diversity in the real world, such as handling various objects and scenes, learning inverse dynamics, and planning in diverse situations.

- **Feature learning:** Back propagation and carefully constructed loss functions allow deep neural networks to learn from data what is important, without the need for explicitly modeling representations or engineering features. It also allows networks to learn multiple representations of similar input data based on the application requirements. This translates into learning the correct distributions to generate actions for a variety of tasks with a single network.

- **Parallelism:** Real-world robotics requires responding at very fast inference speeds on the order of 10Hz or greater. While simpler classical methods are faster than neural nets, when the decision space becomes more complex, search/graph based methods become inefficient from a speed perspective and are not as friendly toward parallelization. Deep learning allows for massive parallelization on hardware accelerators like Tensor Processing Units (TPUs)[46] and Graphics Processing Units (GPUs)[47], which permit millions of matrix multiplications per second in an optimized

CHAPTER 1   INTRODUCTION TO GENERAL PURPOSE ROBOTICS

manner. Network architecture optimizations such as YOLO[11] and EfficientNet[12], discussed in Chapter 2, provide a tradeoff between accuracy and speed based on the application.

To deploy deep learning for robotics, a whole host of infrastructure paradigms are important:

- **Compute:** The improvements in performance from deploying larger and larger AI models is powered by innovations in compute architecture through massive parallelization on TPUs and GPUs. While the theory of deep learning has existed since the 1950s, the current spurt in research and applications was catalyzed by the availability and lowering cost of GPUs in the last decade. Additionally, robot onboard compute capabilities have improved due to platforms like NVIDIA Jetson[13], which allows for fast real-time inference on mobile robots. Over-the-air updating permits the deployment of newer software versions for on-the-field robots, enabling robot software iterations to happen at a rapid pace that pure software companies iterate on.

- **Data:** Deep learning is famously data hungry and its use in robotics requires data harvesting. A good example of this is Tesla's large array of sensor-mounted vehicles on the road that gather data on a wide variety of highly improbable driving scenarios[14]. Scaling also poses questions with respect to the best architecture and training methods. A second example of disruption from large datasets is in visual-language research where large datasets like ImageNet[15], LAION[16],

17

and the Internet have enabled the development of extremely capable neural networks like ChatGPT[17], ResNets[18], and Stable Diffusion[19]. This is covered in greater detail in Chapter 2.

- **Labeling:** Supervised and weakly supervised learning are among the most performant types of deep learning out there. Supervised learning, especially for image data, is made possible by highly streamlined labeling pipelines that generate human feedback and the emergence of labeling companies like Scale AI[20]. Weak supervision from text such as language-image pairs extracted from the Internet also collect very large scale datasets that train neural networks.

- **Simulation:** The widespread availability of simulation engines, such as Gazebo[21], PyBullet[22], and MuJoCo[48], emulate physics in the real world. They unlock new functionalities for robots. For one, simulation allows modeling of and handling emergency/safety critical scenes that are very unlikely in the real world. Learning from simulated data also removes constraints imposed by robot capacity. Doubling your data, if you only rely on real robots, means doubling robot hours and robot capacity, which is costly from a hardware and time perspective, but creating copies of simulated robots to do the same is quite cheap. Additionally, R&D requires iterative development and evaluations and stands to benefit from faster feedback loops that simulation can provide.

Deep Learning Frameworks

In the context of this book, we tend to use artificial intelligence, machine learning, and deep learning interchangeably to refer to data-driven methods. However, strictly speaking, the terms have some distinctions[36], addressed in this section.

Artificial Intelligence (AI) is an overarching term to describe computational techniques capable of performing tasks with human-level intelligence. These tasks include problem-solving, understanding natural language, recognizing patterns, and making decisions based on learnings from model training. AI is the overarching field that consists of ML and DL, as shown in Figure 1-7. Artificial intelligence today is further differentiated into Artificial General Intelligence (AGI), which describes a system that has a wide range of intelligence capabilities useful in everyday life, and Artificial Super Intelligence (ASI), which describes intelligence that far exceeds human level capabilities.

Machine Learning (ML) is a subfield of AI focused on *learning from data*. Instead of being explicitly programmed to perform a task, ML models improve their performance through the data that they are trained on. ML comprises supervised learning (labeled data), unsupervised learning (unlabeled data), and reinforcement learning.

Deep learning (DL) is a subset of ML that uses neural networks that often have many layers *(hence deep)* and uses large scale data to perform complex tasks. These networks help understand images and language and can include more state-of-the-art methods like transformers, GPT, and so on. DL methods often have more layers and parameters and require more data and computational resources than "shallow" ML methods.

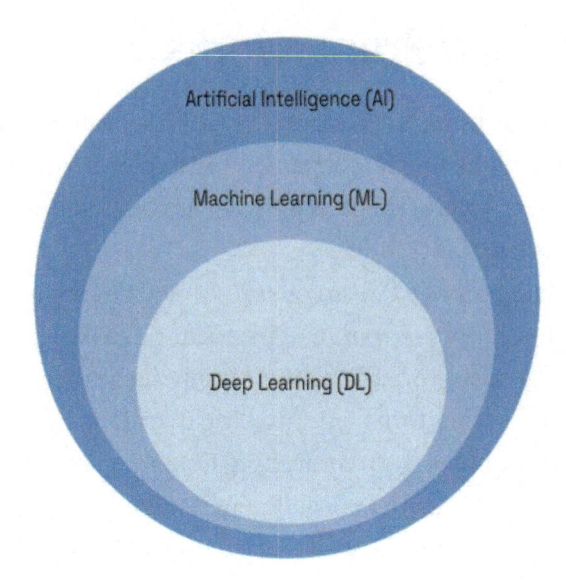

CHAPTER 1　INTRODUCTION TO GENERAL PURPOSE ROBOTICS

Figure 1-7. *Artificial intelligence (AI) is the overarching concept that includes machine learning (ML) and its subset, deep learning (DL). Inspired by [36].*

Within deep learning there are four main types of learning systems: supervised learning[23], unsupervised learning[24], weakly supervised learning[25], and reinforcement learning[26]. In recent years, the lines between these three learning methods have blurred, as combinations or ideas from them can be utilized in a singular system.

1. **Supervised learning** uses datasets with labels of the ground truth that the system should use to predict labels of previously unseen data. This is used for classification, where the output typically consists of discrete classes, or for regression, where predicted outputs are real numbers.

2. **Unsupervised learning** methods attempt to learn useful representations of data without labels. Examples of unsupervised learning methods include clustering, principal component analysis, Gaussian mixture models, auto-encoders, and so on. Unsupervised learning models are typically used for clustering, association, and dimensionality reduction.

3. **Weakly supervised learning** methods use noisy labels in a supervised learning setting. They are used in cases where datasets are expensive to label and aggregating large datasets with weak labels is feasible over a smaller dataset with clean labels. Weakly supervised learning is used to train very powerful models, including CLIP, DALL-E, and so on, using Internet-scale data.

4. **Reinforcement learning** motivates an agent to learn a policy that maximizes a reward function through processing sequences of state-action pairs, observing the achieved rewards, and adapting predictions until it accurately predicts an optimal path, or policy, for the agent. Reinforcement learning provides a framework for robots to autonomously learn through trial-and-error interactions and continuously self-improve with feedback.

These terms are overarching and represent the entire field of AI. The next section looks at frameworks specifically used in robot learning.

CHAPTER 1   INTRODUCTION TO GENERAL PURPOSE ROBOTICS

Robot Learning Frameworks and Objectives

Learning a task from a robotics perspective may be described as generating the distribution of actions given a specific input world model. As mentioned, a key manner in which robotics of the present differs from the past is that we can now design multi-purpose, generalist robots that can do a variety of tasks. Generality, from a deep learning perspective, can be framed in three settings:

1. **Transfer learning:** Given a network trained on task T_i, can we adapt it to learn task T_{i+1}?

2. **Meta-learning:** Given a network that can do task T_1, T_2, and T_{n-1}, can we quickly adapt it to learn task T_n?

3. **Multi-task learning:** Can we train a network on all tasks—T_1, T_2, to T_n—at the same time?

Subsequent chapters address algorithms that explain these learning paradigms in detail.

For a multitask robot, specifying the objective assume various forms. A robot brain can be configured to achieve an objective. But the question of how to convey an objective to a robot remains. This is especially important for deep learning, which is an objective optimization framework. Ideas explained in the book to address this question include the following:

- **Language conditioning:** The practice of specifying targets for robotics using language as an interface. Language has been the natural interface for interaction between humans. The expansion in deep learning for natural language processing has made language interfaces to robots and generative a standard.

- **Goal conditioning:** Goal conditioning often happens on policies where you can train policies to reach a goal state, which is provided as input. For example, goal-conditioned reinforcement learning (GCRL)[27] trains an agent to achieve different goals under particular scenarios.

- **Self-collision:** If the robot is not programmed properly, the robot can collide with itself. The goal of self-collision is to make the robot aware of its body to avoid collision during motion. Another type of collision is avoiding collisions with the environment. For example, suppose a robot experiences an unexpected obstacle while navigating its environment. In that case, it can use collision avoidance to determine the best action and path around the object and continue its task.

- **Hierarchical robot learning:** The goal of hierarchical learning is to break down larger problems into a hierarchy of subproblems. This allows higher-level parent tasks to invoke lower-level child tasks to complete a task.

Toward Embodied General Intelligence

Solving artificial general intelligence (AGI) is one of the most heated and important problems of our generation. Nick Bostrom defines AGI as "an intellect that is much smarter than the best human brains in practically every field"[28]. Wikipedia defines it as "the ability of an intelligent agent to understand or learn any intellectual task that a human being can"[29]. Open Philanthropy describes "transformative AI" with an economic definition, as something that could increase the Gross World Product ten times over[30].

CHAPTER 1   INTRODUCTION TO GENERAL PURPOSE ROBOTICS

But all these specifications mainly focus on digital AI, whose imagined interfaces to the real world are still human. Transforming any industry that predates the Internet—such as manufacturing, construction, driving, logistics, energy, mining, and agriculture—and that encompass large portions of global GDP would require solving *embodied* AI, that is intelligence within an embodiment that has physical reasoning and can manipulate the physical world. Advancements in compute and data have made it much easier for anyone to build and test deep learning models for robotics. But how does one know if an AI is embodied and general? Steve Wozniak has proposed a coffee test for embodied AGI: a machine can learn how to make coffee in an unseen human kitchen[31]. This flies in the face of Moravec's paradox, because AI today can create high-quality graphic images and movies, but making a simple cup of coffee in a generic setting is still impossible. Robotics is hard AI, because it needs to solve computer vision (for understanding the world), language (for interacting with humans and communicating), manipulation and navigation (for acting in the real world), and tool use (for search, embodied reasoning, etc.). Robotics is the hard and over-encompassing version of AI, one that is truly packaged to change the world, and one that inhabits and lives among us, not just behind screens and in data centers. Specific benchmarks for embodied intelligence to measure and track progress toward solving robotics is still ongoing.

Environment Is Deeply Tied to the Definition of Intelligence

What is the north star for embodied intelligence? There is some evolutionary evidence that points to how the solution may look.

Lifetime learning over several episodes has encoded data in our genes, so much that babies understand structured motion and the physics of the world before they understand and comprehend language. What

CHAPTER 1   INTRODUCTION TO GENERAL PURPOSE ROBOTICS

is intelligent is deeply tied to what provides a survival advantage in an environment. For example, aquatic animals have visual systems that are much better[32] at seeing underwater because they've evolved to accommodate for the refraction by water in a way that humans have not. Our sensors that attempt to emulate our visual range, and the data we've collected on that basis, including YouTube videos, suffers from being overfit to our domain of visual capability. In patients who have had their cataracts removed, allowing them to see for the first time, it was seen that despite spending an entire life in a 3D world, they lacked understanding of spatial imagery because their sensors didn't have that input[33]. In essence, environment and agent cannot be removed from the definition of intelligence.

The last decade of AI research has led to the rise of large transformers that are very good at multi-task speech and vision. A language-first AI would be susceptible to the failure modes[34] of a blind agent, beyond the visual context it receives from a training corpus gathered from humans who can see. It logically extends that a visual language model would suffer from an inability to approximate actuator parameters inherent to performing precise control of an embodied agent. Reasoning about the real world requires not just thinking about methodological spaces and language, but also being grounded in a real-world context[35].

In a world built by and designed for humans, an intelligence that is agnostic to sensory-motor dynamics is going to be suboptimal, and superhuman skills beckon physical agency and universal control. **Having physical embodiment is absolutely indispensable to AGI.**

This book explores how to practically reach that goal and build a future with generally intelligent robots.

CHAPTER 1   INTRODUCTION TO GENERAL PURPOSE ROBOTICS

Summary

This chapter covered the following points:

- Recent advancements in AI are moving robots from controlled research labs into real-world applications, thus allowing them to adapt and generalize to dynamic, unpredictable environments. AI allows robots to learn from data, rather than relying on preprogrammed rules, making them versatile across various tasks and industries.

- A robot operates by processing sensor inputs (such as cameras, LiDAR, and IMUs) into actionable outputs via a microprocessor or microcontroller, which then signals actuators (like motors or soft materials) to execute physical movements. This perception-action loop is central to a robot's ability to interact with and manipulate its environment.

- Robots can be classified based on their applications, including industrial robots (for assembly, welding, etc.), service robots (for tasks like cleaning or telepresence), medical robots (for surgeries and diagnostics), military and defense robots (for reconnaissance or bomb disposal), among others. They can also be categorized by their physical structure, such as wheeled, legged, flying, or humanoid forms, depending on their function and environment.

CHAPTER 1 INTRODUCTION TO GENERAL PURPOSE ROBOTICS

- Key design principles include the use of robotic manipulators (arms with joints and links), degrees of freedom (the range of independent movements a robot can perform), the end effector (the tool at the end of a robotic arm used for tasks like gripping), workspace (the area a robot can physically reach), and kinematics (the study of the motion of joints and links, including forward and inverse kinematics for planning).

- The demand for robots is rapidly growing due to their increasing ability to generalize and perform a wide range of tasks in various industries. The development of infrastructure, software frameworks, and systems learning from AI has fueled advancements in the capabilities and adoption of robots.

- Robots learn to perform tasks through four main types of machine learning: supervised learning (training on labeled data), unsupervised learning (discovering patterns in data without labels), weakly supervised learning, and reinforcement learning.

- Various frameworks guide robot learning, including transfer learning, meta-learning, and multi-task learning.

- The ultimate goal of AI in robotics is to build a generally intelligent agent in the physical world.

The next chapter discusses sensors, robot perception, and common neural network and transformer methods that robots use to sense and understand their environment.

CHAPTER 1   INTRODUCTION TO GENERAL PURPOSE ROBOTICS

References

[1] Law, Marcus. "Robotics Reshaping Manufacturing and the Future of Work." Technology Magazine, 31 May 2024, technologymagazine.com/articles/robotics-reshaping-manufacturing-and-the-future-of-work.

[2] Ryan, Melissa, and Karl McLetchie. "How Robots Are Uncovering the Mysteries of the Deep." Oyla Articles: *Ocean Exploration Technology,* Aug. 2022, oceanexplorer.noaa.gov/explainers/technology.html.

[3] https://www.agritecture.com/blog/exploring-the-future-of-agriculture-a-deep-dive-into-robots

[4] https://online-engineering.case.edu/blog/medical-robots-making-a-difference

[5] Tilley, Jonathan. "Automation, Robotics, and the Factory of the Future | McKinsey." McKinsey & Company, 7 Sept. 2017, www.mckinsey.com/capabilities/operations/our-insights/automation-robotics-and-the-factory-of-the-future.

[6] www.mordorintelligence.com/industry-reports/robotics-market.

[7] "Artificial Intelligence (AI) Robots Market Size, Growth, Trend and Forecast to 2023 | MarketsandMarkets." *Markets and Markets,* www.marketsandmarkets.com/Market-Reports/artificial-intelligence-robots-market-120550497.html.

[8] Pinker, Steven. *The Language Instinct: The New Science of Language and Mind.* London, Penguin Books, 1994.

[9] Casse, Bernard. "Council Post: The Demise of Robotics Companies: Learning from Past Mistakes." *Forbes*, 9 July 2021, www.forbes.com/sites/forbesbusinesscouncil/2021/07/09/the-demise-of-robotics-companies-learning-from-past-mistakes/?sh=5b9a1bac2b1d.

[10] Lenz, Ian, et al. "Deep Learning for Detecting Robotic Grasps." *The International Journal of Robotics Research*, vol. 34, no. 4-5, 16 Mar. 2015, pp. 705–724, https://doi.org/10.1177/0278364914549607.

[11] Redmon, Joseph, et al. "You only look once: Unified, real-time object detection." Proceedings of the IEEE Conference on Computer Vision and Pattern Recognition. 2016.

[12] Tan, Mingxing, and Quoc Le. "Efficientnet: Rethinking model scaling for convolutional neural networks." International Conference on Machine Learning. PMLR, 2019.

[13] https://www.nvidia.com/en-us/autonomous-machines/embedded-systems/

[14] https://www.tesla.com/en_ca/support/transitioning-tesla-vision

[15] https://www.image-net.org/

[16] https://laion.ai/

[17] https://openai.com/chatgpt/

[18] He, Kaiming, et al. "Deep residual learning for image recognition." Proceedings of the IEEE Conference on Computer Vision and Pattern Recognition. 2016.

CHAPTER 1 INTRODUCTION TO GENERAL PURPOSE ROBOTICS

[19] Rombach, Robin, et al. "High-resolution image synthesis with latent diffusion models." Proceedings of the IEEE/CVF Conference on Computer Vision and Pattern Recognition. 2022.

[20] https://scale.com/data-engine

[21] https://gazebosim.org/home

[22] https://pybullet.org/wordpress/

[23] Ali, Moez. "Supervised Machine Learning." *DataCamp*, Aug. 2022, www.datacamp.com/blog/supervised-machine-learning.

[24] Pykes, Kurtis. "Introduction to Unsupervised Learning: Types, Applications and Differences from Supervised Learning." *DataCamp*, Jan. 2024, www.datacamp.com/blog/introduction-to-unsupervised-learning.

[25] Kanjilal, Joydip. "An Introduction to Weakly Supervised Learning." *Paperspace Blog,* blog.paperspace.com/an-introduction-to-weakly-supervised-learning/.

[26] https://spinningup.openai.com/en/latest/

[27] Qian, Zhifeng, et al. "Goal-conditioned reinforcement learning with disentanglement-based reachability planning." *IEEE Robotics and Automation Letters* 8.8 (2023): 4721-4728.

[28] Bostrom, Nick. "How long before superintelligence." *International Journal of Futures Studies* 2.1 (1998): 1-9.

[29] https://www.scientificamerican.com/article/what-does-artificial-general-intelligence-actually-mean/

[30] https://docs.google.com/document/d/1IJ6Sr-gPeXdSJugFulwIpvavc0atjHGM82QjIfUSBGQ/edit?fbclid=IwAR3W3XVgKok3caD2TY6zSxqFr2CFSmqpOKX-gObOjup-o5nSJEdWEx2fy3o

[31] https://koopingshung.com/blog/turing-test-is-obsolete-bring-in-coffee-test/

[32] www.wildlifeonline.me.uk/questions/answer/how-can-marine-mammals-see-underwater-but-we-cant#:~:text=Human%20eyes%20have%20evolved%20to

[33] Chatterjee, Rhitu. "Feature: Giving blind people sight illuminates the brain's secrets." *Science Magazine* (2015).

[34] Yang, Zhengyuan, et al. "An empirical study of gpt-3 for few-shot knowledge-based vqa." Proceedings of the AAAI conference on artificial intelligence. Vol. 36. No. 3. 2022.

[35] Ahn, Michael, et al. "Do As I Can, Not As I Say: Grounding language in robotic affordances." *arXiv preprint arXiv:2204.01691* (2022).

[36] https://www.intel.com/content/www/us/en/robotics/artificial-intelligence-robotics.html

[37] Ramesh, Aditya, Mikhail Pavlov, Gabriel Goh, Scott Gray, Chelsea Voss, Alec Radford, Mark Chen, and Ilya Sutskever. "Zero-shot text-to-image generation." In *International conference on machine learning*, pp. 8821-8831. Pmlr, 2021.

[38] Brooks, Peebles, et al., https://openai.com/index/video-generation-models-as-world-simulators/

CHAPTER 1   INTRODUCTION TO GENERAL PURPOSE ROBOTICS

[39] https://www.cnbc.com/2024/09/06/ai-craze-getting-funded-by-tech-giants-distorting-traditional-vcs.html#:~:text=That%20continues%20a%20trend%20from,27%25%20so%20far%20this%20year

[40] Radford, A., Kim, J. W., Xu, T., Brockman, G., McLeavey, C., & Sutskever, I. (2023, July). "Robust speech recognition via large-scale weak supervision." In *International Conference on Machine Learning* (pp. 28492-28518). PMLR.

[41] https://github.com/features/copilot

[42] https://notebooklm.google/

[43] https://google-research.github.io/seanet/musiclm/examples/

[44] Achiam, Josh, Steven Adler, Sandhini Agarwal, Lama Ahmad, Ilge Akkaya, Florencia Leoni Aleman, Diogo Almeida et al. "Gpt-4 technical report." *arXiv preprint arXiv:2303.08774* (2023).

[45] https://www.nytimes.com/2024/07/25/science/ai-math-alphaproof-deepmind.html

[46] https://cloud.google.com/tpu

[47] https://www.intel.com/content/www/us/en/products/docs/processors/what-is-a-gpu.html

[48] https://mujoco.org/

[49] Joseph, Lentin. "Robot Kinematics in a Nutshell." *ROBOCADEMY*, 21 Apr. 2020, robocademy.com/2020/04/21/robot-kinematics-in-a-nutshell/.

[50] Kisacik, Dogan, et al. "Efficiency and survivability of a floating oscillating water column wave energy converter moored to the seabed: an overview of the EsflOWC MaRINET2 database." *Water* 12.4 (2020): 992.

[51] Lässig, Ralph, et al. "Robotics Outlook 2030: How Intelligence and Mobility Will Shape the Future." *BCG Global,* 28 June 2021, `www.bcg.com/publications/2021/how-intelligence-and-mobility-will-shape-the-future-of-the-robotics-industry`.

CHAPTER 2

Robot Perception: Sensors and Image Processing

Humans use five senses to perceive the environment and various cognitive pathways to process this input. This conglomeration of senses, pathways, and the brain forms our perception system, which allows us to detect movement around us, recognize a friend's face, and detect a familiar scent. Similarly, robots need to be aware of and understand what is around them to function in the real world. Perception systems enable robots to accomplish two objectives: to sense their surrounding environment and to comprehend and reason about it.

The first objective of the perception system is achieved by the sensing suite. Sensors act as the eyes and ears of the robot, enabling it to observe and record the physical world around it. Additionally, they allow robots to collect useful data that in turn allows them to evolve and adapt in an environment.

The second objective involves interpreting data from the sensors and extracting relevant information that can aid in completing a robot's objective. Traditionally, this part was done by classical methods that were narrow and specific to a robot and task, making them difficult to generalize to changing environments. Machine learning perception systems, on the

CHAPTER 2 ROBOT PERCEPTION: SENSORS AND IMAGE PROCESSING

other hand, are a lot more robust. They improve with data, and they can be used to map the robot's surroundings, so it can navigate, detect, and track objects.

An overview of this general pipeline is shown in Figure 2-1.

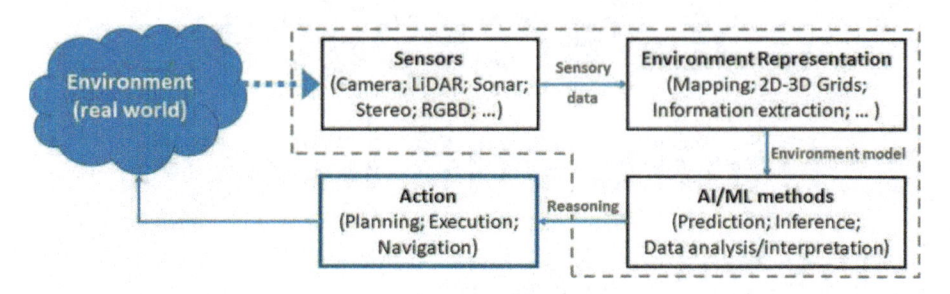

Figure 2-1. A typical robot perception system will include sensors that collect data and ML/AI algorithms for interpretation, planning, and execution of actions. Used with permission, source: `https://www.intechopen.com/chapters/62978`*[39]*

This chapter discusses sensing, explains the types of sensors, and highlights various ML algorithms commonly used for robotic perception.

Sensors

Robots use sensors to receive information about their surroundings and decide how to interact with the world and maneuver around it safely. Cameras take pictures, LiDARs provide point clouds and accurate depth, ultrasonic sensors measure proximity (especially of moving objects), and Inertial Measurement Units (IMUs) give information on orientation and motion. Robots may use singular sensors for object recognition, localization and mapping, collision avoidance, and feedback control. Robots may also use sensor fusion techniques to blend data from various sensors and get more comprehensive information.

CHAPTER 2   ROBOT PERCEPTION: SENSORS AND IMAGE PROCESSING

While robotics applications use a variety of sensors, this book only covers the more popular ones. They fall into four main categories: (1) monocular vision, mainly cameras, (2) depth sensors, (3) range sensors, such as LiDARs and ultrasonic sensors, and (4) inertial measurement units (IMUs).

Vision Sensors (Cameras)

One of a robot system's most important sensors is its camera. Light is captured by the camera sensor, which transforms the light into electrical signals to produce images. They are made up of a variety of pixels or photosensitive components. Every pixel records the amount of light that strikes it.

The two most popular types of image sensors used in robotics are CMOS (Complementary Metal-Oxide Semiconductor) and CCD (Charge-Coupled Device). CCD and CMOS image sensors both convert light into electrons by capturing photons (light particles) with numerous photosites, which are tiny, light-sensitive regions on the sensor that correspond to pixels in the final image. When taking a picture, these photosites collect and store photons as electrical signals. A key aspect that differentiates each sensor is the way accumulated charge (the electrical charge generated by the photosites when they capture photons) of each photosite is transported. CCD sensors transport charges with minimal distortion, resulting in high-quality, sensitive images but consuming significantly more power. CMOS sensors are more flexible and less expensive but tend to be more susceptible to noise and have lower light sensitivity. The exposure time, ISO sensitivity, and aperture size of camera sensors are all tunable parameters, allowing them to adapt to different lighting conditions.

CHAPTER 2   ROBOT PERCEPTION: SENSORS AND IMAGE PROCESSING

Let's outline a few key camera concepts:

1. *Pixel resolution* of camera sensors determines the degree of detail in the captured images. High-resolution sensors include more pixels, making images clearer and with more detail, but they use more memory and processing power.

2. *Channels* refer to color channels used in a camera. A common framework is RGB, where each image has red, green, and blue channels. Most cameras use a variety of color filters—typically the Bayer pattern—on top of the pixels to capture color information. Due to these filters, each pixel may capture red, green, or blue light. The camera sensor generates a full-color image by interpolating the color values from nearby pixels. So, a camera of resolution 480,640 with three channels (red, green, blue) generates an image that can be interpreted as a matrix of size[480,640, 3].

3. *Frame rate* refers to the number of images a camera sensor takes each second. Robotics applications in high speed or dynamic environments can benefit from higher frame rates because they provide real-time perception, thus enabling quicker reaction times.

4. *Post processing*. Raw picture data from cameras may be further processed for noise reduction, white balance correction, color correction, and feature extraction. Image processing algorithms may be then used to improve image quality, find objects,

recognize patterns, and extract depth data. Camera sensors are frequently fused with other sensors, such as LiDAR, radar, or IMUs, in a sensor-fusion strategy to maximize their utility.

Cameras as sensors also have certain limitations:

a. They may struggle in low-light or high-contrast scenes and need a lot of lighting for the best outcome.

b. Cameras may struggle to effectively perceive depth and 3D information without additional sensors.

c. Cameras are also prone to occlusions.

d. Large-scale picture processing can also be computationally demanding.

Key Considerations for Cameras

This section discusses tradeoffs for selecting cameras for your robotic application. As shown in Figure 2-2, three factors mainly determine camera selection.

Frames per second ──────── Resolution ──────── Cost

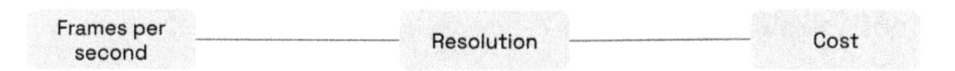

Figure 2-2. *The three main components of a camera you need to consider when selecting it for a robot application are frames per second, resolution, and cost*

Autonomous cars, for instance, require cameras that can capture images at a high frequency due to the speed of the moving car. If a car is moving at 60 miles per hour and passing another car in the opposite direction at the same speed, each frame captured represents a significant range between the two cars. In this scenario, we ideally want to capture

CHAPTER 2 ROBOT PERCEPTION: SENSORS AND IMAGE PROCESSING

and process as many frames per second (fps) as possible. Typically, automotive cameras operate at 30 fps, but this can introduce processing delays for each frame[1]. Therefore, to accurately detect dynamic objects such as pedestrians and predict their path, multiple passes per frame are necessary[1]. If we miss one or a few frames because the camera is not fast enough, the result could be collisions. For instance, an advanced driver-assistance system (ADAS) today might use a camera with an 8-megapixel resolution and a frame rate of 60 fps to ensure reliable detection and understanding of its environment[2].

Cameras used for industrial applications have different specifications. Picking, packing, and grasping objects do not involve high-speed movements and don't necessarily require the high frame rate capture. Usually, a regular 30 fps camera is sufficient.

High frame rate cameras usually tend to have lower resolution, and lower resolution can impact precision of detection, especially with far away objects. Picking cameras both high resolution and high frequency can increase the cost, mainly because they imply higher bandwidth and processing power, and may be just as expensive as a 3D LiDAR. Ultimately, however, the precision requirements and cost constraints of an application drive camera selection.

Event-Based Cameras

Another type of camera is an event-based or neuromorphic camera. It outputs pixel-level changes in brightness. This is in contrast to regular frame cameras, which transmit entire arrays of information of the single frame captured by the shutter at a given time. The data format and output of event-based cameras offer a significant advantage, as the only transmitted data is the individual pixel information that has changed from frame to frame. This allows it to capture objects in high-speed motion with no motion blur. An event camera is shown in Figure 2-3, and the data output of event cameras is shown in Figure 2-4.

CHAPTER 2 ROBOT PERCEPTION: SENSORS AND IMAGE PROCESSING

Compared to regular frame cameras, event-based cameras have some strengths, such as no motion blur, high temporal resolution, and high dynamic range yet low bandwidth. Event cameras are still a novel approach for most solutions, and prices are still not as affordable as regular cameras. However, some applications are very suitable for event-based cameras:

- Due to the low data rate and sparse information provided by event-based sensors, they can effectively track objects with low compute power.

- Frame interpolation, optical flow estimation, and high-speed recording applications benefit from the high temporal resolution and better temporal modeling enabled by event cameras.

Figure 2-3. *A 680 x 480 event camera*

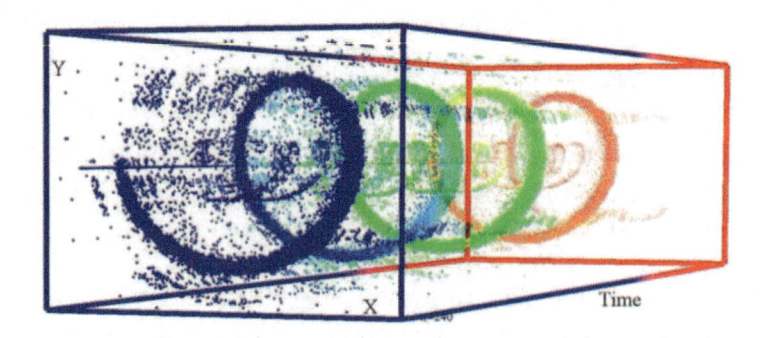

CHAPTER 2 ROBOT PERCEPTION: SENSORS AND IMAGE PROCESSING

Figure 2-4. *Data output of an event-based camera. Used with permission, source:* `https://rpg.ifi.uzh.ch/docs/PAMI17_Gallego.pdf`*[40]*

Research from Davide Scaramuzza's group[3] at The University of Zurich shows the development of a hybrid event- and frame-based object detector. This method combines the high temporal resolution and efficiency of event cameras with the detailed imaging of traditional sensors, significantly reducing perceptual and computational latency while maintaining accuracy. Using a 20 Hz RGB camera with an event camera, the system achieves the same latency as a 5,000 Hz camera with the bandwidth of a 45 Hz camera. For a more detailed look at the latest research in event-based vision and camera, we recommend this[3] resource.

Depth Sensors

Robots also have sensors that measure the depth or distance of objects in their surroundings, such as time-of-flight cameras and structured light sensors. These sensors allow robots to comprehend the three-dimensional structure of their surroundings and carry out tasks like object reconstruction, scene interpretation, and gesture identification because of the exact depth information provided by these sensors. Figure 2-5 shows depth sensors that use time-of-flight (ToF), structured light, and stereo-vision principles to measure depth information.

CHAPTER 2   ROBOT PERCEPTION: SENSORS AND IMAGE PROCESSING

- **Time-of-flight (ToF) sensors**: ToF sensors measure the time it takes for a light signal to bounce back after reflecting off of surrounding objects. The light signal is commonly an infrared (IR) signal. The depth sensor can calculate the time it takes for the round-trip journey to determine the distance to objects.

- **Structured light sensors**: Structured light sensors project a pattern of light onto the scene, such as a grid of infrared dots or a collection of structured patterns. The sensor's infrared camera then records the distorted pattern. The depth sensor may determine depth information based on the distortions generated by object surfaces by examining the deformation of the pattern.

- **Stereo vision sensors**: Similar to human eyes, stereo vision sensors employ a pair of cameras with a known baseline separation. The depth sensor may determine the depth by comparing the disparities or discrepancies between corresponding pixels in the stereo pictures, and each camera captures a slightly different view of the scene. Triangulation techniques are frequently used to determine depth based on discrepancies in pixels.

CHAPTER 2   ROBOT PERCEPTION: SENSORS AND IMAGE PROCESSING

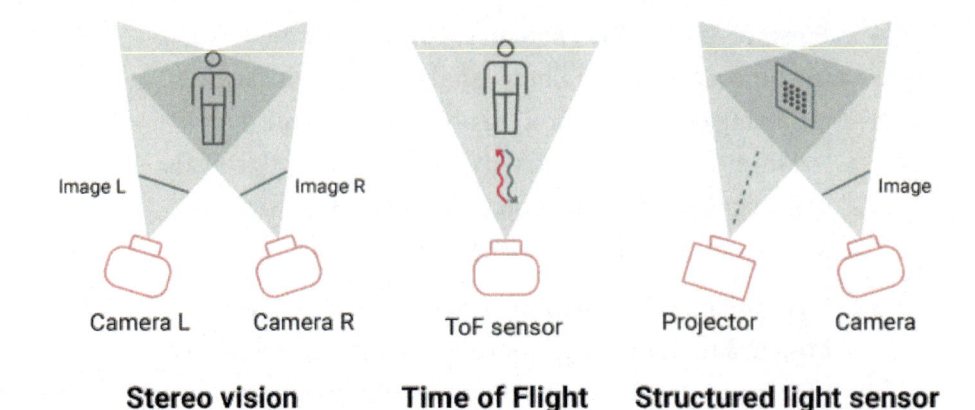

Figure 2-5. *Time-of-flight (ToF), structured light, and stereo vision methods for depth measurement. Used with permission by Wavelength Opto-Electronic, source:* `https://wavelength-oe.com/articles/optics-for-consumer-electronics/`*[41]*

The output from depth sensors is a *point cloud,* a 3D representation of the scene consisting of a collection of 3D points that represent surfaces of the scene, or a depth image, representing the distance of each pixel from the sensor. Many depth sensors such as ToF and structured light will directly output depth images. In some cases, especially for sensors used in robotics applications, the output is a point cloud. A specific location in space and its corresponding color information are represented by each point in the cloud. An example of the 3D point cloud output produced from a depth sensor is shown in Figure 2-6.

CHAPTER 2 ROBOT PERCEPTION: SENSORS AND IMAGE PROCESSING

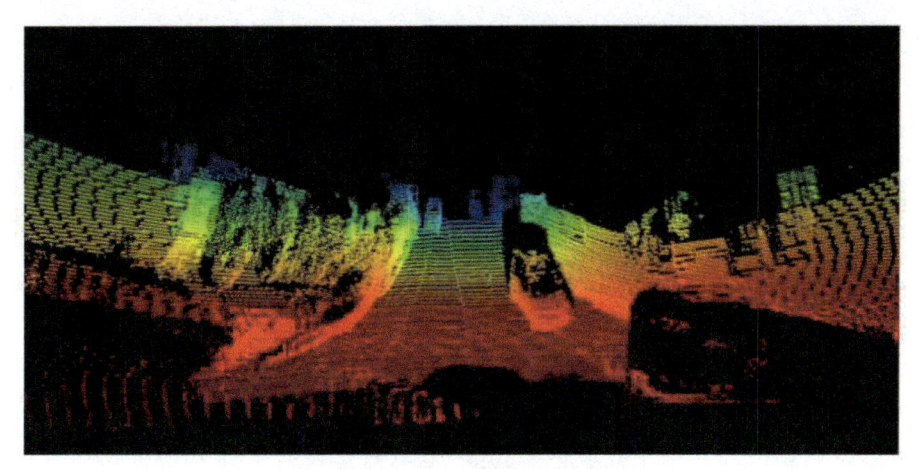

3D Point Cloud Simulation

Figure 2-6. *Example 3D point cloud output from a depth sensor. Used with permission, source:* `https://learnopencv.com/3d-lidar-visualization/`*[42]*

The range, precision, sensitivity to lighting, occlusion issues, and difficulty photographing translucent or reflecting surfaces are drawbacks of depth sensors. But in recent years, improvements in depth sensor technology have improved their performance. For example, they are combined with other sensors, like cameras or inertial measurement units (IMUs), to create a more comprehensive perception system.

Range Sensors

Another important type of sensor in robots are range sensors, such as Light Detection and Ranging (LiDAR), and ultrasonic sensors, which provide information about the distance between the robot and objects in its surroundings. If a range sensor has high fidelity (such as high accuracy and resolution, typically seen in high-density LiDAR), it is useful as a depth sensor for accurately measuring distances from the robot in question to other objects. The following section delves into how LiDAR achieves this feat. Overall, range sensors help robots navigate and avoid collisions by allowing 3D perception.

45

CHAPTER 2 ROBOT PERCEPTION: SENSORS AND IMAGE PROCESSING

LiDAR

LiDAR (Light Detection and Ranging) sensors emit laser pulses and track how long it takes for the pulses to reach nearby objects and return to the sensor. The LiDAR sensor then determines the distances to objects by calculating the round-trip time, based on the speed of light.

Although some LiDARs also employ visible light, laser pulses primarily emitted by LiDARs are infrared (IR) light. Depending on the particular sensor, laser pulses are either released in brief bursts or as continuous beams. Time-of-flight (ToF) is used here along with high-speed electronics and exact timing mechanisms to measure the round-trip time of light more precisely.

An overview of the LiDAR reflection process is shown in Figure 2-7.

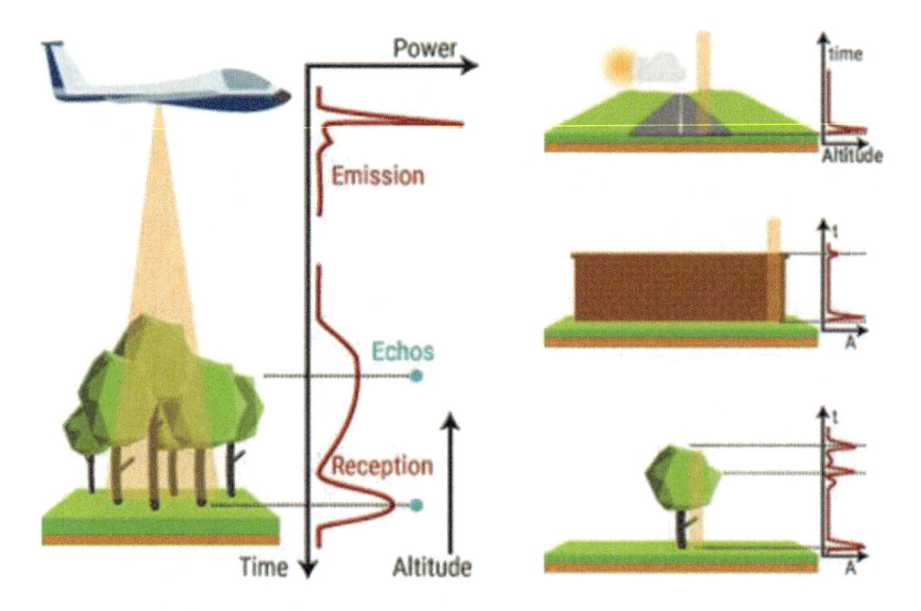

Figure 2-7. Overview of LiDAR sensor. The LiDAR sensor receives the laser pulses as they are reflected back when they strike objects. A receiver on the LiDAR sensor picks up the returning laser pulses and calculates the appropriate time of flight. Used with permission, source: `https://www.yellowscan.com/knowledge/how-does-lidar-work/[43]`

CHAPTER 2   ROBOT PERCEPTION: SENSORS AND IMAGE PROCESSING

LiDAR sensors provide a point cloud image of the surroundings by integrating the distance readings from several laser pulses. Each point in the point cloud represents a distinct position in 3D space, together with the relevant distance details. Since LiDAR sensors provide point clouds as their raw data output, additional processing is necessary to build high-resolution 3D maps, eliminate outliers, segment objects, and filter out noise.

A high-density LiDAR is one where more beams are emitted synchronously. A wide LiDAR is one where the opening angle of the beam is larger, such that if it is mounted on top of a self-driving car, it can get coverage even around the car itself. This is useful in situations such as when riders get on and off, to determine whether the curb is clear, and so on.

Most LiDAR sensors use one of these popular scanning techniques:

- **Mechanical scanning:** Mechanical scanning LiDARs employ a *spinning* mirror or prism that directs laser pulses in various directions, enabling the LiDAR sensor to record a 360-degree image of its surroundings.

- **Solid-state scanning:** Solid-state LiDARs do not use moving elements, but steer laser pulses at various angles by using electronically controlled solid-state emitters and receivers such as phased arrays or optical phased arrays. Due to the lack of moving parts, they're more energy efficient, more durable, and tend to be smaller. The downside is that there is a lower range and field of view.

- **Hybrid solid-state LiDAR:** This technique combines elements of mechanical and solid-state LiDAR, often using a rotating mirror to achieve a wider field of view while employing solid-state components for beam steering within that field of view. It offers a compromise between the wide field of view of mechanical LiDARs and the durability and compact size of solid-state LiDARs.

47

CHAPTER 2　ROBOT PERCEPTION: SENSORS AND IMAGE PROCESSING

2D LiDARs operate on a single horizontal plane and emit beams in a fan-like pattern. They're very low resolution but are enough in many simple autonomous robots to do simultaneous localization and mapping, such as in indoor environments like offices, warehouses, and so on. An example of a 2D LiDAR can be seen in Figure 2-8. In most scenarios, this data is combined with a depth sensor like an Intel RealSense to capture objects, obstacles, and blind spots for the LiDAR.

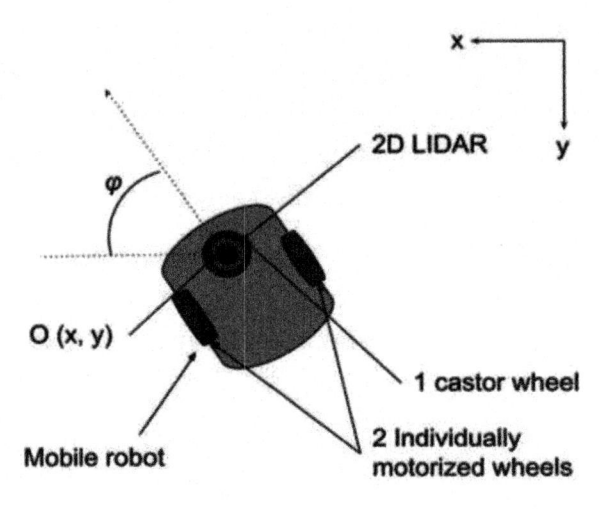

Figure 2-8. Setup of an indoor mobile robot with a 2D LiDAR for navigation and mapping. The robot has two individually motorized wheels and one castor wheel. Its position is tracked as O(x,y) with orientation ϕ, and the 2D LiDAR sensor scans the environment to detect obstacles and surroundings. Used with permission, source: https://www.mdpi.com/1424-8220/23/5/2534[44]

Advanced autonomous robots implement 3D LiDARs for the perception stack due to the benefit of 360-degree coverage and more precise and larger mapping capabilities. The output from 3D LIDARs is shown in Figure 2-9.

CHAPTER 2   ROBOT PERCEPTION: SENSORS AND IMAGE PROCESSING

Figure 2-9. *Output data of a 360 degree, 3D LiDAR. Used with permission, source:* `https://ieeexplore.ieee.org/stamp/stamp.jsp?tp=&arnumber=5980322[45]`

LiDAR sensors provide precise 3D perception, long-range sensing, a 360-degree field of view, resistance to lighting conditions, high data density, and obstacle detection capabilities in robotics applications. However, they can be expensive, use a lot of power, have low resolution for small objects, have problems with shiny or transparent surfaces, work worse than cameras in bad weather like snow/rain, and need large computational resources for processing. LiDAR sensors should be carefully weighed against these advantages and disadvantages to see if they are appropriate for particular robotic jobs and settings.

Ultrasonic Sensors

For tasks involving obstacle identification, collision avoidance, presence detection, distance measuring, and navigation, ultrasonic sensors are frequently used in robotics. They use sound waves to interact with the

CHAPTER 2   ROBOT PERCEPTION: SENSORS AND IMAGE PROCESSING

environment. In particular, ultrasonic sensors produce high-frequency sound waves often inaudible to humans (20 kHz and above). A piezoelectric transducer built into the sensor transforms electrical energy into ultrasonic sound waves.

Once the sound waves are created, they spread outward like a cone-shaped beam. The sound waves go in a straight line until they come into contact with a surface or object. The object's surface characteristics cause the produced sound waves to reflect or bounce back when they collide with it. The ultrasonic sensor's receiver detects these reflected sound waves. The time taken for the sound waves to travel to an object and back to the sensor is measured and used to determine the distance between the sensor and the object using the speed of sound in the medium, usually air. Figure 2-10 shows an overview of this process.

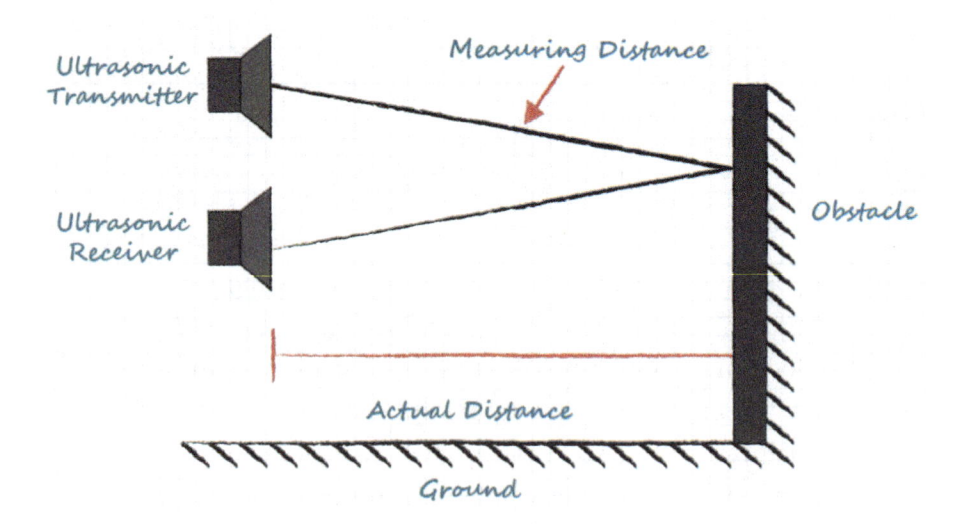

Figure 2-10. Overview of ultrasonic sensors showing how the transmitter emits sound waves that bounce off an obstacle and are received by the receiver, with the distance calculated based on the time taken for the sound to return. Used with permission, source: https://www.cuidevices.com/blog/the-basics-of-ultrasonic-sensors *[46]*

CHAPTER 2   ROBOT PERCEPTION: SENSORS AND IMAGE PROCESSING

The sensor's design, the frequency of the sound waves, and the surrounding environment are only a few of the variables that affect the range and accuracy of ultrasonic sensors. In general, ultrasonic sensors have an accuracy range of a few millimeters to a few centimeters and can detect things within a few centimeters to several meters.

However, there are several restrictions on ultrasonic sensors. For example, they can have trouble detecting items with uneven surfaces or ones that are smaller than the sound waves' wavelength. They can also be impacted by background noise and echoes, impairing how accurately they calculate distance. These are important factors to consider when determining whether ultrasonic sensors are the best for your robotics application.

Inertial Measurement Units (IMUs)

Another sensor worth noting are the IMUs that give robots basic motion-measuring capabilities. IMUs record data on acceleration, angular velocity, and magnetic fields, thereby allowing robots to understand *their own* motion and allow feedback to correct their motions.

The three main sensors that make up an IMU are accelerometers, gyroscopes, and occasionally magnetometers.

- **Accelerometers:** Accelerometers measure linear acceleration along three orthogonal axes, often x, y, and z. They use the inertia principle, which states that acceleration produces an electrical signal when a mass is moved. Accelerometers calculate the object's acceleration by analyzing the electrical output.

- **Gyroscopes:** Gyroscopes calculate the rotational rate or angular velocity around each of the three axes. To detect orientation changes, they rely on the angular momentum principles. Gyroscopes track the Coriolis effect as the item rotates and produce a signal corresponding to the rotation rate.

51

CHAPTER 2 ROBOT PERCEPTION: SENSORS AND IMAGE PROCESSING

- **Magnetometers**: Magnetometers are not always a part of an IMU, but they are occasionally integrated to give information about the object's orientation with respect to the Earth's magnetic field. They can be used to determine the magnetic field's strength and direction.

Figure 2-11 shows accelerometers and gyroscopes with the movement along three axes.

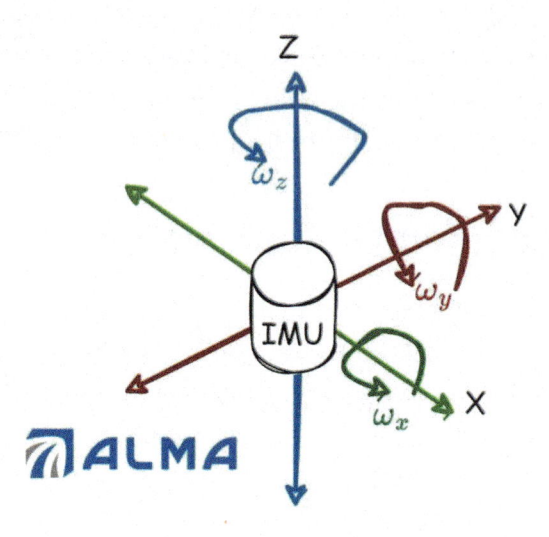

Figure 2-11. The accelerometer and gyroscope within an Inertial Measurement Unit (IMU) detect angular velocity along the X, Y, and Z axes, positioned at 90° to each other. Used with permission, source: https://towardsdatascience.com/what-is-imu-9565e55b44c, created by Dr Barak Or[47]

IMUs use sensor-fusion techniques to understand the object's motion thoroughly. The IMU estimates an object's position, orientation, velocity, and acceleration by combining data from accelerometers, gyroscopes, and magnetometers and frequently utilizing techniques like Kalman filters or complementary filters.

IMUs are essential for tracking and managing robot movements. Thanks to IMUs, which continually measure and update orientation, velocity, and acceleration, robots can retain stability, modify their movements, and react to outside influences. This data is used for balancing humanoid robots, operating robotic arms, and enabling quick navigation in self-driving cars, among other things.

Inertial navigation uses IMUs to determine the position and trajectory of the robot by integrating the data from accelerometers and gyroscopes over time. This method is particularly helpful for navigation inside buildings or in difficult circumstances where other external localization systems, like GPS, are not available or dependable.

IMUs enable accurate motion tracking and control by providing real-time, high-frequency data regarding the robot's orientation, acceleration, and velocity. However, they are prone to accumulating mistakes over time due to sensor drift, which can reduce their accuracy. They are similarly unable to provide information on absolute positions without outside references. Additionally, magnetic field interference can cause problems for the magnetometers inside IMUs, reducing their dependability in some circumstances. In general, IMUs are useful sensors for robotic motion detection and control, but in applications where high accuracy and absolute position data are essential, these limitations should be accounted for.

Problems in Perception

Now that you know the different types of sensors used in robotics and how they work, you need to understand how sensory data can be useful for robotics. The most common type of input data that robots use are images from cameras. Images help robots perceive their environment, such as identifying objects and their locations, and then carry out tasks accordingly. Common perception tasks for robotics include classification, semantic segmentation, instance segmentation, and object detection. These essential tasks are the foundation for how many robots see, reason, and manipulate objects.

Classification

For robots to manipulate objects, they need to have a visual understanding of the object and its surroundings. Image classification allows images to be labeled based on a fixed set of categories. For example, if a kitchen robot is tasked with picking up a cup, it must first identify a cup in an image and differentiate it from other categories of objects such as a glass, plate, or spoon. The robot does this by taking in an array of pixels (the image) and assigning a class to it. Say that you have a dataset of N images, each labeled with one of the K classes. The goal is to use this dataset, pass it through a model, and learn what every one of the classes looks like. You can then evaluate how well your model learned the classes of objects by making predictions on a new set of images.

Segmentation

Imagine an example of an autonomous vehicle trying to navigate the road and understand its surroundings on a busy street with pavement, a car, and a bus in the foreground. In the background, there is a building, a tree, and the sky. Image segmentation aims to assign each pixel in the image that the car sees to the object to which it belongs. It can separate the foreground from the background, identify the precise location of any cars on the road, identify pedestrians, and mark where the road is.

The pixel-level understanding that image segmentation provides can help robots understand how they can navigate their environment. For example, image segmentation can be used to extract an object that you may be grasping from a bin of objects or discover the safe areas to drive for a self-driving car. An encoder-decoder structure is often used for segmentation, as shown in Figure 2-12.

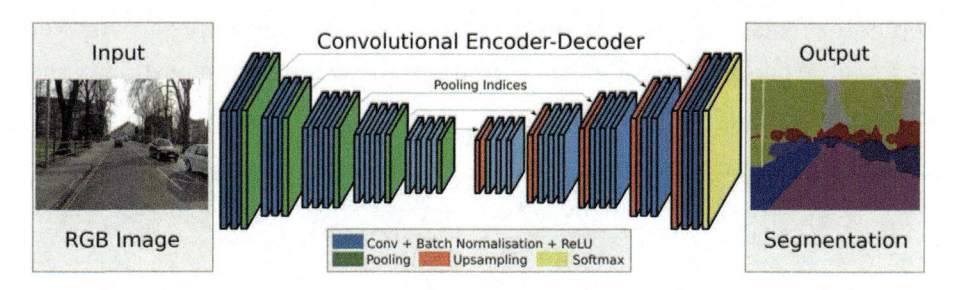

Figure 2-12. *A convolutional encoder-decoder architecture for image segmentation. The RGB input image is processed through layers of convolution, pooling, and upsampling, followed by a softmax function to generate the segmented output. Used with permission, source:* https://arxiv.org/pdf/1511.00561 *[48]*

The CNN section dives more into what each of these layers means. Essentially, the encoder comprises a pre-trained classification network—which means it was already trained on large quantities of data—commonly a ResNet, paired with a decoder network. The encoder downsamples the image to learn a compact representation, and the decoder upsamples this representation to reconstruct the high-resolution pixel space. Using the low-resolution spatial tensor generated by the encoder that encapsulates condensed high-level information, the decoder generates high-resolution segmented outputs.

Simply stacking the encoder and decoder layers results in a loss of low-level information that is important for the model to have in later layers. To make up for the missing information, skip connections are used to allow the decoder to access the low-level features generated by the encoder layers. This allows it to capture both low-level and high-level features in later parts of the model. The primary concept behind skip connections is that the intermediate outputs from the encoder are merged with the inputs to the decoder's intermediate layers at specific points.

There are two major image segmentation types: semantic and instance. With semantic segmentation, all objects that are of the same kind (a person) are marked using one class label, while similar objects get

CHAPTER 2   ROBOT PERCEPTION: SENSORS AND IMAGE PROCESSING

separate labels in instance segmentation (i.e., different people are marked as separate labels). The difference between these two segmentation types is shown in Figure 2-13.

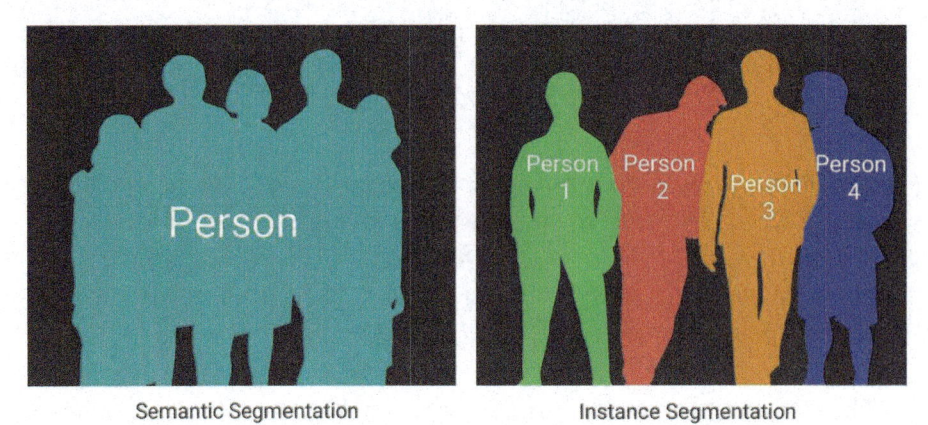

Figure 2-13. Semantic segmentation (left) and instance segmentation (right). Used with permission, source: `https://www.taus.net/resources/blog/introduction-to-image-annotation-for-ml-and-ai`*[49]*

Semantic Segmentation

Semantic segmentation takes in an image and produces a segmentation map where each pixel is assigned a class label, thereby categorizing objects into predefined categories. It aggregates multiple objects of the same category (i.e., different people) into a single entity (humans). For instance, in an autonomous car's street scene, segmentation might classify general categories of objects a car sees, such as pedestrians, bikes, vehicles, and so on. The output segmentation map matches the input image's dimensions (width and height), with channels corresponding to predicted classes. Each channel has a binary mask that labels pixels by class, identifying where each class is in the input image.

CHAPTER 2   ROBOT PERCEPTION: SENSORS AND IMAGE PROCESSING

Once you get the output predictions, how do you know how good the predictions are? A commonly used metric for segmentation is Intersection over Union (IoU), which calculates the percentage of overlap in the predicted image and the ground truth by dividing the number of pixels that are identical in the target and prediction masks by the total number of pixels present in either mask[4]. Figure 2-14 shows an example of this calculation.

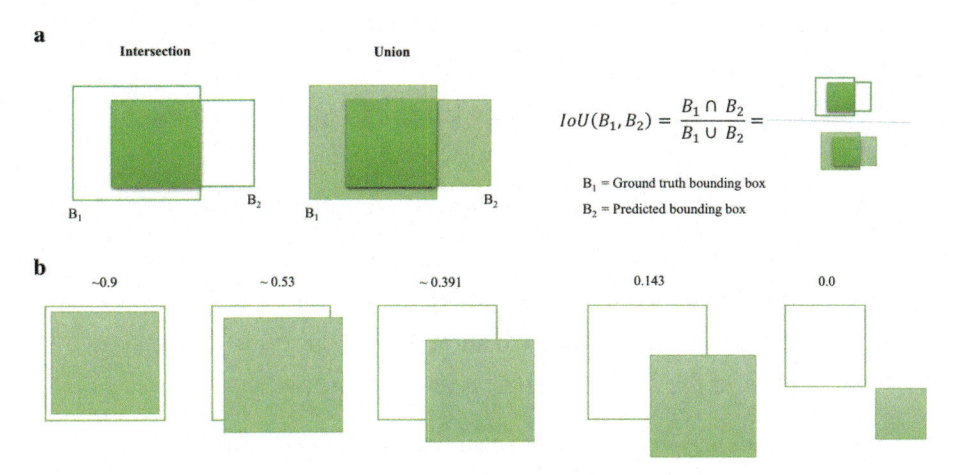

Figure 2-14. Example of Intersection over Union (IoU) calculation for object detection. (a) IoU is calculated as the ratio of the intersection area between the ground truth bounding box and the predicted bounding box to their union. (b) Various IoU values are shown for different degrees of overlap between the predicted and ground truth boxes. Used with permission, source: `https://www.researchgate.net/publication/335127265_AI-powered_banana_diseases_and_pest_detection`*[50]*

Instance Segmentation

Instance segmentation identifies individual objects within categories of objects, such as people, cars, houses, and so on. For instance, categories such as animals might be further segmented into dogs, cats, and birds, and

57

categories like dogs may be segmented into dog 1, dog 2, and so on. This is done by clustering pixels that belong to a single instance of a dog against others. Instance segmentation can be very important for self-driving cars, where you want to have a detailed understanding of your surroundings, such as complex streets with many pedestrians and moving objects like cars.

How do you evaluate predictions you get from an instance segmentation? One way is by using mean average overlap (mAP)[5]. Before looking at mAP, it's useful to explain the concepts of precision and recall. To grasp precision and recall, it can be helpful to break down the confusion matrix, as shown in Table 2-1:

- A **true positive** means that the prediction and target mask pair has an IoU score greater than some threshold you've set (usually 0.5 or more). This means the model has successfully predicted the segmentation or detected the object based on ground truth.

- A **true negative** occurs when the model does not predict a label that it shouldn't (not in ground truth).

- A **false positive** is when the model predicts an object that doesn't exist.

- A **false negative** is when the model fails to identify an object that it should have based on the ground truth.

Table 2-1. *Confusion Matrix Illustrating the Differences Between Precision and Recall. Used with permission, source:* `https://octave-jkh.medium.com/theory-behind-confusion-matrix-bccabd3ad7d7`*[51]*

		Actual	
		Positive	Negative
Predicted	Positive	True Positive	False Positive
	Negative	False Negative	True Negative

Recall measures the number of objects that were correctly identified by taking the number of correctly identified objects and dividing by the number of actual objects. *Precision* measures how many of the identified objects were correct by taking the number of correctly predicted objects and dividing by the total number of predicted objects.

Recall = True positives / # of ground truths

Precision = True positives / # predictions

Average precision builds on top of these concepts by calculating the area under a precision-recall curve to evaluate how well a segmentation or object-detection model performs. As an example, imagine a self-driving car that is detecting cars on a road. Let's discuss this in the context of confidence intervals, where a high confidence interval means that there is a greater certainty that true values lie in that range. At a higher confidence interval, a model might detect only a few cars on the road (high precision, low recall). On the other hand, at a lower confidence interval, more cars may be detected but many of them could be false positives (not actual cars), referring to a lower precision but higher recall.

AP helps you understand the tradeoff that exists between precision and recall in many real-world scenarios. mAP builds on top of average precision (AP), which represents the area beneath the precision-recall curve (PR curve). mAP is simply all the AP values averaged over classes/categories and is useful when you are detecting multiple classes or objects. Harshit Kumar created a good resource[6] if you are interested in diving deeper into these metrics and understanding how they are calculated.

Object Detection

Object detection[7] classifies objects in an image and then specifies where they are in the image using bounding boxes. Each bounding box is characterized by a point, width, height, and associated class labels, sometimes identifying multiple objects within the image. Robots in industrial settings and self-driving cars have to do various tasks such as navigating, picking, and placing, based on their ability to recognize objects. Most commonly, AP and mAP are used as metrics to evaluate different object-detection models.

For example, take a self-driving car, where you want to detect other cars on the highway. The output of the model's prediction is shown in red boxes in Figure 2-15, whereas the green boxes are the ground truth detections. Each prediction is evaluated based on the IoU score, which measures how well the predicted bounding box overlaps with the ground truth bounding box. These IoU scores help determine true positives, false positives, and false negatives, which are then used to calculate precision and recall. By using the precision and recall values at different thresholds, you can calculate AP and mAP across different classes.

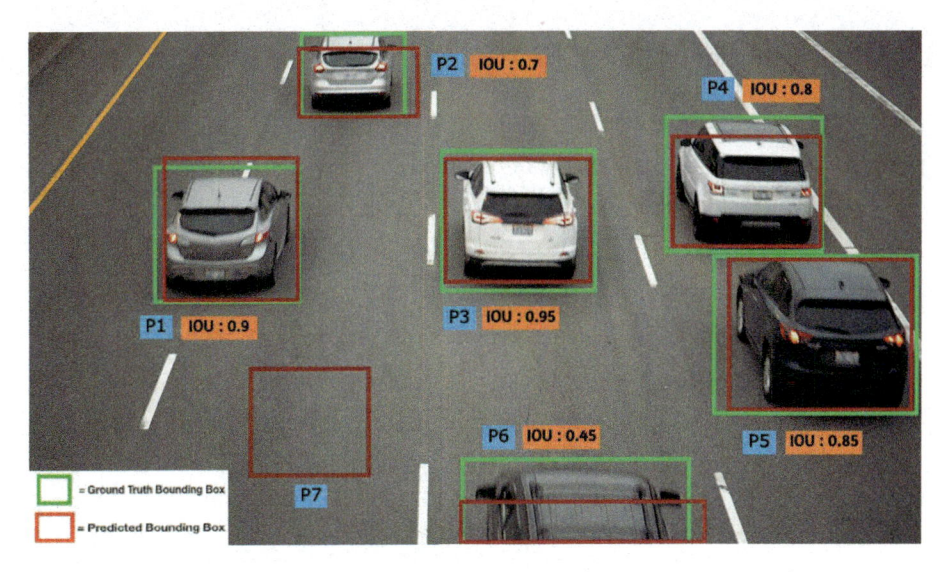

CHAPTER 2   ROBOT PERCEPTION: SENSORS AND IMAGE PROCESSING

Figure 2-15. *Object detection on cars, showing predicted bounding boxes (red) and ground truth bounding boxes (green) with Intersection over Union (IoU) scores for each detected car (P1 to P7). Used with permission, source:* https://www.v7labs.com/blog/mean-average-precision#:~:text=let%27s%20dive%20in!-,What%20is%20Mean%20Average%20Precision%20(mAP)%3F,values%20from%200%20to%201 *[5]*

Convolutional Neural Nets Overview

One approach to image classification, segmentation, and object detection is to use convolutional neural networks (CNNs)[8]. For a deeper dive into how CNNs work, we recommend going through Stanford's CS231n[9] course notes and assignments. This section briefly overviews CNNs before jumping into how CNNs are used for perception in robotics.

CNNs are deep neural networks that can classify and identify specific features from data, such as images collected using sensors on robots. Key parts of a CNN are the convolutional layers, pooling layers, and fully-connected (FC) layers.

61

CHAPTER 2   ROBOT PERCEPTION: SENSORS AND IMAGE PROCESSING

Convolutional Layers

A convolutional layer extracts image features by applying small filters (kernels) over smaller regions of the input data. For example, a 3x3 filter scans over a 32x32x3 input image (height, width, and color channels) and captures patterns like edges and textures within those small sections. As the filter moves across the image, it creates a 30x30x1 feature map (assuming no padding and a stride of 1; we discuss these parameters soon) that learn features like corners and edges at each spatial location. Figure 2-16 shows a filter being used on a two-dimensional input to generate a feature map.

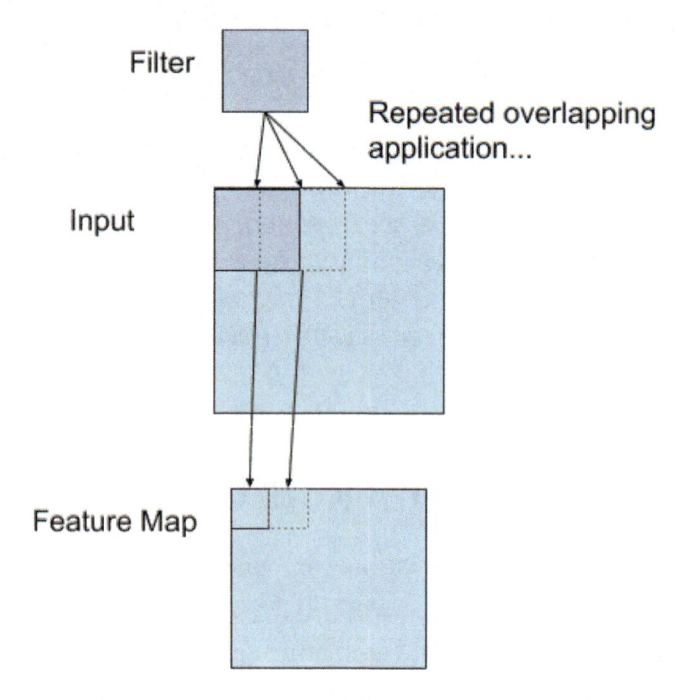

Figure 2-16.   A feature map is generated by applying a filter to an input 2D image. The filter moves across the image, capturing important features like edges and textures at each location to produce the final feature map. Used with permission, source article "Building a Convolutional Neural Network in PyTorch": `https:// machinelearningmastery.com/building-a-convolutional- neural-network-in-pytorch/`*[52]*

CHAPTER 2   ROBOT PERCEPTION: SENSORS AND IMAGE PROCESSING

An image and a filter, both represented as matrices, are essentially multiplied at corresponding values and then summed to produce an output. This output, known as a feature map, is then fed into subsequent layers of the network to help learn features in the input image. Figure 2-17 provides a visual overview of this process.

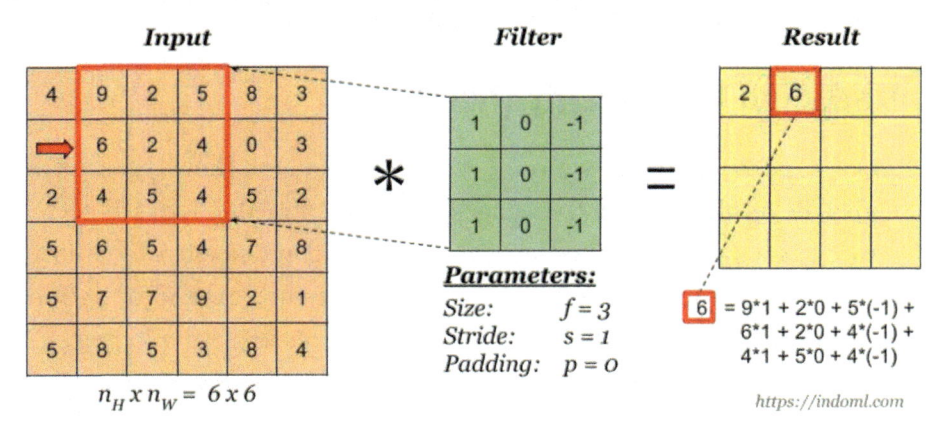

Figure 2-17. *Example of a convolution operation where a 3x3 filter is applied to a 6x6 input image with a stride of 1. Used with permission, source:* `https://www.ijsrp.org/research-paper-1019/ijsrp-p9420.pdf`*[53]*

Three parameters determine the size of the feature map:

- **Depth:** Indicates how many filters are used during the convolution process.

- **Stride:** Indicates how many pixels the filter matrix moves across the input matrix in each step. For example, a stride of 1 means the filter shifts one pixel at a time. Larger strides lead to smaller feature maps.

- **Padding:** Zero-pad the edges of the input matrix so that the filter can be applied to elements at the border of the input image and these features can be captured.

You can differ these values in the filter matrix to generate different feature maps for an input image. By stacking these convolutional layers, they can detect a variety of visual patterns in a hierarchical way. The earlier layers create feature maps that identify simple patterns, edges, and corners. The later layers start to discern more complex objects like animals, furniture, facial features, and landscapes.

Introducing Nonlinearity (ReLU)

After every convolution operation, Rectified Linear Unit (ReLU) is used to increase the nonlinearity in the images. ReLU as a function looks like f(x) = max(x, 0, such that negative elements are set to 0 and positive elements stay the same. Broadly speaking, ReLU is used to manage the information that moves forward through the network. ReLU does this by replacing all negative pixel values in the feature map with 0. An overview of this process is shown in Figure 2-18. Other types of nonlinearity that are used as alternatives for ReLU are Tanh and Sigmoid. In the context of CNNs, ReLU has been found to perform better in most situations.

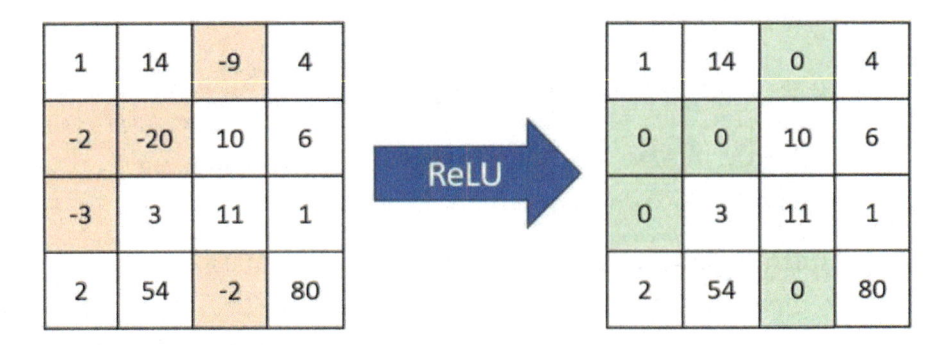

Figure 2-18. Example of the ReLU (Rectified Linear Unit) activation function applied to a matrix. Negative values in the input are replaced with 0, while positive values remain unchanged. ReLU introduces nonlinearity into the model. Used with permission, source: `https://www.linkedin.com/pulse/cnn-activation-functions-global-average-pooling-softmax-n-bhatt/`[54]

Pooling Layers

The pooling layer is often placed after a convolutional layer, with its primary role being to reduce the dimensions of the previously created feature map. Reducing the feature map size cuts down on parameters and computational needs. The network also becomes more robust to minor changes in the input image, such as distortions and shifts, because it extracts the maximum or average values within a specific area. Pooling can be performed in two primary ways: max pooling, where the largest element from the feature map section is selected, and average pooling, which computes the mean of each section of a predetermined size in the image.

Fully Connected Layers

The outputs from the final pooling and convolutional layers are transformed into a flattened vector by unrolling the 3D matrix into a single dimension. This vector typically encapsulates high-level features of the input image, which then serve as inputs to the fully connected (FC) layer. The role of the FC layer is to leverage these features to classify the input image into various categories, as determined by the training data. FC layers are generally placed just before the output layer.

CNNs for Perception

This section explains a simple example of a cleaning robot that uses a CNN to detect objects on a table and then clean the table. As shown in Figure 2-19, each object in the image has been located and identified with a certain level of accuracy.

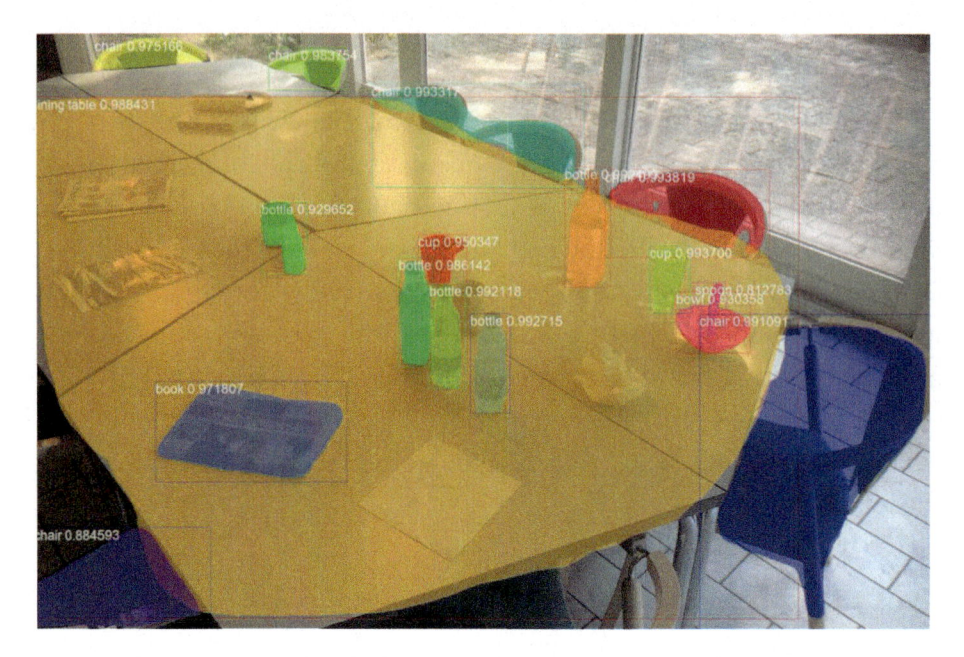

CHAPTER 2   ROBOT PERCEPTION: SENSORS AND IMAGE PROCESSING

Figure 2-19. *Example of a CNN detecting various objects on a table. Used with permission, source:* `https://www.ubiqisense.com/news/robot-to-clean-canteens`*[55]*

One general way to solve this problem using a CNN is as follows:

1. Take an image of the table with the items we want to clean/organize using a sensor.

2. Input this image into the CNN.

3. Divide the image into various smaller sections, known as regions or patches, and treat each region as a separate image.

4. Pass all the regions to the CNN and classify the contents of each region into various classes.

5. After assigning each divided region to its respective class, all these regions can be merged to re-create the original image, now annotated with the identified objects.

Although this method might initially seem effective, it faces challenges due to objects in images having varying aspect ratios and locations. For instance, some objects might fill most of the image, while others appear as small parts. This variability means a significant number of regions would be needed, greatly increasing computational demands. To address this, region-based CNN (R-CNN)[10] can be used, which employs a proposal method to select fewer, but more relevant, regions. The next section explains how R-CNN works to streamline this process.

R-CNN

Rather than examining many regions, R-CNN[10] suggests potential regions within an image to determine if they contain objects. This method uses selective search, which recursively groups similar regions based on color, texture, and size, or edge boxes, which generate object bounding boxes based on edges in the image to identify regions of interest (RoI). Features from each RoI are then extracted using a pre-trained CNN. Following this, a classifier is used to classify objects for each RoI.

One significant limitation of the R-CNN model is that it's slow and computationally intensive. It generates and processes thousands (~2,000) of distinct regions per image. For applications like a cleaning robot, having the R-CNN take tens of seconds to analyze a new image would be impractically slow for real-time implementation.

Thankfully, there's a model, called Fast R-CNN, that aims to solve some of the problems with the R-CNN model.

CHAPTER 2   ROBOT PERCEPTION: SENSORS AND IMAGE PROCESSING

Fast R-CNN

As mentioned, a key limitation affecting the performance of an R-CNN is the computation that is done for each region proposal. In many real-world scenarios, regions often overlap and independently extracting features for each one can result in redundant computations[11]. For example, in a robotics application where a robot navigates a cluttered environment, multiple overlapping regions might contain parts of various objects like boxes, furniture, people, and so on. Fast R-CNN[12] addresses the inefficiencies of R-CNN by processing the entire image at once for feature extraction, instead of handling individual region proposals separately.

Here's how it works:

- The entire image is input into a CNN, generating convolutional feature maps. You can pass in a 512x512 image from the robot's camera, which then produces feature maps with a lower spatial resolution. These capture key information from the images, like edges, patterns, and shapes, that are needed for object detection or segmentation.

- These maps are used to identify RoIs, which are areas where objects are likely present. Say the network finds ROIs that contain different boxes, furniture items, and so on. Each RoI is then resized through an RoI pooling layer into a fixed size that can be passed into a fully connected network. These feature maps are then fed into a fully connected neural network where the objects in each RoI are classified and bounding box regression is used to refine location of the bounding boxes.

- By using the same feature map for all object proposals, Fast R-CNN significantly improves the efficiency of object detection, unlike the original R-CNN, which processes each region independently.

Although Fast R-CNN performs better than R-CNN, it still relies on selective search to find the RoI, which is a slow and time-consuming process. A cleaning robot would likely be working with a large dataset and would need a model that could detect each item on a table very quickly. To solve this problem, another iteration of R-CNN, called Faster R-CNN[13], was developed.

Faster R-CNN

Faster R-CNN[13] improves the speed of object detection compared to Fast R-CNN. The key feature distinguishing Faster R-CNN from other models is its inclusion of a Region Proposal Network (RPN). Placed after the final convolutional layer, the RPN generates object proposals directly, eliminating the need for selective search. It slides a window, such as a 3x3 convolutional layer, across the feature maps from the CNN, creating k anchor boxes of varying shapes and sizes at each position. The RPN then estimates the likelihood of each anchor containing an object and refining coordinates to better fit the detected object. An overview of how RPN works is shown in Figure 2-20.

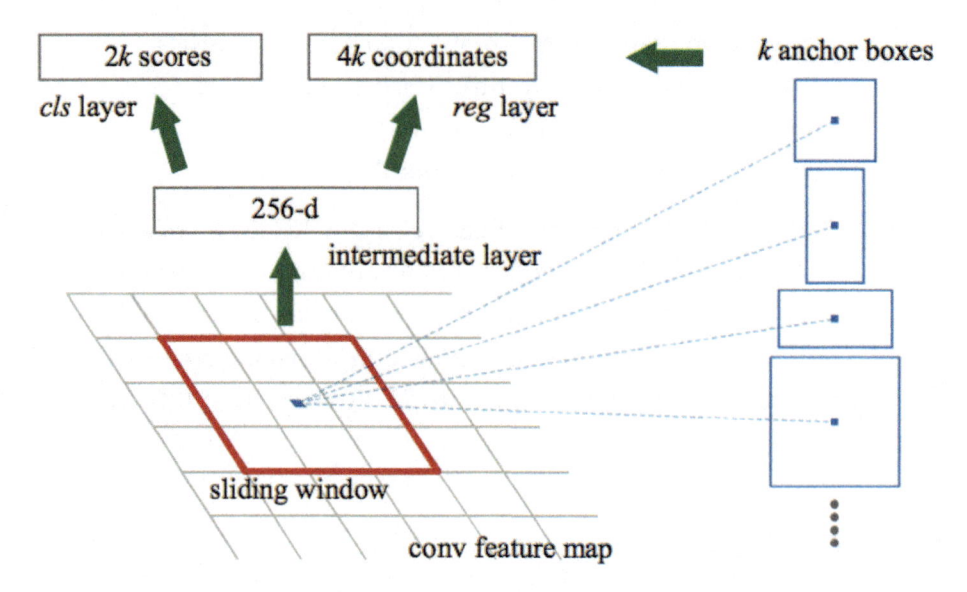

CHAPTER 2   ROBOT PERCEPTION: SENSORS AND IMAGE PROCESSING

Figure 2-20. *Process for how RPN works. Used with permission, source: https://arxiv.org/abs/1506.01497[13]*

Overall, Faster R-CNN is approximately ten times faster than Fast R-CNN while maintaining similar accuracy on datasets like VOC-2007[14]. This makes Faster R-CNN a preferred algorithm for object detection. Table 2-2 provides a quick comparison of these versions of R-CNN.

Table 2-2. *Comparison of R-CNN, Fast R-CNN, and Faster R-CNN on Speed and Test Time Per Image. Used with permission, source: https://cv-tricks.com/object-detection/faster-r-cnn-yolo-ssd/[14]*

	R-CNN	Fast R-CNN	Faster R-CNN
Test Time per Image	50 Seconds	2 Seconds	0.2 Seconds
Speed Up	1x	25x	250x

Mask R-CNN

Mask R-CNN[15] is an extension of the R-CNN models that allows for pixel-level classification, making it useful for image-segmentation tasks. While other R-CNN models—like R-CNN, Fast R-CNN, and Faster R-CNN—are primarily designed for object detection and bounding box predictions, Mask R-CNN adds a branch for predicting segmentation masks. This means it not only detects objects but also provides a detailed pixel-by-pixel mask for each detected object. It has been applied in various domains, such as object segmentation, distance measurement for robot grasping, and vehicle detection in self-driving cars. This could also be useful for the cleaning robot if you wanted to get more granular and segment out all of the objects it would be cleaning from the image.

Mask R-CNN extends Faster R-CNN by adding a segmentation branch, a fully convolutional network that generates a binary mask for each RoI, which predicts the exact pixels that belong to each object. A CNN is used for generating feature maps from the input image. The RPN then identifies RoIs that are most likely to contain objects. RoI Align is used to match the RoIs from the image with the feature maps. Having precise matching can ensure the exact location and shape of objects is preserved, which can be useful for generating segmentation masks. Each ROI is then processed by three branches. The classification branch identifies the objects in each RoI. The bounding box regression branch is used to refine the coordinates of the bounding box. Finally, a mask branch generates a binary mask for each RoI. An illustration of Mask R-CNN is presented in Figure 2-21.

The loss function used in Mask R-CNN encompasses three elements: classification loss, which assesses the accuracy of the predictions relative to the actual class; bounding box loss, which evaluates the model's localization accuracy; and mask prediction loss, which is determined by calculating the binary cross-entropy between the predicted mask and the ground truth.

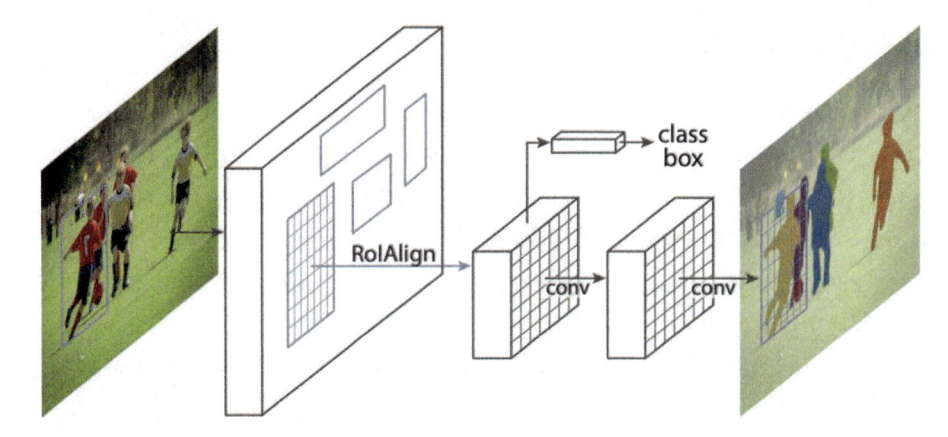

CHAPTER 2   ROBOT PERCEPTION: SENSORS AND IMAGE PROCESSING

Figure 2-21. *Main components of Mask R-CNN. Used with permission, source:* https://arxiv.org/pdf/1703.06870[15]

ResNet

You've read about how it's important to optimize the architecture of these models to make perception tasks faster to compute. But what about increasing the accuracy of these models to perform perception tasks better? For example, if a robot is used in environments like surgeries, it needs to be able to detect or classify objects with extremely high precision.

A key question in deep learning is focused on understanding whether simply stacking more layers will improve model performance. The rationale is that these added layers progressively capture more complex features. However, it has been observed that beyond a certain network depth, accuracy plateaus and then quickly deteriorates. Traditional CNN models have a maximum effective depth, beyond which they suffer from the vanishing gradient problem—during back-propagation, the gradients can become exceedingly small due to repeated multiplication, nearly diminishing them.

CHAPTER 2   ROBOT PERCEPTION: SENSORS AND IMAGE PROCESSING

Residual Neural Networks (ResNet)[16] were proposed by He et al. in 2015 as a breakthrough that solved the vanishing gradient problem and allowed people to build larger CNNs that had improved performance in perception tasks. For this reason, ResNets are used as the backbone today for many computer vision tasks. They are not only limited to image classification but can also solve a wide range of problems in image segmentation and object detection—even in robotics.

Skip Connection: The Strength of ResNet

A key aspect of ResNets lies in the skip connection, which has become a fundamental component in many convolutional architectures. Skip connections offer an alternative route for gradient flow in backpropagation, enabling the output of a layer to feed into subsequent layers. This is done through vector addition, where the output of a layer is added directly to the output of a deeper layer. This addition creates a shortcut for the gradient during backpropagation, allowing it to bypass certain layers. By multiplying the gradient by 1 (identity function), the skip connection ensures that the gradient does not diminish and maintains its original value. ResNets are constructed by stacking these residual blocks that incorporate skip connections. Figure 2-22 illustrates a single residual block with a skip connection.

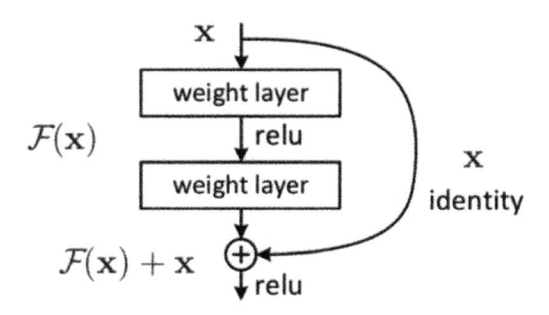

Figure 2-22.   A residual block. Used with permission, source: https://arxiv.org/pdf/1512.03385[16]

U-Net

The concept of skip connections has influenced the design of architectures such as U-Net[17], which was originally developed by Ronneberger et al. for biomedical image segmentation. U-Nets are now being applied to other image segmentation tasks, including robotics perception. The architecture is an encoder-decoder structure with skip connections, known for its efficiency with fewer training images and highly accurate segmentations. U-Nets can segment images up to 512x512 pixels using a modern GPU (according to 2015) in under a second.

As illustrated in Figure 2-23, a distinctive aspect of U-Net is the replacement of pooling operations with upsampling operators, which increase the resolution of the output. The encoder uses convolutional and pooling layers to capture features and reduce the spatial dimensions of the input image, creating a compressed representation of the input image. The decoder then uses upsampling operators to increase the resolution of the output from the encoder, refining the features captured by the encoder into a high-resolution, segmented output. U-Net also incorporates skip connections, which directly transfer feature maps from the encoder to the corresponding layers in the decoder. These skip connections help retain important spatial information that might otherwise be lost during the downsampling process.

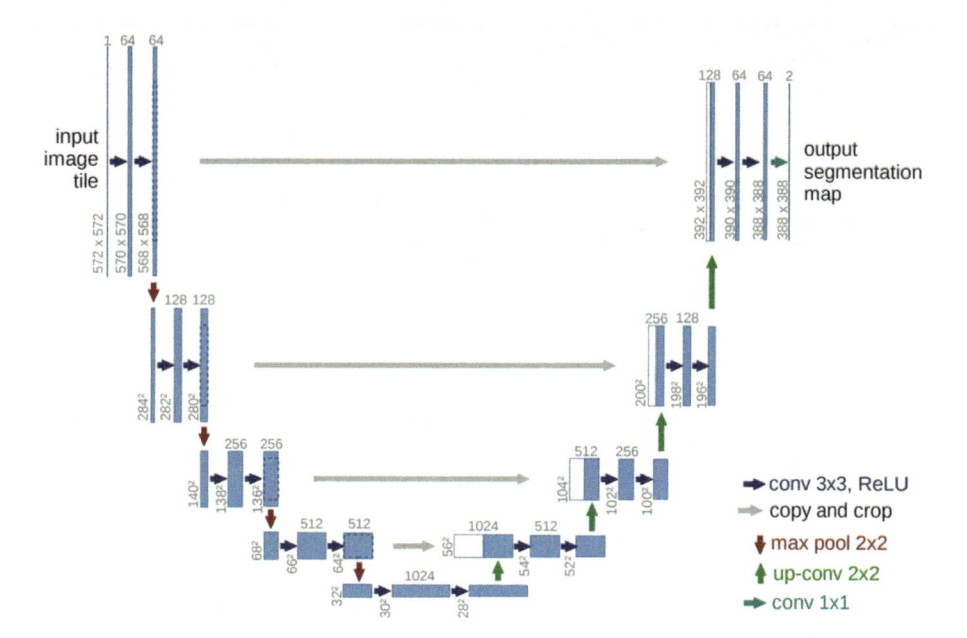

CHAPTER 2   ROBOT PERCEPTION: SENSORS AND IMAGE PROCESSING

Figure 2-23. U-Net architecture. Used with permission, source: https://arxiv.org/pdf/1505.04597 [17]

U-Net has been used commonly in segmentation tasks for robotics given its speed and the ability of the encoder-decoder structure to learn accurate spatial information. For example, it has been used in research for robotic instrument segmentation for robotic surgery, lane detection for self-driving cars, and for robot grasping.

U-Net is notable in that it forms the backbone of several early diffusion models and even Stable Diffusion. An important property of U-Net is that the input and output are the same size, meaning the model maintains the spatial dimensions of the input throughout the network. This design allows U-Net to learn and encode detailed image features (latent encodings) within the encoder, which can then be used by some decoder or another model. Recent architectures use transformers in place of CNNs to make U-Nets that run faster. Despite the change in modules, the fundamental concepts of U-Nets transfer.

CHAPTER 2   ROBOT PERCEPTION: SENSORS AND IMAGE PROCESSING

EfficientNet

One way that CNNs can be scaled up is by adding more layers. Oftentimes, you can add more layers such as to ResNet and then manually make the network deeper and wider, or use higher-resolution images. EfficientNet[18] uses a compound scaling method that adjusts all three dimensions simultaneously based on a predefined scaling factor.

By using a user-defined coefficient and fixed scaling constants, this method ensures that there is proportional growth across all three factors. For example, if you want to double the computational resources, you might increase the depth 1.2 times, the width 1.1 times, and the resolution 1.15 times[18]. These empirically determined parameters help improve the features that CNNs capture by allowing them to learn more complex patterns (depth), capture more granular details (width), and detect smaller objects (resolution). Figure 2-24 shows a comparison of the scaling aspects. They also maintain computational efficiency by holistically optimizing the use of all available resources and features. This approach allows EfficientNet to achieve higher accuracy and efficiency, making the network up to ten times smaller yet faster compared to traditional scaling methods.

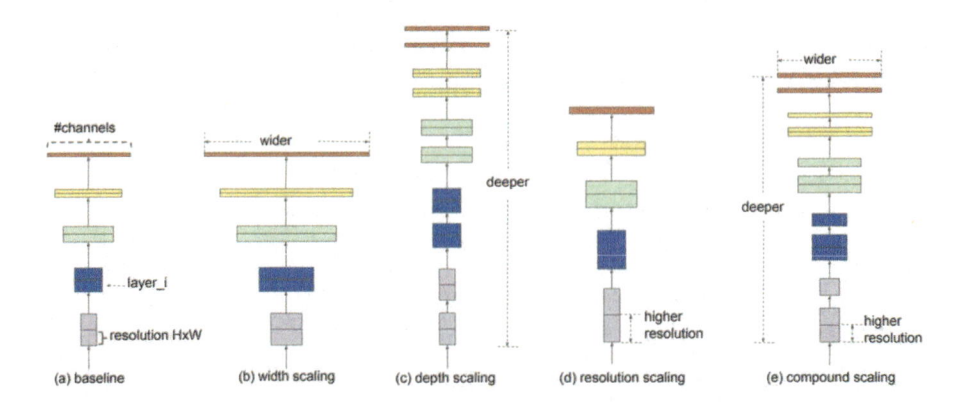

Figure 2-24.   Comparison of scaling methods. Used with permission, source: https://arxiv.org/pdf/1905.11946[18]

One-Stage Detectors

Many of the object-detection algorithms mentioned so far handle object detection as a classification problem by first generating object proposals and then sending these proposals to classification/regression or segmentation heads. What if there was a way to look at the complete image all at once rather than looking at only the generated region proposals? These methods would be able to compute extremely quickly and could be run real time.

A method called You Only Look Once (YOLO)[19] aims to do this by making predictions of bounding boxes and class probabilities simultaneously.

YOLO

YOLO (You Only Look Once) has been commonly applied in robotics for object detection applications due to its simplicity and speed. For example, a more recent version of YOLO, called YOLO v10, achieves an impressive 54.4 percent AP for real-time object detection. YOLO has been used for identifying objects using mobile robots, robotic grasping, and other autonomous navigation applications.

YOLO works by dividing the input image into a grid of cells. Each cell in the grid is responsible for predicting the bounding boxes and determining the confidence scores for objects within that cell. However, this can result in multiple overlapping boxes for the same object. YOLO addresses this issue using Non-Maximal Suppression, which keeps the box with the highest confidence score and removes the redundant overlapping boxes. The YOLO architecture is shown in Figure 2-25.

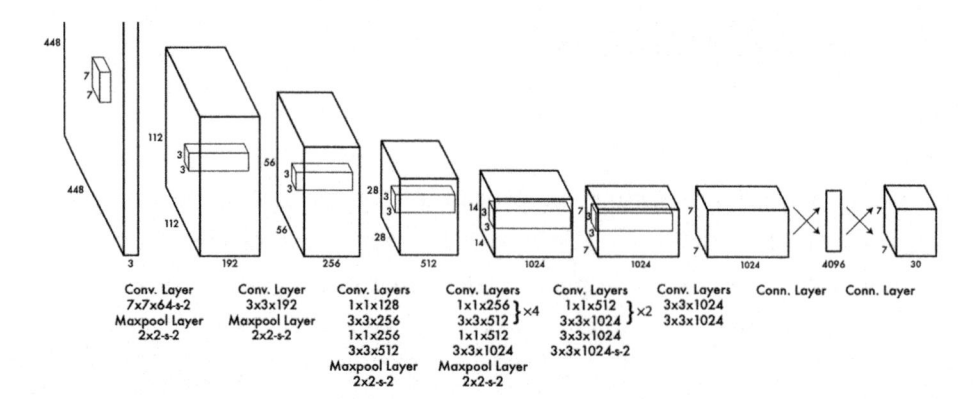

CHAPTER 2 ROBOT PERCEPTION: SENSORS AND IMAGE PROCESSING

Figure 2-25. *YOLO architecture. Used with permission, source: https://arxiv.org/pdf/1506.02640[19]*

SSD

Similar to YOLO, the single-shot detector (SSD)[20] detects objects in a single pass using a pretrained base network (like VGG16) followed by convolutional layers that generate multiple feature maps at different scales. These maps allow SSD to detect objects of different sizes, where early layers capture small objects and deeper layers detect larger ones. SSD uses receptive fields, which are specific regions of the input image that each feature map covers, to handle objects of different sizes. Similar to YOLO, SSD applies Non-Maximal Suppression to remove redundant bounding boxes and keep the most confident predictions.

A visual comparison of the SSD and YOLO architectures is provided in Figure 2-26.

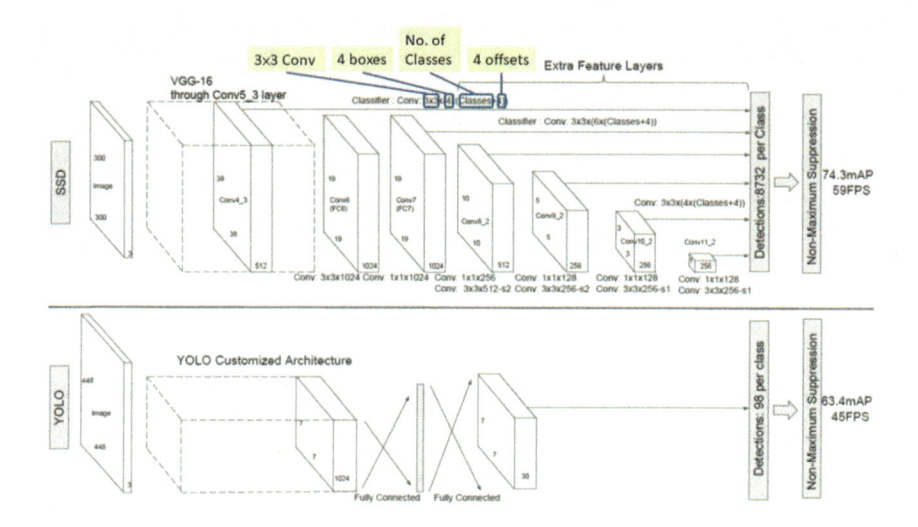

CHAPTER 2   ROBOT PERCEPTION: SENSORS AND IMAGE PROCESSING

Figure 2-26. *SSD (top) and YOLO (bottom) architecture. Used with permission, source:* https://towardsdatascience.com/review-ssd-single-shot-detector-object-detection-851a94607d11 *[56]*

YOLO and SSD are both known to be fast. However, SSD is known to provide better accuracy for objects of varying sizes because of its multiscale feature maps and receptive fields. A comparison of the performance of SSD, YOLO, and Faster R-CNN is shown in Figure 2-27.

Model Comparison

Choosing the right model is crucial for robotics and depends on the problem you are trying to solve and how you set it up. The results of Faster R-CNN, YOLO, and SSD based on mAP and FPS on Pascal VOC2007 are shown[20] in Figure 2-27. The input images from these datasets are tested with different resolutions to compare results. It is important to note that newer versions of YOLO (YOLO v9, v10), which have an improved architecture, have much higher mAP and FPS, but we've provided results from this study as a benchmark.

CHAPTER 2   ROBOT PERCEPTION: SENSORS AND IMAGE PROCESSING

Method	mAP	FPS	batch size	# Boxes	Input resolution
Faster R-CNN (VGG16)	73.2	7	1	~ 6000	~ 1000 × 600
Fast YOLO	52.7	155	1	98	448 × 448
YOLO (VGG16)	66.4	21	1	98	448 × 448
SSD300	74.3	46	1	8732	300 × 300
SSD512	76.8	19	1	24564	512 × 512
SSD300	74.3	59	8	8732	300 × 300
SSD512	76.8	22	8	24564	512 × 512

Figure 2-27. Comparison of Faster R-CNN, YOLO, and SSD based on average precision (mAP) and FPS, on the Pascal VOC2007 dataset. Used with permission, source: `https://arxiv.org/pdf/1512.02325`*[20]*

In another example, a paper[21] by Google Research looked at various object-detection models and the tradeoffs between speed (GPU time) and accuracy (mAP). They generally found that R-FCN and SSD models are faster, as they require less GPU time per image, but they tend to have lower accuracy (mAP). On the other hand, Faster R-CNN models tend to be slower but achieve higher accuracy. For Faster R-CNN models, the speed can also be improved by reducing the number of regions proposed during detection. It's important to note that different combinations of feature extractors (e.g., Inception Resnet V2, MobileNet) and architectures (e.g., Faster R-CNN, SSD, and so on) result in different performance benefits, as shown in Figure 2-28. This tradeoff will differ depending on the problem you are working on and your dataset. This tradeoff is essential to consider, especially with real-time detection applications in robotics.

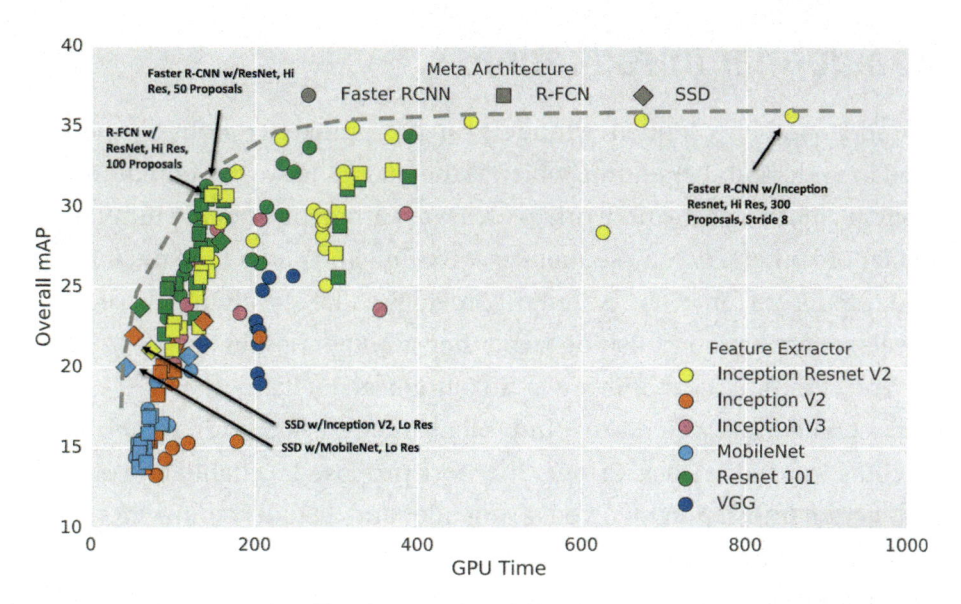

CHAPTER 2   ROBOT PERCEPTION: SENSORS AND IMAGE PROCESSING

Figure 2-28. *Tradeoffs of various object-detection models when it comes to GPU time vs mAP. Used with permission, source: https://arxiv.org/pdf/1611.10012 [57]*

Transformers for Perception

The previous section covered CNN-based vision techniques. However, in recent times, transformer-based[23] vision techniques have provided a strong alternative to solving vision problems discussed in this chapter. Transformers, as opposed to CNNs, rely on self-attention mechanisms to model global connections among visual elements, thereby improving the model's understanding of contextual information. In robotics, applications like object detection, image segmentation, and visual reasoning—where global context and fine-grained interactions are essential for precise decision-making—are ideally suited to transformers.

CHAPTER 2   ROBOT PERCEPTION: SENSORS AND IMAGE PROCESSING

Transformer Introduction

To understand how sequence models can be applied to robotics, you first need to understand attention, which is the neural network element that explicitly motivates the network to focus on certain parts of the input data and ignore others. In long sequences, unrolling the net in time means older inputs are forgotten. Attention provides a direct path to older inputs, thereby reducing the vanishing/exploding gradient problem.

The attention mechanism was first proposed by Bahdanau et al. [22] as embedding weights to jointly align and translate during a neural machine translation task. Prior to the work proposed by Bahdanau et al. [22], neural translation involved an encoder and decoder setup where an encoder converted inputs to an embedding representation and a decoder converted the embedding back into tokens in a target language. In Bahdanau et al.[22], a simple feedforward network was used to calculate alignment scores between the input and output tokens. This score was used to weigh the context vector of the RNN decoder. This would later become known as *additive attention* since feedforward applies a linear, additive function on inputs.

$$c_i = \sum_{j=1}^{T_x} \alpha_{ij} h_j.$$

Equation 2-1

In Equation 2-1, a_ij is a learned weight given to the jth input to calculate the ith output. This mechanism was soft attention. Since then, several attention mechanisms have been proposed:

- **Content-based attention**[35]: This method uses the cosine similarity between the target state and the source hidden state.

- **Additive attention**[22]: Using a trainable weight vector, followed by a tanh activation over the combination or concatenation of the target state and the source hidden state.

- **General attention**[36]: A trainable weight matrix is applied to the source hidden state and the dot product is taken with the target state.

- **Dot-product attention**[36]: The dot product of the target state and the source hidden state is computed.

- **Scaled dot-product attention**[23]: Similar to dot-product attention, but the score is scaled by the square root of the dimension n of the source hidden state.

- **Cross-attention**[38]: The attention scores are computed as the dot product between the queries Q (from one input sequence) and the keys K (from a different input sequence), scaled by the square root of the dimension of the keys divided by the number of heads, h. After applying the softmax function to these scores, they are used to weigh the values V. This technique can be used for conditioning on multiple input streams.

- **Flash attention**[37]: For Flash attention, tiling breaks the large attention matrix into smaller blocks that fit within fast memory (SRAM). Each block is processed independently, keeping intermediate data local to avoid frequent reads/writes to slower memory (HBM). This reduces memory overhead and speeds up computation, while making sure the correct softmax output is computed across the sequence.

CHAPTER 2   ROBOT PERCEPTION: SENSORS AND IMAGE PROCESSING

The Transformer

An important neural network architecture that bases itself on the ability to utilize attention over long sequences is the transformer. Introduced in Vaswani et al.[23], titled "Attention Is All You Need," transformers have changed the terrain of deep learning by providing better than human performance in speech and vision. The best language models in the world, including GPT-3[24] and PaLM[25], are transformers.

Outside of being applied to tasks such as machine translation, text generation, and language understanding, transformers can be utilized for object detection and tracking in robotic vision systems. Using self-attention mechanisms, transformers can capture global dependencies and spatial relationships between different regions, leading to improved object recognition and tracking capabilities. Other use cases in robotics include path planning and navigation tasks in robotics where transformers can be used to capture long-range dependencies and use contextual information, improving the robot's ability to navigate complex and dynamic environments.

Vaswani et al.[23] categorizes an attention network generically as mapping a set of queries and key-value pairs to outputs where they are all vectors. Figure 2-29 shows that the output is a weighted sum of all the values, and the weight is calculated as a dot product of the query and key vectors.

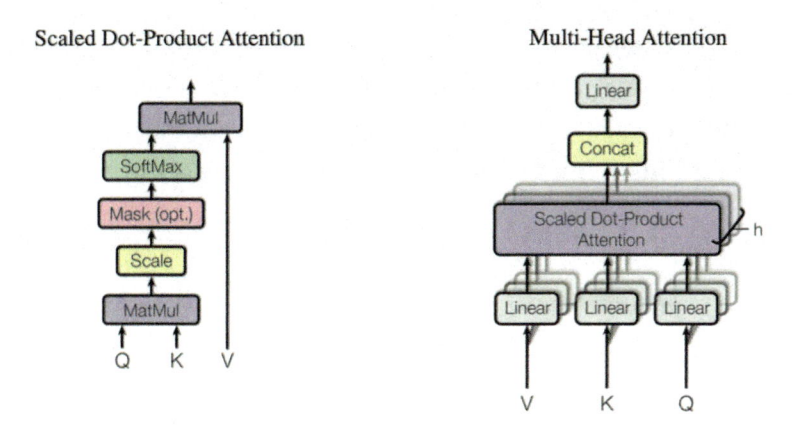

Figure 2-29. *Scaled dot product attention and multi-head attention. Used with permission, source:* https://papers.nips.cc/paper_files/paper/2017/hash/3f5ee243547dee91fbd053c1c4a845aa-Abstract.html *[23]*

In scaled dot product attention, the attention is calculated using Equation 2-2:

$$\text{Attention}(Q, K, V) = \text{softmax}(\frac{QK^T}{\sqrt{d_k}})V$$

Equation 2-2

where Q is a query, K is the key, V are value vectors, and d_k is the dimension of the key vector. Compared to additive attention introduced in Bahdanau et al.[22], dot product attention is faster to compute and space-efficient. In scaled dot product attention, the 1/ sqrt(d_k) keeps the input value of the softmax value normalized, since for large values of d_k, the dot product is pushed to areas where the softmax function has small gradients.

The original transformer paper uses both scaled dot product attention and multi-head attention. Multi-head attention is used to project query, key values into multiple heads/representation subspaces to apply scaled dot product attention over them. In multi-headed attention, you

CHAPTER 2 ROBOT PERCEPTION: SENSORS AND IMAGE PROCESSING

parallelize by first projecting values, keys, and queries to the d_v, d_k, and d_q dimensional arrays, respectively, then applying scaled dot product attention, as shown in Equation 2-3. This allows the model to attend to different information from different embedding subspaces jointly.

$$\text{MultiHead}(Q, K, V) = \text{Concat}(\text{head}_1, ..., \text{head}_h)W^O$$
$$\text{where head}_i = \text{Attention}(QW_i^Q, KW_i^K, VW_i^V)$$

Equation 2-3

The transformer (as shown in Figure 2-30) employs multi-headed attention with an encoder-decoder setup in three places:

1. In the encoder as self-attention where the keys, values, and queries come from the encoder, such that the encoder can attend to all its positions.

2. Between the encoder and decoder, where keys and values come from the encoder and queries come from the previous output of the decoder.

3. In the decoder as self-attention, where the keys, values, and queries come from the decoder, such that the decoder can attend to all its positions.

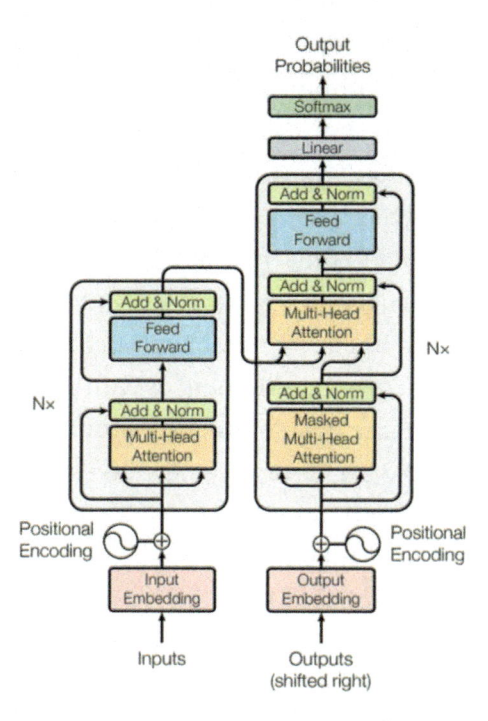

CHAPTER 2 ROBOT PERCEPTION: SENSORS AND IMAGE PROCESSING

Figure 2-30. *The transformer. Used with permission, source:*
https://papers.nips.cc/paper_files/paper/2017/hash/3f5
ee243547dee91fbd053c1c4a845aa-Abstract.html[23]

Transformers for Vision

This section looks at methods that allow you to apply transformers to perception problems mentioned in this chapter, namely classification, detection, segmentation, and captioning.

Image Classification with Vision Transformer (ViT)

One of the first applications of transformers to computer vision that could be applied for robotic perception was published by Dosovitskiy et al.[26] in the form of the vision transformer (ViT). By utilizing its ability to interpret visual data and capture contextual relationships, the vision transformer

CHAPTER 2   ROBOT PERCEPTION: SENSORS AND IMAGE PROCESSING

(ViT) can be used for image classification. The ViT is useful when you need a thorough grasp of the environment because of its attention mechanism, which enables it to model long-range dependencies and capture fine-grained information. A robot's perception abilities can be improved by modifying the ViT architecture and training it on domain-specific data. This allows the robot to move around, interact, and make better judgements based on visual input.

The ViT tokenizes an image into patches. Then, it feeds it into a transformer as if the patches were sequences with a position embedding corresponding to the patch's position in the picture. The architecture of this is shown in Figure 2-31. This architecture, in essence, unifies natural language research and computer vision research, because you can now treat images as language by tokenizing an image into a sequence of patch tokens, just like how a sentence/document is a sequence of language tokens. Prior to this paper, convolutional neural nets dominated vision processing and transformers dominated language processing, but ViT set the foundations for transformers dominating vision and eventually visual-language or multimodal processing.

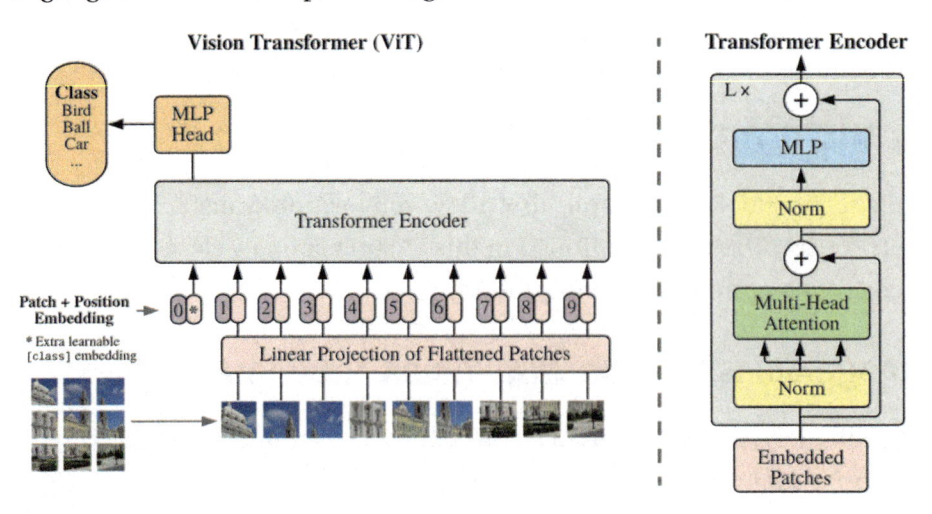

Figure 2-31. Vision transformer architecture. Used with permission, source: https://arxiv.org/abs/2010.11929 [26]

While breaking a picture up into multiple patches seems counterintuitive to the purpose of retaining geometric correlation across patches, it was found that when the model is pretrained on a very large dataset (> 100M images) and then fine-tuned to a classification task, it learned the relationships between the position embeddings and could extract features across patches. With smaller dataset sizes, a ResNet-based model was still dominant given that convolution neural nets preserve inductive biases about translational equivariance and locality. A closer inspection found that the initial layers learn to attend to features in the patches, preserving its low dimensional structure. Once position embedding is added, there are similarities in the embeddings between close patches and those in the same row/column, meaning that the network learns the larger geometrical context of the image. From that perspective, the attention distance of the ViT is comparable to the receptive field of ResNets.

The ViT handles images of higher resolution by increasing the sequence length and 2D interpolation of pretrained position embeddings corresponding to the new patch positions.

The paper also found that ViTs outperform ResNets on compute vs performance, requiring two to four times less memory for the same performance. Interestingly, the model didn't saturate, indicating the possibility of higher performance via scaling.

Scaling Vision Transformers

Although ViTs are state of the art for many computer vision tasks, scaling them like regular dense transformers had proven to be infeasible due to training instabilities. The paper, "Scaling Vision Transformers to 22 Billion Parameters,"[27] dives into why the traditional method of training ViTs produces instability during scaling and how to modify the architecture to prevent it. As a result of these investigations and modifications, they introduce the ViT 22B model, composed of 22 billion parameters, which

CHAPTER 2　ROBOT PERCEPTION: SENSORS AND IMAGE PROCESSING

at the time was 5.5x larger than the previous vision backbone ViT-e of four billion parameters. ViT 22B deeply inspired PaLM-e[28], which combines ViT and a large language model to create an early multimodal robot foundation model. Chapter 4 discusses this more in depth.

The authors apply three main modifications to traditional ViT to enable scaling:

1. **Parallel layers:** ViT 22B applies attention and MLP blocks in parallel, rather than sequentially as in a traditional transformer. This enables grouping some operations such as parallelization of linear projections of the MLP and attention blocks. This technique results in a 15 percent increase in speed during training without affecting performance.

$$y' = \text{LayerNorm}(x),$$
$$y = x + \text{MLP}(y') + \text{Attention}(y').$$

2. **QK normalization:** At around 8B parameters, it was noticed that attention logits become really large values and exhibit very low entropy, leading to a divergence in training loss that makes training unstable. Applying a LayerNorm to queries and keys before scaled dot product attention fixes this issue, as shown in Figure 2-32. Attention weights are then calculated as

$$\text{softmax}\left[\tfrac{1}{\sqrt{d}}\text{LN}(XW^Q)(\text{LN}(XW^K))^T\right],$$

where X is input, and W^Q and W^K correspond to query and key weight matrices, respectively.

CHAPTER 2   ROBOT PERCEPTION: SENSORS AND IMAGE PROCESSING

3. **Omitting bias terms:** Removing the bias from QKV projections and LayerNorms was first found useful in the PaLM 540B paper[25] and was repeated in the ViT-22B paper as well, since it improves accelerator utilization without compromising performance. Contrary to PaLM 540B, ViT 22B preserves bias terms in MLP dense layers because it was found to not harm utilization.

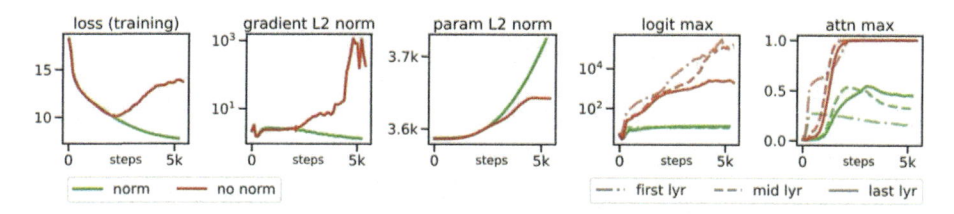

Figure 2-32. Improvement in training metrics due to query/key normalization (green curve). Used with permission, source: https://arxiv.org/pdf/2302.05442[27]

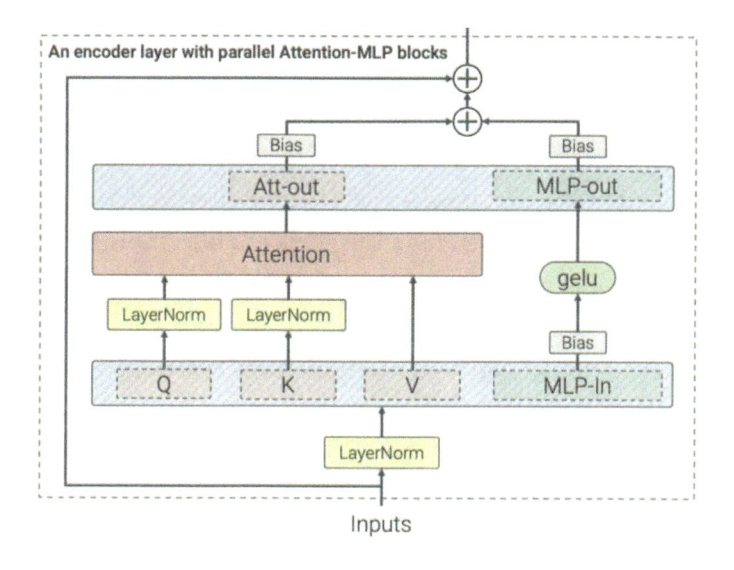

Figure 2-33. Parallel ViT-22B layer with QK normalization. Used with permission, source: https://arxiv.org/pdf/2302.05442[27]

CHAPTER 2   ROBOT PERCEPTION: SENSORS AND IMAGE PROCESSING

Figure 2-33 shows the modified ViT encoder with the modifications applied in ViT-22B. It is trained on an extended JFT dataset comprising around 4B images. The practical implementation of ViT 22B was done on JAX with model and data parallelism; it incorporates measures to optimize for throughput by improving computation (per device) and communication (between devices). ViT 22B approaches SoTA results on several perception tasks, including classification, semantic segmentation, and monocular depth estimation.

Learning Joint Image-Language Features

Applying language with computer vision such as using CLIP, introduced by Radford et al.[29], allows you to use a ViT-like transformer to get richer visual features. The idea of CLIP is to jointly learn embeddings from a large corpus of image-text pairs in order to use zero-shot image classification. Instead of training an image encoder and classifier, you jointly train an image encoder and text encoder, which are fed batches of (image, text) pairs with a label indicating whether they match. There are a few strong reasons for doing this.

It removes the requirement of labeling the data in the format intended for classification tasks. Even ImageNet only labels 1,000 classes, which is far fewer than object descriptors required for generalized vision. Compared to it, natural language supervision trains on descriptive text, without explicit and formatted labeling. This enables training from the large number of image/text descriptions on the Internet, compared to previous benchmarks, as well learning attributes of the image such as object types, aesthetics, style, and other features that people are likely to write online about.

Contrastive objectives are also found to be better than predictive objectives at learning representations. In a contrastive setting, the model is trained to predict which text is paired with which image rather than predicting the exact words of that text. This is because the jointly trained image encoder learns the representations required to describe the same image in many ways rather than using a singular label. An overview of CLIP is shown in Figure 2-34.

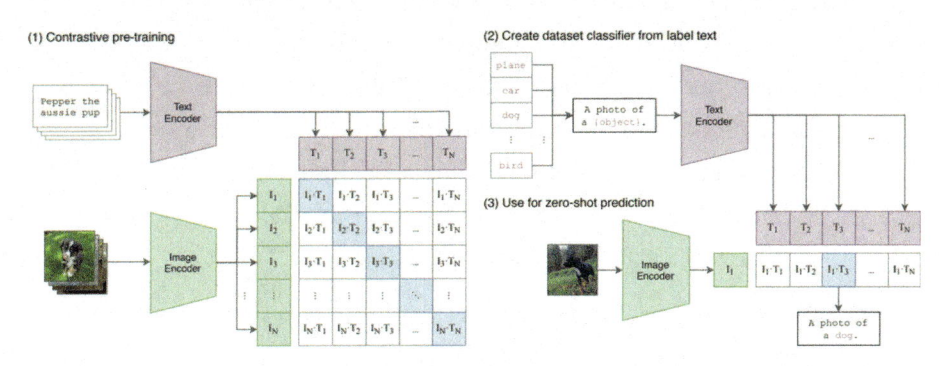

Figure 2-34. Summary of CLIP. Used with permission, source: https://arxiv.org/pdf/2103.00020v1[29]

The image encoder backbone could be a standard ResNet or the vision transformer and the text encoder could be another text transformer such as in Vaswani et al.[23]

Robotic perception can be improved by richer image understanding made possible by CLIP features. Language allows more flexible querying of objects than is allowed by pure classification/detection methods. CLIP's skill in correlating image and text features can help robots make inferences based on language instruction from a human to relevant object/image features in the environment to complete challenging visual language tasks. Chapter 4 discusses the applications of language in robotics in better depth.

Open Vocabulary Object Detection with Transformers

While the original ViT models perform detection on ImageNet benchmarks[30], OWL-ViT[31] combines the techniques of ViT and CLIP to perform open vocabulary object detection. They train a large CLIP network on billions of Internet-scale, paired image-text data and fine tune it to a smaller detection dataset with millions of examples.

How does OWL-ViT modify CLIP?

1. OWL-ViT removes the token pooling and projection layer of CLIP and instead projects the output token embeddings to get per object detection labels. For each object query, a probability of how much the query relates to the image and a bounding box is predicted. At this point, closed class object detection becomes a special case, where every class label is applied as a query to every image.

2. While early fusion is generally useful to extract the right image features, early fusion can make things slower in OWL-ViT, since you would need to parse through the entire image for every query. Instead, the queries are independently passed through a text encoder, allowing the use of thousands of queries per image and boosting inference efficiency.

3. Since language and image are fused late, you could also perform object detection conditioned on image embeddings instead of text embeddings, enabling image-conditioned object detection.

4. The model is trained using focal sigmoid cross-entropy loss, which fits datasets with a large number of classes, where labels are disjoint and each image has several examples of positive (present) and negative (absent) labels.

OWL-ViT gets really good performance on open vocabulary object detection, by a large margin from prior work. It also shows zero-shot object detection, signaling the transfer of knowledge and representations from Internet-scale data to object detection tasks. OWL-ViT is fast enough to run open vocabulary detection within 100ms, which means it can be widely used as an auto-labeling system. Figure 2-35 shows the process of pretraining an image and text encoder using CLIP and the OWL-VIT used for open-vocabulary detection.

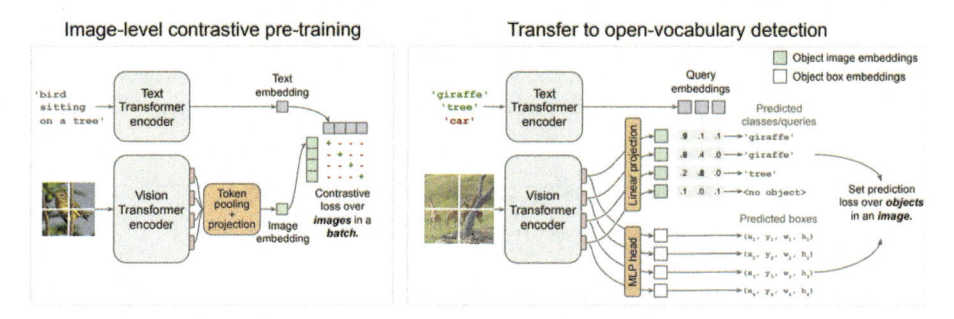

Figure 2-35. Left: CLIP pretraining. Right: OWL-VIT for detection. Used with permission, source: https://arxiv.org/pdf/2205.06230[58]

Promptable Open Vocabulary Segmentation

Transformers have also been applied to solve the image segmentation problem with a method called Segment Anything Model (SAM)[32]developed by Meta Research in 2023. They followed this up with Segment Anything 2[33] in 2024, which applies segmentation across video frames. SAM defines promptable segmentation as the task of generating a valid segmentation mask for any prompt specifying textual or spatial information to look for in an image, in a manner that is flexible, ambiguity-aware, and in real time. SAM was trained on 11 million images and 1.1 billion segmentation masks[34], becoming an extremely valuable foundation model in computer vision.

CHAPTER 2   ROBOT PERCEPTION: SENSORS AND IMAGE PROCESSING

SAM has three main components:

1. An **image encoder** that encodes an input image into an embedding to extract high level features and semantic representation, in this case a ViT

2. A **prompt encoder** to encode different types of prompts (sparse and dense) into embedding vectors

3. A **mask decoder** that uses the image and prompt embeddings to predict segmentation masks. It updates these embeddings using self-attention and cross-attention between the prompt and image embeddings

Figure 2-36 shows an overview of how these three components interact in the Segment Anything framework. SAM has an end-to-end latency of 50ms from input to mask generation and is very fast.

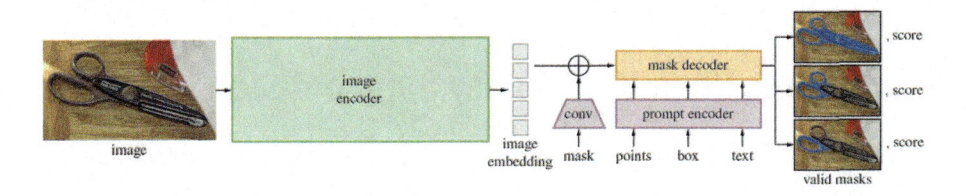

Figure 2-36. *The Segment Anything processing pipeline. Used with permission, source:* https://arxiv.org/pdf/2304.02643 *[32]*

To extend SAM to video segmentation, SAM2 introduces a memory attention, a memory encoder, and a memory bank. SAM2 extends segmentation in the temporal dimension by allowing a prompt on one or more images of a video, and expecting a segmented output across frames. The prompt could be a click, a text description, and so on. The image encoder in SAM2 produces one embedding per frame of video. The memory attention conditions the output segmentation mask prediction on past images, past predictions, and any new prompts by applying self-attention on current images and cross-attention to memories of frames

CHAPTER 2 ROBOT PERCEPTION: SENSORS AND IMAGE PROCESSING

and past predictions. The memory encoder generates memories by downsampling the output. The memory bank is a first-in, first-out (FIFO) queue that stores as spatial feature maps M memories for frames and N memories for prompted frames as well as object pointers for the target object to be segmented as vector embedding.

SAM2 is trained jointly on image and video data with interactive prompting and shows remarkable performance for zero-shot open vocabulary segmentation in videos.

Summary

The chapter covered the following concepts:

- Robots use various sensors to perceive their environment, including vision sensors like cameras for capturing images, depth sensors for measuring distance, range sensors like LiDAR and ultrasonic sensors for navigation, and IMUs to track motion and orientation.

- Robots perform tasks like classification, semantic segmentation, instance segmentation, and object detection. These tasks help robots navigate, manipulate objects, and interpret environments.

- CNNs can be used to process images through layers like convolutional layers, ReLU, and pooling layers. The chapter discussed different models, including R-CNN, Fast/Faster R-CNN, YOLO, and SSD, including how such models balance speed and accuracy for tasks like object detection and segmentation, each with tradeoffs in performance.

- Transformers treat images as sequences, allowing them to capture global context and dependencies within an image. Models like vision transformers (ViT) and detection transformers (DETR) can be used for object detection and segmentation, and they outperform traditional CNNs in certain scenarios. The chapter also explored how transformers can be scaled efficiently, and how methods like ViT (for classification), CLIP (for joint image-text understanding), and SAM (for segmentation) can be useful in robotics.

This chapter focused mainly on 2D image-processing techniques. The next chapter discusses 3D image-processing methods, multimodal perception, and sensor fusion.

References

[1] Johannesson, Pär-Olof. "What's the Difference between Frame- and Event-Based Lidar?" *Electronic Design*, 20 Jan. 2021, www.electronicdesign.com/markets/automotive/article/21152870/terranet-whats-the-difference-between-frame-and-event-based-lidar.

[2] https://www.micron.com/about/blog/storage/ssd/adas-camera-requirements-driving-memory-needs

[3] rpg.ifi.uzh.ch/research_dvs.html

[4] Rosebrock, Adrian. "Intersection over Union (IOU) for Object Detection." *PyImageSearch*, 7 Nov. 2016, pyimagesearch.com/2016/11/07/intersection-over-union-iou-for-object-detection/.

CHAPTER 2   ROBOT PERCEPTION: SENSORS AND IMAGE PROCESSING

[5] Shah, Deval. "Mean Average Precision (MAP) Explained: Everything You Need to Know." V7 Labs, 7 Mar. 2022, www.v7labs.com/blog/mean-average-precision.

[6] Kumar, Harshit. "Evaluation Metrics for Object Detection and Segmentation: mAP." Harshit Kumar, 20 Sept. 2019, kharshit.github.io/blog/2019/09/20/evaluation-metrics-for-object-detection-and-segmentation.

[7] teamraft.com/2020/05/01/object-detection-in-a-nutshell/

[8] Kollu, SaiSumanth. "Introductory Note on Deep Learning." *Analytics Vidhya,* 28 Nov 2022, www.analyticsvidhya.com/blog/2022/01/introductory-note-on-deep-learning/?utm_source=related_WP&utm_medium=www.analyticsvidhya.com/blog/2021/06/deep-dive-into-time-series-data-with-single-neuron/.

[9] "CS231n Convolutional Neural Networks for Visual Recognition." cs231n.github.io/.

[10] Girshick, Ross, et al. "Rich feature hierarchies for accurate object detection and semantic segmentation." Proceedings of the IEEE Conference on Computer Vision and Pattern Recognition. 2014.

[11] d2l.ai/chapter_computer-vision/rcnn.html.

[12] Girshick, Ross. "Fast r-cnn." Proceedings of the IEEE International Conference on Computer Vision. 2015.

[13] Ren, Shaoqing, et al. "Faster R-CNN: Towards real-time object detection with region proposal networks." IEEE Transactions on Pattern Analysis and Machine Intelligence 39.6 (2016): 1137-1149.

[14] Sachan, Ankit. "Zero to Hero: Guide to Object Detection Using Deep Learning: Faster R-CNN,Yolo,SSD." CV Tricks, cv-tricks.com/object-detection/faster-r-cnn-yolo-ssd/.

[15] He, Kaiming, et al. "Mask r-cnn." Proceedings of the IEEE International Conference on Computer Vision. 2017.

[16] He, Kaiming, et al. "Deep residual learning for image recognition." Proceedings of the IEEE Conference on Computer Vision and Pattern Recognition. 2016.

[17] Ronneberger, Olaf, Philipp Fischer, and Thomas Brox. "U-net: Convolutional networks for biomedical image segmentation." Medical image computing and computer-assisted intervention–MICCAI 2015: 18th international conference, Munich, Germany, October 5-9, 2015, proceedings, part III 18. Springer International Publishing, 2015.

[18] Tan, Mingxing, and Quoc Le. "EfficientNet: Rethinking model scaling for convolutional neural networks." International Conference on Machine Learning. PMLR, 2019.

[19] Redmon, Joseph, et al. "You only look once: Unified, real-time object detection." Proceedings of the IEEE Conference on Computer Vision and Pattern Recognition. 2016.

[20] Liu, Wei, et al. "Ssd: Single shot multibox detector." Computer Vision–ECCV 2016: 14th European Conference, Amsterdam, The Netherlands, October 11–14, 2016, Proceedings, Part I 14. Springer International Publishing, 2016.

[21] Kim, Jeong-ah, Ju-Yeong Sung, and Se-ho Park. "Comparison of Faster-RCNN, YOLO, and SSD for real-time vehicle type recognition." 2020 IEEE International Conference on Consumer Electronics-Asia (ICCE-Asia). IEEE, 2020.

[22] Bahdanau, Dzmitry, Kyunghyun Cho, and Yoshua Bengio. "Neural machine translation by jointly learning to align and translate." *arXiv preprint arXiv*:1409.0473 (2014).

[23] Vaswani, Ashish. "Attention is all you need." *arXiv preprint arXiv*:1706.03762 (2017).

[24] Brown, Tom B. "Language models are few-shot learners." *arXiv preprint ArXiv*:2005.14165 (2020).

[25] Chowdhery, Aakanksha, et al. "Palm: Scaling language modeling with pathways." *Journal of Machine Learning Research* 24.240 (2023): 1-113.

[26] Dosovitskiy, Alexey, et al. "An image is worth 16x16 words: Transformers for image recognition at scale." *arXiv preprint arXiv*:2010.11929 (2020).

[27] Dehghani, Mostafa, et al. "Scaling vision transformers to 22 billion parameters." International Conference on Machine Learning. PMLR, 2023.

[28] Driess, Danny, et al. "Palm-e: An embodied multimodal language model." *arXiv preprint arXiv*:2303.03378 (2023).

[29] Radford, Alec, et al. "Learning transferable visual models from natural language supervision." International Conference on Machine Learning. PMLR, 2021.

[30] `https://paperswithcode.com/sota/image-classification-on-imagenet`

CHAPTER 2   ROBOT PERCEPTION: SENSORS AND IMAGE PROCESSING

[31] Minderer, Matthias, et al. "Simple open-vocabulary object detection." European Conference on Computer Vision. Cham: Springer Nature Switzerland, 2022.

[32] Kirillov, Alexander, et al. "Segment anything." Proceedings of the IEEE/CVF International Conference on Computer Vision. 2023.

[33] Ravi, Nikhila, et al. "Sam 2: Segment anything in images and videos." *arXiv preprint arXiv*:2408.00714 (2024).

[34] Buhl, Nikolaj. "Meta AI's Segment Anything Model (Sam) Explained: The Ultimate Guide." *Encord*, 6 Apr. 2023, encord.com/blog/segment-anything-model-explained/.

[35] Graves, Alex. "Neural Turing Machines." *arXiv preprint arXiv*:1410.5401 (2014).

[36] Luong, Minh-Thang. "Effective approaches to attention-based neural machine translation." *arXiv preprint arXiv*:1508.04025 (2015).

[37] Dao, Tri, et al. "Flashattention: Fast and memory-efficient exact attention with io-awareness." *Advances in Neural Information Processing Systems* 35 (2022): 16344-16359.

[38] Lin, Hezheng, et al. "Cat: Cross attention in vision transformer." 2022 IEEE International Conference on Multimedia and Expo (ICME). IEEE, 2022.

[39] Premebida, Cristiano, Rares Ambrus, and Zoltan-Csaba Marton. "Intelligent robotic perception systems." *Applications of Mobile Robots* (2018): 111-127.

CHAPTER 2   ROBOT PERCEPTION: SENSORS AND IMAGE PROCESSING

[40] Gallego, Guillermo, et al. "Event-based, 6-DOF camera tracking from photometric depth maps." IEEE Transactions on Pattern Analysis and Machine Intelligence 40.10 (2017): 2402-2412.

[41] Ng, Bryan, et al. "Understanding the Role of Optics for Consumer Electronics in 2024." Wavelength Opto-Electronic Singapore, 27 Nov. 2023, wavelength-oe.com/articles/optics-for-consumer-electronics/.

[42] Durai, Pranav. "3D LiDAR Visualization: Case Study on 2D KITTI Depth Frames." Learnopencv.com, 5 Dec. 2023, learnopencv.com/3d-lidar-visualization/.

[43] "What is LiDAR and how does it work ?" *YellowScan*, 25 Aug 2023, www.yellowscan.com/knowledge/how-does-lidar-work/.

[44] Bouazizi, Mondher, Alejandro Lorite Mora, and Tomoaki Ohtsuki. "A 2D-Lidar-equipped unmanned robot-based approach for indoor human activity detection." *Sensors* 23.5 (2023): 2534.

[45] Pandey, Gaurav, et al. "Visually bootstrapped generalized ICP." 2011 IEEE International Conference on Robotics and Automation. IEEE, 2011.

[46] Smoot, Jeff. "The Basics of Ultrasonic Sensors." *Same Sky*, Apr. 2021, www.sameskydevices.com/blog/the-basics-of-ultrasonic-sensors.

[47] Or, Barak. "What Is IMU?" *Medium*, 31 Jul 2021, towardsdatascience.com/what-is-imu-9565e55b44c.

[48] Badrinarayanan, Vijay, Alex Kendall, and Roberto Cipolla. "SegNet: A deep convolutional encoder-decoder architecture for image segmentation." IEEE Transactions on Pattern Analysis and Machine Intelligence 39.12 (2017): 2481-2495.

[49] Sayedi, Husna. "Introduction to Image Annotation for ML and AI." Taus.net, 19 April 2021, www.taus.net/resources/blog/introduction-to-image-annotation-for-ml-and-ai.

[50] Selvaraj, Michael Gomez, et al. "AI-powered banana diseases and pest detection." *Plant Methods* 15 (2019): 1-11.

[51] Keells, John. "Theory behind Confusion Matrix - OCTAVE - John Keells Group - Medium." *Medium*, 26 Nov. 2023, octave-jkh.medium.com/theory-behind-confusion-matrix-bccabd3ad7d7.

[52] Tam, Adrian. "Building a Convolutional Neural Network in PyTorch" MachineLearningMastery.com, 8 April 2023, machinelearningmastery.com/building-a-convolutional-neural-network-in-pytorch/.

[53] Tammina, Srikanth. "Transfer learning using vgg-16 with deep convolutional neural network for classifying images." International Journal of Scientific and Research Publications (IJSRP) 9.10 (2019): 143-150.

[54] Bhatt, Anil. "CNN - Activation Functions, Global Average Pooling, Softmax, Negative Likelihood Loss" Linkedin.com, 19 Mar. 2020, www.linkedin.com/pulse/cnn-activation-functions-global-average-pooling-softmax-n-bhatt/.

[55] "IoT Sensors Enable Robot to Clean through Object Detection." Ubiqisense.com, 26 Jan 2021, www.ubiqisense.com/news/robot-to-clean-canteens.

[56] Tsang, Sik-Ho. "Review: SSD — Single Shot Detector (Object Detection)." Medium, Towards Data Science, 3 Nov. 2018, towardsdatascience.com/review-ssd-single-shot-detector-object-detection-851a94607d11.

[57] Huang, Jonathan, et al. "Speed/accuracy trade-offs for modern convolutional object detectors." Proceedings of the IEEE conference on computer vision and pattern recognition. 2017.

[58] Minderer, Matthias, et al. "Simple open-vocabulary object detection with vision transformers. ArXiv abs/2205.06230 (2022)." (2022).

CHAPTER 3

Robot Perception: 3D Data and Sensor Fusion

3D sensor data collected from LiDAR or depth cameras is critical for real-world robotic perception, as it allows robots to perceive the three-dimensional structure of objects around them. 3D sensor data is often collected and used by industrial and consumer robots, but one of the most relevant applications is in the development of autonomous vehicles. The key force behind autonomous vehicles is 3D sensor data collected from LiDAR or depth cameras. From 3D sensor data, autonomous vehicles can accurately perceive the three-dimensional structure of objects, pedestrians, and other vehicles around them. This information enables robust object detection, precise localization, and reliable path planning, leading to improved safety, situational awareness, and more efficient decision-making in real-time driving scenarios. Some autonomous vehicle companies also rely on 3D data processing to create detailed maps, aiding in navigation and ensuring smooth and reliable autonomous driving experiences.

Overall, 3D data processing is extremely important for robotic perception, mapping, and navigation. By processing 3D sensor data, such as point clouds and depth maps, robots can perceive their surroundings in three dimensions, detect obstacles, and plan collision-free paths to reach their destinations. This chapter addresses methods to process data with depth perception such as 3D maps as well as sensor fusion. It also covers techniques used to compress data from multimodal sensors such as radar, LiDAR, and cameras.

3D Data Processing

To understand 3D data processing, you first need to study how 3D data is represented in the framework of machine learning. Then you can understand the methods and learn from the data representations that the model generates.

Data Representation

3D data, typically from a LiDAR, which is a stereo vision camera with depth or infrared sensor, is generally represented in the form of point clouds, voxels, or meshes.

Let's consider each of these terms.

- A **point cloud** is a set of point coordinates in space, generally represented as a cartesian (x, y, z). A multiple return LiDAR sensor can provide a point cloud.

- A **voxel** or a volumetric pixel is a 3D pixel. It is again represented as a pixelized (x, y, z) point where the coordinates are limited by resolution of the voxel grid. A sparse voxel is a memory efficient representation where only cells containing information are stored. Octrees are a variant of sparse voxels where adjacent cells that are identical are aggregated, allowing for data compression.

CHAPTER 3   ROBOT PERCEPTION: 3D DATA AND SENSOR FUSION

- A **mesh** is a polygon representation of 3D data where the data is represented as a composition of polygon surfaces, generally triangles.

A depth map is typically the output of a stereo vision camera where, for every (x,y) pixel, there's a depth (z). Note that this is a 2.5D representation, as you cannot have multiple values of depth for the same pixel.

Parametric modeling refers to representing objects in terms of surfaces and volumes that share parameters, such that interlinked attributes change together. An example is a CAD model of an object, where height and width can change with scale. The difference between a 3D mesh image and a voxelized image is shown in Figure 3-1.

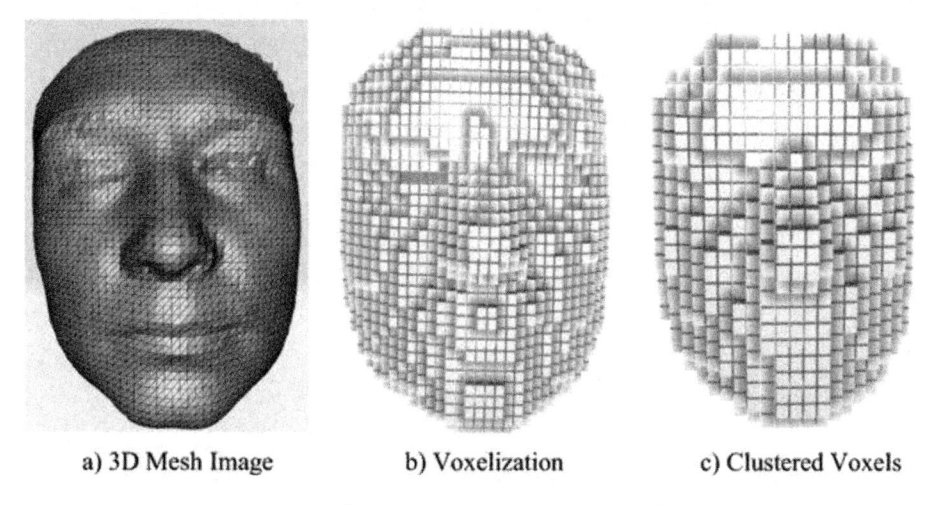

a) 3D Mesh Image   b) Voxelization   c) Clustered Voxels

Figure 3-1. *3D mesh and voxel representation of a human face. Used with permission, source:* `https://link.springer.com/article/10.1007/s11042-020-08688-x[29]`

With voxels, the size of the dataset increases cubically with space, while the data stays ordered. With point clouds, you can represent varied spatial density and resolutions effectively, but a point set is permutation-invariant and models that use them need to factor that property.

CHAPTER 3 ROBOT PERCEPTION: 3D DATA AND SENSOR FUSION

For the purpose of robotics, this chapter focuses on processing point clouds, since, following the advent of LiDARs, they have been more widely adopted in research and industry.

Processing Point Clouds

Point clouds are fundamentally irregular in representation. They are also permutation-invariant, meaning changing the order of the points does not change the dataset. In the past, point clouds were processed after their voxelization, such as with VoxNet[1] and 3D ShapeNet[2]. However, this increases memory usage and introduces quantization artifacts, such as removing information in high point density clusters.

PointNet[3], shown in Figure 3-2, is a pioneering architecture for object classification and segmentation on point clouds. The network takes a set of *n* unordered points and converts them to transform invariant feature vectors using multilayer perceptrons.

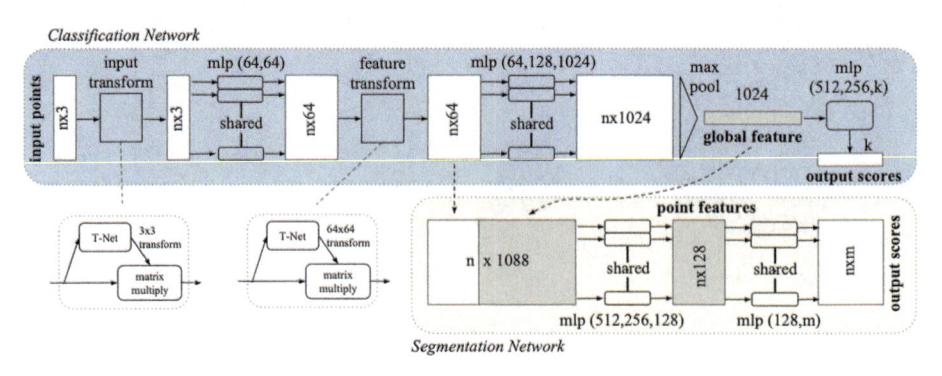

Figure 3-2. PointNet architecture. Used with permission, source: https://arxiv.org/abs/1612.00593[3]

The inputs are aligned to a canonical space by calculating an affine transformation using a T-net, which is a mini-network resembling the bigger network, shown in Figure 3-2, that contains point-independent feature extraction, max pooling, and fully connected layers. The feature

CHAPTER 3   ROBOT PERCEPTION: 3D DATA AND SENSOR FUSION

vectors are concatenated to original points and these joint point features are aggregated using multiple MLP[4] and max pooling layers[5] to output per point scores. At the time of its release, PointNet was the state of the art in semantic segmentation for point clouds. However, since max pooling is an aggregation and doesn't preserve local structures, PointNet is not robust with fine grained patterns and highly complex scenes.

The drawback concerning local structures was improved upon by PointNet++[6], shown in Figure 3-3. PointNet++ attempts to solve this issue by first partitioning the set of points into overlapping local regions using a Euclidean distance metric for 3D space. Then, instead of using a single max pooling operator on the entire point cloud, PointNet++ builds a hierarchical grouping of these points, extracts local features using PointNet, and processes them in hierarchical groups to get higher level features in an iterative fashion. Essentially, they recursively apply PointNet on a nested partition of the input set. The overlapping partitions are generated by defining balls in the input space with varying sizes and centers, and the centroids of these clusters/balls are generated using the farthest point sampling (FPS) algorithm[7]. In comparison to volumetric CNNs[8], which always have the same receptive field, applying PointNet to these Euclidean kernels of varying sizes adapts to the variation of point cloud density and variation in feature scales. This is achieved during training by random input dropout such that the network learns to adaptively combine features extracted from multiscale neighborhoods based on input data.

CHAPTER 3   ROBOT PERCEPTION: 3D DATA AND SENSOR FUSION

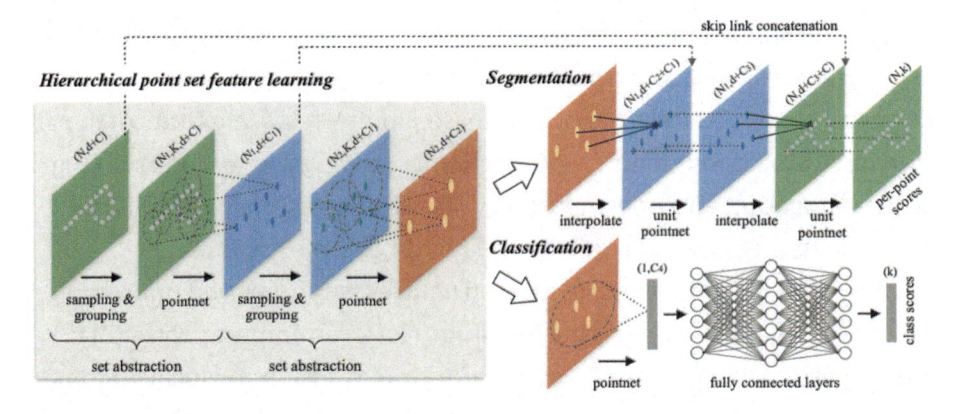

Figure 3-3. PointNet++ architecture. Used with permission, source: https://arxiv.org/pdf/1706.02413 [6]

The interpolation operation shown in the segmentation head of the network in Figure 3-3 is achieved by interpolating features at point j using an inverse distance weighted averaging over k nearest neighbors (see Equation 3-1).

$$f^{(j)}(x) = \frac{\sum_{i=1}^{k} w_i(x) f_i^{(j)}}{\sum_{i=1}^{k} w_i(x)} \quad \text{where} \quad w_i(x) = \frac{1}{d(x, x_i)^p}, \; j = 1, ..., C$$

Equation 3-1, source: https://arxiv.org/pdf/1706.02413 [6]

These are then concatenated with skip level point features from the abstraction layer and passed through unit PointNets (close to 1*1 convolution). These are then propagated back until the original point set is labeled.

However, PointNet++ still has room to improve. Feature extraction using PointNet that uses max pooling loses spatial information.

PointCNN[9], shown in Figure 3-5, aims to use the ability of traditional CNNs to capture local and hierarchical features with increasing receptive fields on point clouds. One way to apply CNNs on point clouds is by voxelizing the data and then applying CNNs on the voxels. However, this

CHAPTER 3   ROBOT PERCEPTION: 3D DATA AND SENSOR FUSION

is not an efficient representation of data and voxel CNNs take up too much memory due to dimensionality. PointCNN solves this problem by introducing an $\mathcal{X}$-conv operator that transforms the point cloud into a latent and canonical space. The algorithm for this is shown in Figure 3-4.

ALGORITHM 1: $\mathcal{X}$-Conv Operator
Input : $\mathbf{K}, p, \mathbf{P}, \mathbf{F}$
Output : $\mathbf{F}_p$ ▷ Features "projected", or "aggregated", into representative point p
1: $\mathbf{P}' \leftarrow \mathbf{P} - p$ ▷ Move $\mathbf{P}$ to local coordinate system of p
2: $\mathbf{F}_\delta \leftarrow MLP_\delta(\mathbf{P}')$ ▷ **Individually** lift each point into C_δ dimensional space
3: $\mathbf{F}_* \leftarrow [\mathbf{F}_\delta, \mathbf{F}]$ ▷ Concatenate $\mathbf{F}_\delta$ and $\mathbf{F}$, $\mathbf{F}_*$ is a $K \times (C_\delta + C_1)$ matrix
4: $\mathcal{X} \leftarrow MLP(\mathbf{P}')$ ▷ Learn the $K \times K$ $\mathcal{X}$-transformation matrix
5: $\mathbf{F}_\mathcal{X} \leftarrow \mathcal{X} \times \mathbf{F}_*$ ▷ Weight and permute $\mathbf{F}_*$ with the learnt $\mathcal{X}$
6: $\mathbf{F}_p \leftarrow \mathrm{Conv}(\mathbf{K}, \mathbf{F}_\mathcal{X})$ ▷ Finally, typical convolution between $\mathbf{K}$ and $\mathbf{F}_\mathcal{X}$

Figure 3-4. *X-conv algorithm. Used with permission, source:* https://arxiv.org/pdf/1801.07791 [9]

If there are N points in the point set, the $\mathcal{X}$-conv operation samples K neighborhood points for every point in input set, which constitutes its local neighborhood. Each local neighborhood is brought to its own coordinate system. The $\mathcal{X}$-conv operation is completely differentiable and converts the points to a feature space that is deeper, with smaller spatial resolution.

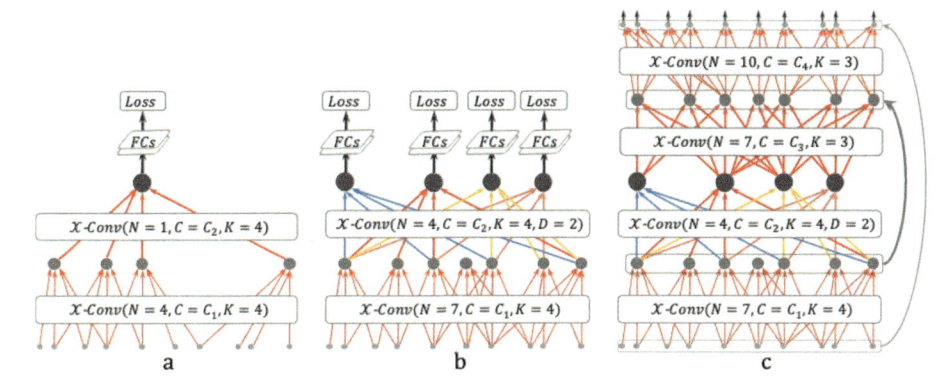

Figure 3-5. *Point CNN architecture for classification (a and b) and for segmentation (c). Used with permission, source:* https://arxiv.org/pdf/1801.07791 [9]

113

CHAPTER 3 ROBOT PERCEPTION: 3D DATA AND SENSOR FUSION

Each X-conv layer reduces the number of points with increasing receptive fields, with the final point seeing all the points. For detection, the final X-conv layer is paired with a fully connected layer and loss function to train the network. For segmentation, X-conv is applied in a fashion similar to the U-Net, where there are successive d-conv operations until the original points have propagated features. PointCNN is found to perform better than PointNet++ in several classification tasks.

Dynamic Graph CNN (DGCNN)[10], shown in Figure 3-6, is an algorithm that aims to improve upon the fact that PointNets cannot capture features at local scale due to permutation invariance of points. It does so by introducing an edge convolution operation.

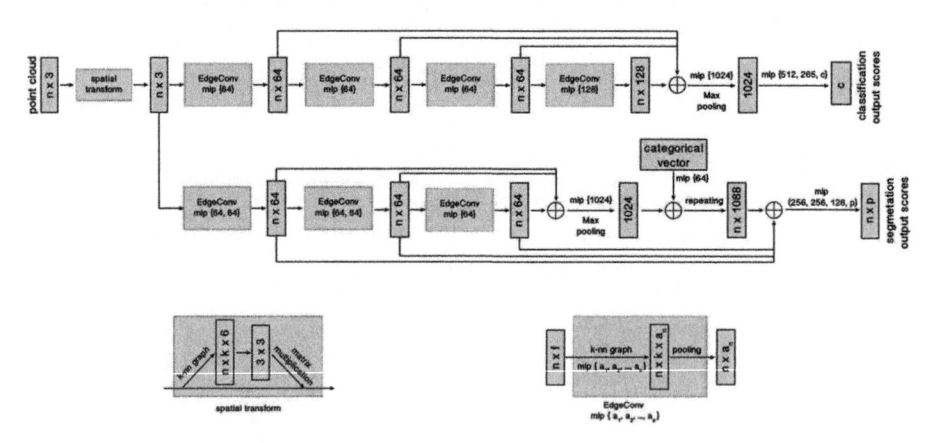

Figure 3-6. DGCNN for classification (top path) and segmentation (bottom path). Used with permission, source: https://arxiv.org/pdf/1801.07829 [10]

DGCNN constructs a local neighborhood graph of k nearest points and updates the graph at each layer by adding convolution-like operations on edges, thus connecting pairs of points in a neighborhood. The neighbors change in each layer and the graph is recomputed, hence it is dynamic.

The edgeconv operation takes an $n*f$ tensor (where n is the number of points and f is input feature size to that layer), applies a multilayer perceptron, and calculates edge features for each point. For classification, the last edgeconv layer is aggregated to get a global 1D descriptor, which is used to arrive at a class label for the point set. For segmentation, the 1D descriptor is concatenated with edge features from each edgeconv layer to provide global and local features for each point. This allows it to arrive at per point segmentation output scores.

DGCNN performs close to PointCNN on classification and segmentation tasks (+-3%). However, the fixed size of edge features can limit its performance at different scales and resolutions. Given that input features contribute differently to the nodes, attention mechanisms could further improve performance by looking at the relevant features when variable sized input is involved.

Research Opportunities

1. **Exploiting context:** Most point cloud based models treat points at a local scale as independent in order to maintain permutation invariance. But this means they're unable to extract finer features and exploit the relationship between a point and its neighbors.

2. **Real-time performance:** Despite the abundance of models for point clouds, most robotic systems have onboard compute and need to respond quickly. This calls for lighter models with lower inference time.

3. **The pitfall of supervised learning:** Most existing point cloud models utilize supervised learning, which is not robust to unseen scenarios and requires large amounts of data to train and generalize. Weakly supervised or unsupervised frameworks can improve these limitations.

Multimodal Perception and Sensor Fusion

The goal of multimodal perception is to use a combination of different sensing modalities to get a robot to accomplish a task. Sensor fusion is combining input from more than one sensor to gain a deeper understanding of the surrounding environment. Multi-modal sensor input can also make perception more robust in the case of failure modes in any of the sensors—this is an important safety mechanism. Sensor fusion is often used in applications like self-driving cars to fuse some combination of LiDARs, radars, and cameras to exploit their various strengths and improve performance of the overall combined system. For example, LiDARs have good depth perception but don't provide color information. While cameras don't provide information about an object's depth, they can provide color information. Radars, on the other hand, provide direct speed measurements of obstacles and complement LiDAR and cameras.

The utility and specifics of sensor fusion is ultimately warranted by the cost and precision required by a robot application. One downside of multisensor perception is the added system complexity of synchronizing information and inputs of sensor fusion. For sensor fusion to work, the sensors need to be calibrated, registered with respect to the other, and need to use a common time reference. Given various sensing modalities, you need a way to fuse them to achieve an accurate 3D representation of the world for the robot to act in. The following sections cover some common strategies to accomplish this.

Fusion Strategies

Recent advances in deep learning have led to two main sensor fusion strategies:

1. **Late fusion:** Processes each sensing modality independently until the very end. For example, it will fuse bounding boxes from LiDAR and from the camera that were processed separately.

CHAPTER 3   ROBOT PERCEPTION: 3D DATA AND SENSOR FUSION

2. **Early fusion:** Leverages recent advances in deep learning to fuse the raw sensor reading in the early stages. For example, it will fuse pixels and point clouds directly.

This section focuses on integrating LiDAR and camera data to leverage the camera's high resolution and its ability to classify objects, along with LiDAR's capability to measure distances and perceive the world in 3D. You can see how both of these work together by looking at fusing point clouds and pixels in the early fusion mode and fusing bounding boxes in the late fusion mode.

To illustrate the advantages of sensor fusion, different sensors and their main uses in autonomous vehicles are shown in Figure 3-7.

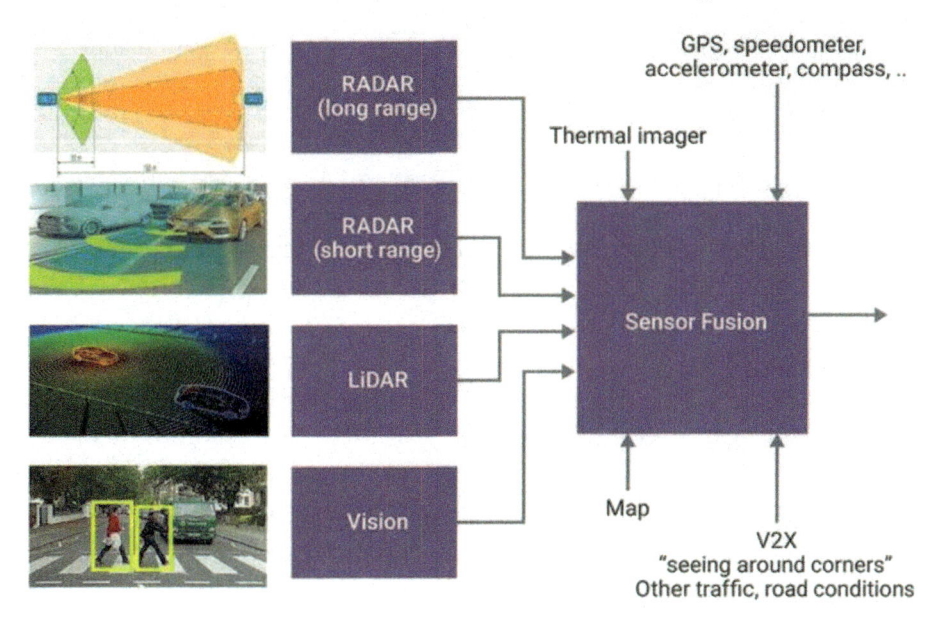

Figure 3-7. Overview of sensor fusion in autonomous vehicles, combining data from long-range and short-range RADAR, LiDAR, vision cameras, thermal imagers, GPS, and other sensors. Used with permission, source: https://semiengineering.com/a-dsp-for-implementing-high-performance-sensor-fusion-on-an-embedded-budget/ by Synopsys[30]

CHAPTER 3　ROBOT PERCEPTION: 3D DATA AND SENSOR FUSION

Fusing Raw Data: Early Fusion

In early fusion, raw data like point clouds from LiDAR and camera images are fused together directly[11]. The goal is to create a 2D projection of the 3D point cloud within the camera frame, where each point's location is now directly comparable to objects and features in the 2D camera image. This three-step process is shown in Figure 3-8.

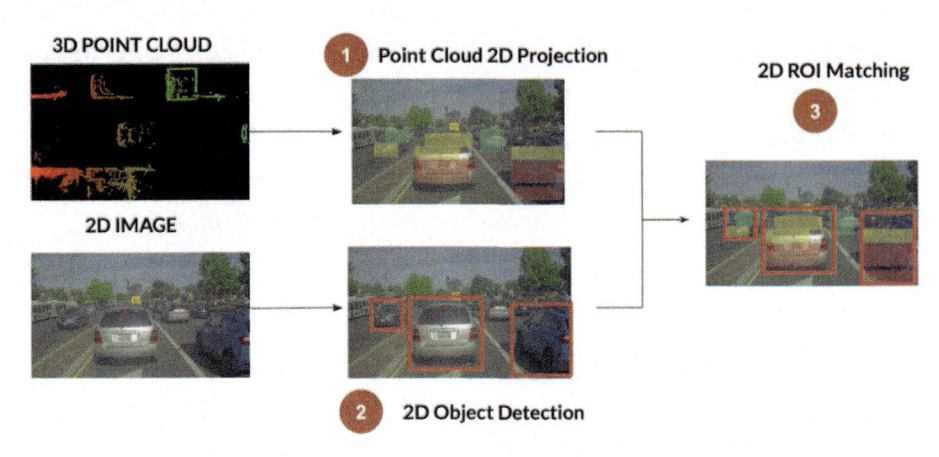

Figure 3-8. Pipeline for early sensor fusion. Used with permission, source: https://www.thinkautonomous.ai/blog/lidar-and-camera-sensor-fusion-in-self-driving-cars/ *by Think Autonomous[11]*

Step 1: Projection of Point Cloud

The first step translates the 3D point clouds captured by the LiDAR into a format that can be used given the 2D data type for the camera. This involves a few steps:

- Convert each 3D point from the LiDAR data into homogeneous coordinates. This involves converting the standard Euclidean coordinates (x, y, z) and adding an additional dimension, making the point (x, y, z, w).

Specifically, each point in the LiDAR's 3D point cloud (x, y, z) is converted to homogeneous coordinates (x, y, z, 1).This extra coordinate allows for different transformations using matrix multiplication and simplifies aligning 3D data with 2D data.

- A transformation matrix (including rotation and translation) aligns the LiDAR coordinate system with the camera coordinate system. This is useful for extrinsic (relative to the LiDAR, the position/orientation of the camera) camera calibration.

- Another transformation matrix projects the 3D points onto the 2D image plane of the camera. This matrix accounts for intrinsic (camera-specific properties like focal length and optical center) calibration. This data is useful to accurately translate and rotate the LiDAR points to align them with the camera's point of view.

- After projection, the points are transformed back from homogeneous to standard Euclidean coordinates. 2D points can now be directly compared with objects and features in the 2D camera image.

If interested, this course[12] talks more about what homogeneous coordinates are and how these specific projections and rotations are done. The result from this step is shown in Figure 3-9.

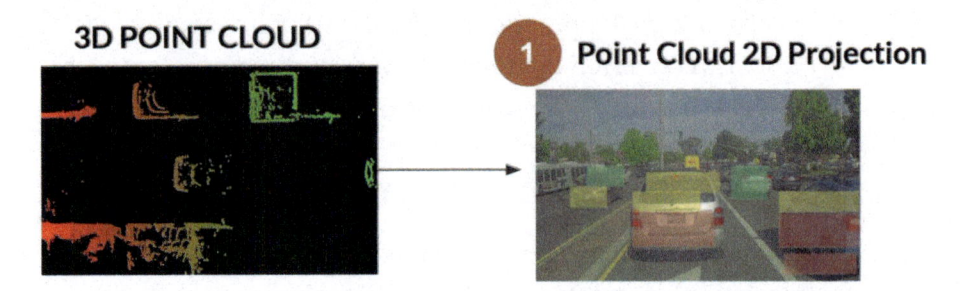

CHAPTER 3   ROBOT PERCEPTION: 3D DATA AND SENSOR FUSION

Figure 3-9. Going from a 3D point cloud to a point cloud projection in 2D. Used with permission, source: https://www.thinkautonomous.ai/blog/lidar-and-camera-sensor-fusion-in-self-driving-cars/ by Think Autonomous[11]

Step 2: Object Detection

Now that the projected point cloud image is aligned with the image, various models such as YOLOv4[13] can be used to do object detection. You can learn more about how these object-detection models work in Chapter 2.

Step 3: Region of Interest (RoI) Matching

Once the objects are detected in the camera image using 2D bounding boxes, the next step is to match the 3D points from the LiDAR data with these detected objects. This process is called *Region of Interest* (RoI) matching.

- The 3D points from the LiDAR, which were projected onto the 2D image plane in the previous steps, are now compared against the 2D bounding boxes generated by the object detection model.
- For each 2D bounding box, each 3D point is associated with a detected object.

CHAPTER 3   ROBOT PERCEPTION: 3D DATA AND SENSOR FUSION

- Once the points are associated with specific objects, each 3D point is labeled according to the detected object it belongs to. For example, points in a bounding box for a car are labeled as part of that car.

Fusing Outputs: Late Fusion

In late fusion, the data from different sensors is processed separately before combining the results[11]. With this approach, independent object detections are done on each sensor's data, resulting in either 2D or 3D bounding boxes, which are then fused. An overview of how the process works in 3D is shown in Figure 3-10.

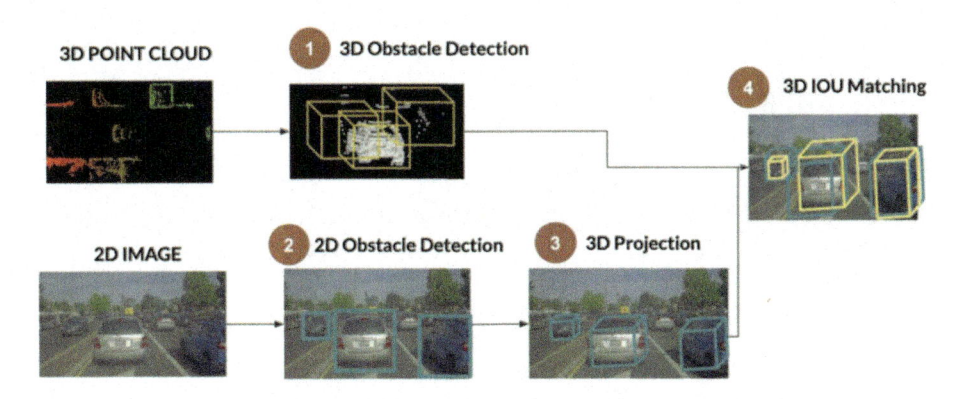

Figure 3-10.   Late fusion pipeline using a 3D point cloud and 2D image as input. Used with permission, source: https://www.thinkautonomous.ai/blog/lidar-and-camera-sensor-fusion-in-self-driving-cars/ by Think Autonomous[11]

Steps 1 and 2: LiDAR 3D Object Detection and Camera 3D Object Detection

There are many LiDAR deep learning models that can be used for 3D object detection, such as PointNet[3], PointNet++[6], VoxelNet[14], and SECOND (a spatially sparse convolutional network)[15], some of which are covered in this chapter. These models generate 3D bounding boxes for objects that are detected and include their positions, dimensions, and orientations within the LiDAR's coordinate system. For 3D camera detection, some object detection methods used in 2D can be altered so that they work with 3D data, such as adding a depth estimation network to YOLO[16]. It's important to note that the bounding boxes are generated as (x, y, z length, width, height, rotation) within the coordinate system of your data (LiDAR or camera).

From here, the 3D bounding boxes from one coordinate system are converted to the other. For example, the bounding boxes can be converted from a LiDAR coordinate system to the camera coordinate system. This allows both sets of the bounding boxes to be aligned in the same frame of reference.

Step 3: IOU Matching

The next step is to match these bounding boxes to identify the same objects detected by both sensors. This process involves matching in space or in time.

- **Matching in space**: For each pair of 3D bounding boxes in LiDAR and camera data, the IOU score is computed. An IOU score above a certain threshold indicates that the bounding boxes likely correspond to the same object.

- **Matching in time**: This extends the matching process over time. Using techniques like the Kalman Filter[17] and the Hungarian Algorithm[18], objects can be tracked across different frames. If a bounding box from one frame overlaps with one in a subsequent frame (based on IOU), the object is considered the same.

- The Kalman Filter operates by first predicting a vehicle's future position using prior knowledge and kinematic equations, then measuring the actual position with sensor data to compare against the prediction, and finally updating the prediction to improve accuracy based on the new information. This helps provide a refined estimation of the object's trajectory. We won't be focusing too much on classical methods for sensor fusion, but if interested you can learn more about Kalman filters here[19].

Some information from this section was referenced from[11] and we recommend it as a good source.

LiDAR-Camera Fusion

Now that you understand the main frameworks for sensor fusion, let's dive into deep learning models that are deployed for LiDAR-camera fusion[25][20][21].

Proposal-level Fusion Methods

Combining LiDAR and camera, proposal-level fusion is where "proposals" or ROIs are generated from one type of sensor data and then refined and augmented using data from another sensor.

The overall pipeline for proposal-level fusion methods is:

- Initial guesses or proposals are provided for where objects might be located. For example, a 2D CNN can be used to generate 2D bounding boxes around where objects might be located.

CHAPTER 3   ROBOT PERCEPTION: 3D DATA AND SENSOR FUSION

- 2D proposals can be expanded to 3D proposals using the projection matrix of the camera. This is called a 3D frustum, representing the volume that goes from the 2D bounding box space to the 3D space that the 2D bounding box might occupy.

- Data from LiDAR can be integrated into these 3D proposals by seeing which point clouds fall within each frustum, essentially allowing you to add depth information to the initial 2D proposals.

- Within each frustum, the combined data is used to further refine the detection and classification of objects. Finally, the exact position, dimensions, and orientation of the objects within the 3D space are determined to estimate precise 3D bounding boxes for the detected objects.

Frustum PointNets[22] use RGB-D data for 3D object detection. This starts by using a 2D CNN to detect objects in the camera image, generating 2D bounding boxes. These 2D bounding boxes are projected into 3D space, creating frustums. The points within the frustums create a frustum point cloud. A detailed depiction of the Frustum PointNet is shown in Figure 3-11.

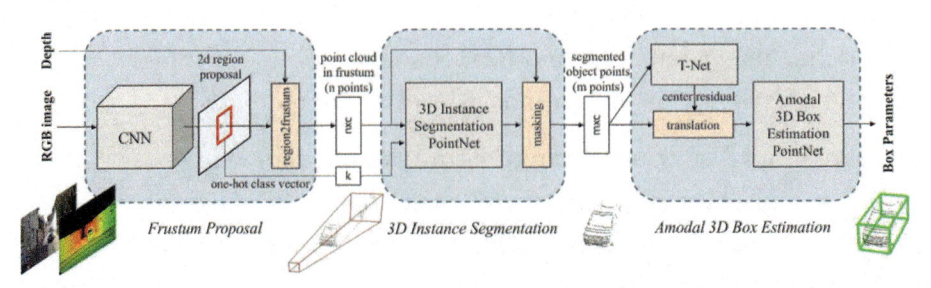

Figure 3-11. *Main components of Frustum PointNet. Used with permission, source: https://arxiv.org/pdf/1711.08488[22]*

CHAPTER 3 ROBOT PERCEPTION: 3D DATA AND SENSOR FUSION

LiDAR points within each frustum are extracted and processed by a network to classify whether each point belongs to the object or the background. A T-Net then refines these points by shifting their centroid to align it more closely with the object's true center. Another network estimates the precise 3D bounding box, including the object's position, dimensions, and orientation.

These steps and their results are illustrated in Figure 3-12, showing how these transformations are applied to derive the final object detection and positioning in 3D space.

(a) camera coordinate   (b) frustum coordinate   (c) 3D mask coordinate   (d) 3D object coordinate

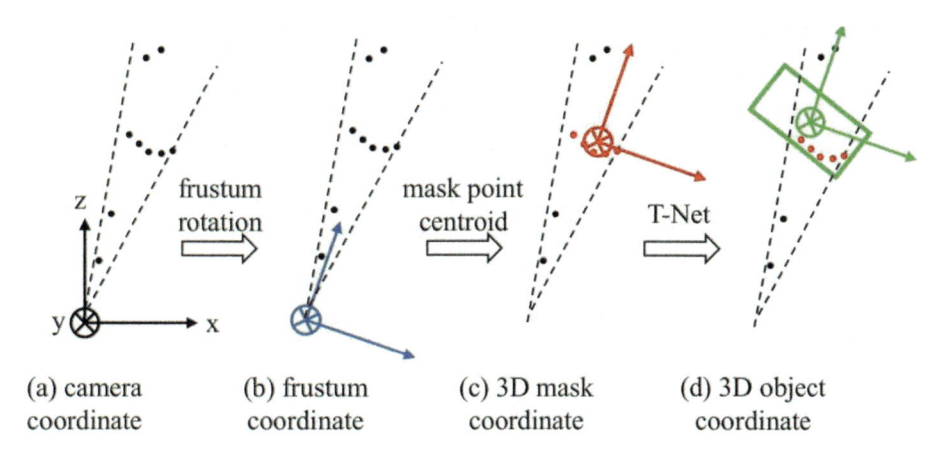

Figure 3-12. *Transformations in Frustum PointNets. Default camera coordinate system(a). Frustum coordinate system is obtained after rotating the frustum to center the view (b). 3D mask coordinate aligns the mask point's centroid at the origin (c). 3D object coordinate is predicted using T-Net (d). Used with permission, source:* https://arxiv.org/pdf/1711.08488 *[22]*

By generating initial proposals with one sensor and refining them with another, proposal-level fusion has better accuracy and reliability of object detection and classification, which can be especially important in complex environments like those encountered by self-driving cars.

Point-level Fusion Methods

On the other hand, point-level fusion techniques augment the LiDAR point cloud data with the camera features.

The overall steps for point-level fusion methods include:

- Sensor calibration, which includes determining the internal parameters (focal length, optical center, etc.) and external parameters (relative position and orientation between the camera and LiDAR). This can help align the camera and LiDAR to be in a similar coordinate system.

- A CNN can be used to extract main features from the camera data.

- Each LiDAR point is projected onto the camera image plane using the calibration parameters. This helps map each LiDAR point to each camera point.

- The LiDAR points can be combined with their corresponding camera features and methods, such as PointNet++ and VoxelNet, and can be used to perform tasks like object detection, segmentation, and classification.

In some cases, point-level fusion offers a more detailed and accurate representation because it combines visual features with each LiDAR point. This can help improve the model's ability to detect and classify objects.

Input-Level Fusion: PointPainting and PointAugmenting

PointPainting[23] is an example of a point-level fusion method that consists of three main stages:

CHAPTER 3　ROBOT PERCEPTION: 3D DATA AND SENSOR FUSION

- Initially, images are segmented, providing you with a segmentation mask that scores each pixel by categories.

- LiDAR points are mapped to their corresponding locations in the segmentation mask, and each point is assigned the semantic scores from the image, adding contextual visual information.

- The final stage utilizes the point clouds with the additional semantic information to identify and localize objects in 3D space. This final point cloud, which now contains both geometric and semantic information, can be used for more accurate detection of objects than just LiDAR or camera data detection.

An illustration of this process can be viewed in Figure 3-13, which showcases how PointPainting integrates image and LiDAR data to increase accuracy when doing object detection for autonomous driving systems.

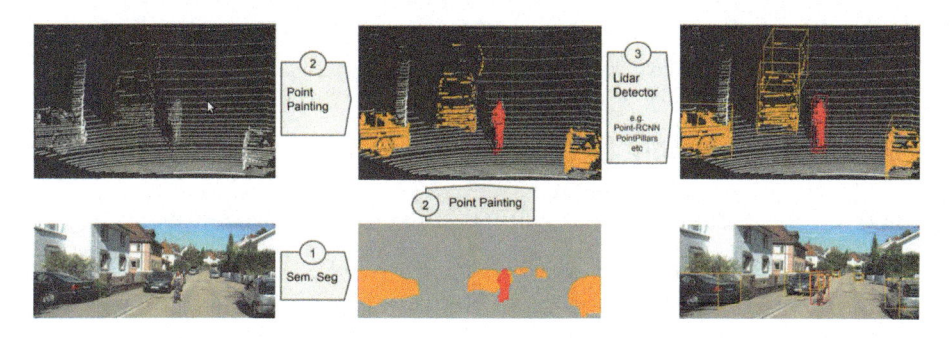

Figure 3-13. Overview of PointPainting and its three main stages: (1) An image-based semantic segmentation network, (2) PointPainting (fusion), (3) A LiDAR-based object detector. Used with permission, source: https://arxiv.org/pdf/1911.10150 [23]

127

PointAugmenting[24] builds on the PointPainting method, but instead of using simple categorical scores from the segmentation mask, features extracted by a CNN are mapped onto the LiDAR points. This is called a cross-modal fusion strategy. It combines visual features with depth data, which can improve the accuracy of 3D object detection.

PointAugmenting is shown in Figure 3-14, which illustrates how this technique combines and utilizes data from LiDAR and camera sensors to improve detection in autonomous vehicles.

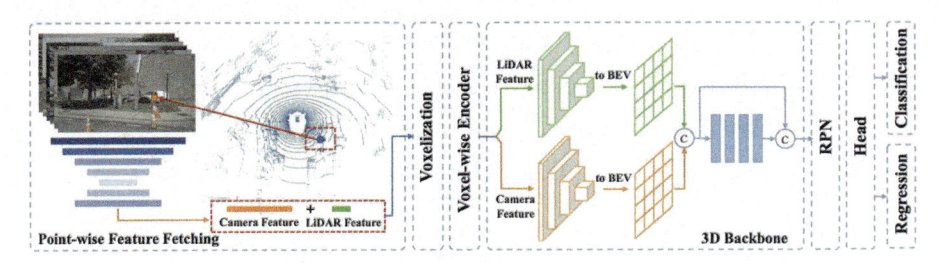

Figure 3-14. Overview of PointAugmenting. Used with permission, source: https://openaccess.thecvf.com/content/CVPR2021/ papers/Wang_PointAugmenting_Cross-Modal_Augmentation_ for_3D_Object_Detection_CVPR_2021_paper.pdf[24]

Feature-level Fusion: DeepFusion

An effective mechanism for aligning camera and LiDAR features is a critical component that is missing from existing work like PointPainting and PointAugmenting. As shown in Figure 3-15, DeepFusion[25] is a model that does fusion at the feature level, meaning it combines features extracted from both LiDAR and camera data, with the goal of improving the alignment to increase performance on downstream tasks.

CHAPTER 3   ROBOT PERCEPTION: 3D DATA AND SENSOR FUSION

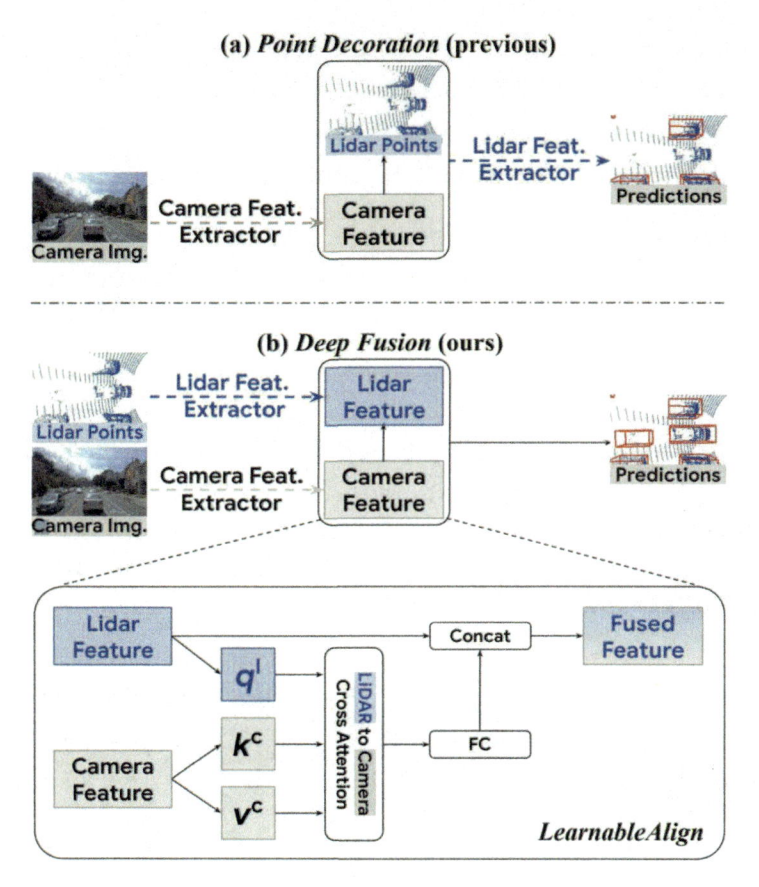

Figure 3-15. *Overview of DeepFusion (b). Used with permission, source: https://arxiv.org/pdf/2203.08195 [25]*

A key component of DeepFusion is Inverse Augmentation (InverseAug) [26], which reverses the augmentations applied during training before the data fusion step. The main idea is that when the LiDAR point cloud is rotated or transformed during training, it becomes difficult to match 3D points with their correct 2D locations in the camera image. InverseAug solves this by storing the applied transformations and reversing them

CHAPTER 3   ROBOT PERCEPTION: 3D DATA AND SENSOR FUSION

before combining the data, as shown in Figure 3-16. This ensures that the 3D points return to their original positions, making it easier to map them to the 2D camera view. This process is illustrated in Figure 3-17, showing alignments both with and without the application of InverseAug.

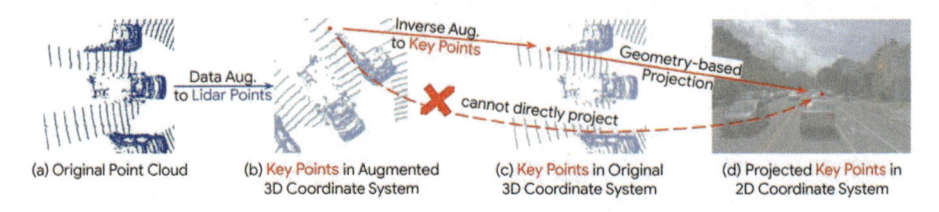

Figure 3-16. *Overview of the InverseAug method. The original LiDAR point cloud (a). After applying augmentation to LiDAR points (b). Key points in the original 3D coordinate system (c). Projected points in the 2D coordinate system (d). Used with permission, source:* https://arxiv.org/pdf/2203.08195 [25]

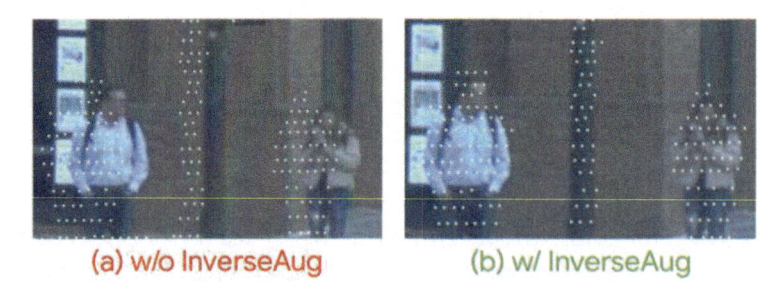

Figure 3-17. *Alignment without InverseAug (a). Alignment with InverseAug (b). Used with permission, source:* https://arxiv.org/pdf/2203.08195 [25]

To further improve the precision of alignment between different types of data, DeepFusion introduces a method called Learnable Alignment (LearnableAlign), as illustrated in Figure 3-18. This is mainly a cross-modality attention mechanism. This mechanism processes the features from LiDAR data (each voxel) and the corresponding camera data. It then calculates an output that is a weighted sum of the camera features,

CHAPTER 3   ROBOT PERCEPTION: 3D DATA AND SENSOR FUSION

meaning that it gives more importance to the most relevant camera features. By focusing on the most relevant features from the LiDAR and camera data through weighted attention, LearnableAlign ensures that the feature alignment is more precise, which helps improve the accuracy of object detection and classification tasks.

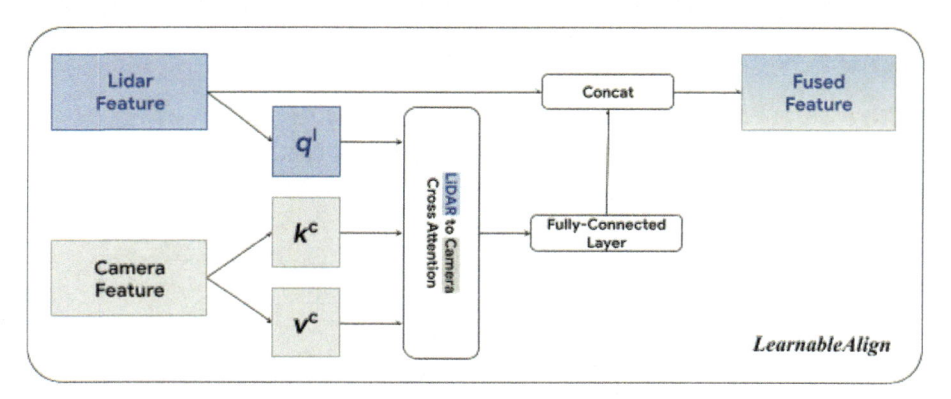

Figure 3-18. *Process of Learnable Alignment in DeepFusion where LiDAR and camera features are aligned. Used with permission, source:* *https://arxiv.org/pdf/2203.08195 [25]*

Putting this all together, DeepFusion[25] improves object detection by combining LiDAR and camera data. First, it transforms LiDAR points into useful features and extracts features from camera images using ResNet. These features are then aligned and fused using InverseAug, which reverses transformations for better alignment, and LearnableAlign, which uses an attention mechanism to focus on the most relevant features. The fused data is processed by a 3D detection model, including the backbone and detection head of the PointPillars system, to produce the final detection results. The entire workflow of this process, from initial data input to the production of detection outputs, is shown in Figure 3-19.

CHAPTER 3   ROBOT PERCEPTION: 3D DATA AND SENSOR FUSION

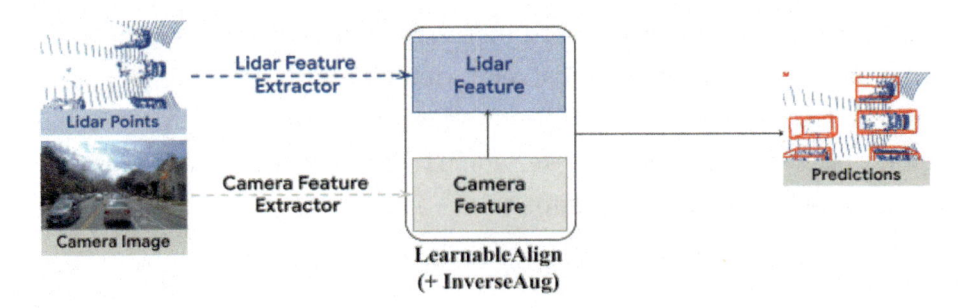

Figure 3-19. *High-level overview of the workflow for DeepFusion. Used with permission, source:* `https://arxiv.org/pdf/2203.08195`*[25]*

BEVFusion

Building on the idea of multisensor fusion, BEVFusion[27][28] takes a different approach to 3D perception tasks. It converts the input from both sensors into features and transforms them into a bird's-eye view (BEV) space, which is a top-down perspective of the environment. This unified view makes it easier to integrate the data. A BEV encoder then processes these combined features, which are used by specialized task-specific components to perform various 3D perception tasks, such as object detection and tracking, as illustrated in Figure 3-20.

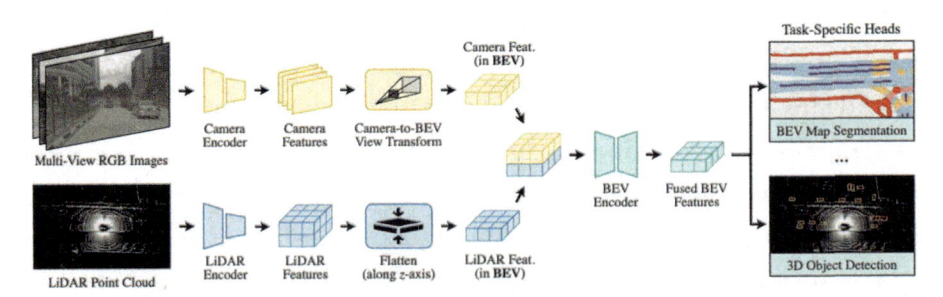

Figure 3-20. *BEVFusion process of extracting features from various inputs and turning them into a shared BEV space. Used with permission, source:* `https://arxiv.org/pdf/2205.13542`*[27]*

If you want to learn more about this approach, we recommend reading the original paper[27]. However, the main benefit of BEV is that it provides a consistent and unified top-down perspective that simplifies the fusion of the LiDAR and camera data. Since the data is in a shared space, there is improved spatial alignment that can improve the accuracy of 3D perception tasks.

Overall, sensor fusion can combine data from many sensors, such as cameras and LiDAR, to leverage their complementary strengths. This integration can improve the performance of common robotic perception tasks such as detection, segmentation, and tracking.

Summary

In this chapter you learned:

- How 3D data can be represented as voxels (3D pixels in a grid), point clouds (sets of 3D points), and meshes (polygonal surfaces). Point clouds can be processed using methods like PointNet, PointNet++, PointCNN, and Dynamic Graph CNN (DGCNN).

- Two main fusion strategies exist: early fusion (combining raw sensor data like point clouds and images directly) and late fusion (combining processed data such as bounding boxes from different sensors).

- LiDAR-camera fusion methods combine data from LiDARs and cameras. Techniques like Frustum PointNets, along with point-level fusion methods such as PointPainting and PointAugmenting, improve detection by projecting LiDAR points onto camera images

- Feature-level fusion approaches like DeepFusion align camera and LiDAR features using techniques such as InverseAug and LearnableAlign. Additionally, BEVFusion converts sensor data into a unified bird's-eye view (BEV), therefore improving 3D object detection and tracking.

The next chapter discusses how large language models (LLMs) are applied to robotic planning, control, and mapping, including the use of foundation models, multimodal approaches, and end-to-end robot control methods and diffusion models.

References

[1] Maturana, Daniel, and Sebastian Scherer. "VoxNet: A 3D Convolutional Neural Network for Real-Time Object Recognition." 2015 IEEE/RSJ International Conference on Intelligent Robots and Systems (IROS). IEEE, 2015.

[2] Chang, Angel X., et al. "ShapeNet: An Information-Rich 3D Model Repository." *arXiv preprint arXiv*:1512.03012 (2015).

[3] Qi, Charles R., et al. "PointNet: Deep learning on point sets for 3D classification and segmentation." Proceedings of the IEEE Conference on computer Vision and Pattern Recognition. 2017.

[4] Bento, Carolina. "Multilayer Perceptron Explained with a Real-Life Example and Python Code: Sentiment Analysis." *Medium*, 21 Sept. 2021, towardsdatascience.com/multilayer-perceptron-explained-with-a-real-life-example-and-python-code-sentiment-analysis-cb408ee93141.

[5] https://machinelearningmastery.com/pooling-layers-for-convolutional-neural-networks/

[6] Qi, Charles Ruizhongtai, et al. "PointNet++: Deep hierarchical feature learning on point sets in a metric space." *Advances in Neural Information Processing Systems* 30 (2017).

[7] Hu, Jordan. "Farthest Point Sampling in 3D Object Detection." Jordan Hu, 20 Sept. 2020, jskhu.github.io/fps/3d/object/detection/2020/09/20/farthest-point-sampling.html.

[8] Qi, Charles R., et al. "Volumetric and multi-view CNNs for object classification on 3D data." Proceedings of the IEEE Conference on Computer Vision and Pattern Recognition. 2016.

[9] Li, Yangyan, et al. "PointCNN: Convolution on x-transformed points." *Advances in Neural Information Processing Systems* 31 (2018).

[10] Wang, Yue, et al. "Dynamic graph CNN for learning on point clouds." *ACM Transactions on Graphics* (tog) 38.5 (2019): 1-12.

[11] "LiDAR and Camera Sensor Fusion in Self-Driving Cars." *Think Autonomous,* 14 May 2021, www.thinkautonomous.ai/blog/lidar-and-camera-sensor-fusion-in-self-driving-cars/.

[12] https://courses.thinkautonomous.ai/stereo-vision

[13] Bochkovskiy, Alexey, Chien-Yao Wang, and Hong-Yuan Mark Liao. "Yolov4: Optimal speed and accuracy of object detection." *arXiv preprint arXiv:*2004.10934 (2020).

[14] Zhou, Yin, and Oncel Tuzel. "VoxelNet: End-to-end learning for point cloud based 3D object detection." Proceedings of the IEEE Conference on Computer Vision and Pattern Recognition. 2018.

[15] Yan, Yan, Yuxing Mao, and Bo Li. "Second: Sparsely embedded convolutional detection." *Sensors* 18.10 (2018): 3337.

[16] Redmon, Joseph, et al. "You only look once: Unified, real-time object detection." Proceedings of the IEEE Conference on Computer Vision and Pattern Recognition. 2016.

[17] https://thekalmanfilter.com/kalman-filter-explained-simply/

[18] https://www.thinkautonomous.ai/blog/hungarian-algorithm/

[19] https://www.udacity.com/course/sensor-fusion-engineer-nanodegree--nd313

[20] https://medium.com/@mohit_gaikwad/deepfusion-lidar-camera-deep-fusion-for-multi-modal-3d-object-detection-c7db1e25670d

[21] Liu, H., Wu, C. & Wang, H. Real time object detection using LiDAR and camera fusion for autonomous driving. Sci Rep 13, 8056 (2023). https://doi.org/10.1038/s41598-023-35170-z

[22] Qi, Charles R., et al. "Frustum Pointnets for 3D Object Detection from rgb-d Data." Proceedings of the IEEE Conference on Computer Vision and Pattern Recognition. 2018.

CHAPTER 3   ROBOT PERCEPTION: 3D DATA AND SENSOR FUSION

[23] Vora, Sourabh, et al. "PointPainting: Sequential fusion for 3D object detection." Proceedings of the IEEE/CVF Conference on Computer Vision and Pattern Recognition. 2020.

[24] Wang, Chunwei, et al. "PointAugmenting: Cross-modal augmentation for 3D object detection." Proceedings of the IEEE/CVF Conference on Computer Vision and Pattern Recognition. 2021.

[25] Li, Yingwei, et al. "DeepFusion: LiDAR-camera deep fusion for multi-modal 3D object detection." Proceedings of the IEEE/CVF Conference on Computer Vision and Pattern Recognition. 2022.

[26] "LiDAR-Camera Deep Fusion for Multi-Modal 3D Detection." *Google Research,* 12 April 2022, `research.google/blog/lidar-camera-deep-fusion-for-multi-modal-3d-detection/?hl=ru&m=1`.

[27] Liu, Zhijian, et al. "BEVFusion: Multi-task multi-sensor fusion with unified bird's-eye view representation." 2023 IEEE International Conference on Robotics and Automation (ICRA). IEEE, 2023.

[28] D, Eric. "BEVFusion - Eric D - Medium." *Medium,* 16 June 2023, `medium.com/@yunding.eric/bevfusion-b9397afa0401`.

[29] Sharma, Sahil, and Vijay Kumar. "Voxel-based 3D face reconstruction and its application to face recognition using sequential deep learning." *Multimedia Tools and Applications* 79.25 (2020): 17303-17330.

[30] Willems, Markus. "A DSP for Implementing High-Performance Sensor Fusion on an Embedded Budget." *Semiconductor Engineering,* 11 Nov. 2021, `semiengineering.com/a-dsp-for-implementing-high-performance-sensor-fusion-on-an-embedded-budget/`.

CHAPTER 4

Foundation Models in Robotics

Foundation models developed for language research are now boldly taking on many fields, including robotics. Many of these advancements are fueled by making information look like language, i.e., making information resemble sequences of tokens. You learned in Chapter 2 how vision research is now fully dominated by transformers[43] by tokenizing the inputs (the images and outputs) to look like tokens. Foundation models and their rise in robotics naturally grew as an outcome of using language in robotics.

This chapter first explains, very briefly, how large foundation models are trained. Then it delves into how language became increasingly used as a connective tissue in robotics and how that research evolved into end-to-end robot control with large transformers. Finally, the chapter covers the rise of diffusion models and their applications in robot control.

Large Foundation Models

Language research has been leading the way in machine learning for the last decade. Several recent language models based on large-scale transformers have made the news, including GPT-4[1], Gemini[2], and

CHAPTER 4   FOUNDATION MODELS IN ROBOTICS

Llama[3]. They're products of a long line of research starting from Bert[4], earlier GPTs[5][6][7], PaLM[8], and more. This section discusses how large multimodal models are trained with Llama[9] as an example.

Developing a large multimodal model involves two main steps:

1. **Pretraining:** In this stage the model is trained on vast amounts of multilingual data from the Internet. In the case of Llama 3[10], a 405 B model was trained on 15.6 T tokens, with a token vocabulary of 128K, where 28K vocab tokens support non-English languages. Llama transformer uses grouped query attention[11] and key value caches[12] to improve inference speed (refer to the resources to learn more about these topics).

2. **Post-training:** In this stage, the model is improved on instruction following by using a smaller set of high-quality data coming from human annotations and synthetic data.

Llama 3 uses a compositional approach to multimodality and has separate vision and speech encoders.

1. **Image encoder and adapters:** An image encoder is trained on a large corpus of image text pairs, akin to CLIP[13], covered in Chapter 2, to generate representations that are aware of visual content and its language representations. Two separate adapters are trained for images and videos on image-text and video-text pairs respectively, during the training of which the pretrained language model is kept frozen, while the weights of the pretrained image encoder are edited. These adapters use cross-attention to feed pretrained image encoder representations into the pretrained language model.

CHAPTER 4   FOUNDATION MODELS IN ROBOTICS

2. **Speech encoder and image adapter:** Akin to the image setting, a separate speech encoder is trained and a speech adapter is trained with cross-attention to align the embeddings of the pretrained speech encoder and the large, pretrained language model.

Pretraining has three major components:

1. **Web data curation:** Involves creating a large high-quality dataset while applying techniques for *deduplication*, removal of personally identifiable information and adult content via both heuristic and learned filtering.

2. **Data mix determination:** Scaling law experiments are done to predict the performance of a large model by interpolating the performances of several smaller models in order to arrive at an optimal data mix, and to understand how different datasets contribute/affect the large model mixture. Ultimately, they arrive at 50 percent general knowledge tokens, 25 percent math and reasoning tokens, 17 percent code tokens, and 8 percent multilingual data.

3. **Initial pretraining:** Llama pretraining uses a cosine learning rate schedule with decay and warm up, and starts with a smaller batch size and sequence length, which is gradually increased as pretraining progresses.

4. **Long context pretraining:** Pretraining on long contexts is introduced during the final stages of pretraining, since compute in self-attention layers grows quadratically with context length. Supported

141

CHAPTER 4　FOUNDATION MODELS IN ROBOTICS

context length is increased in increments, with adaptation to increased context lengths measured by recovery of performance on shorter context lengths and needle-in-a-haystack[14] tests for the increased context length.

5. **Annealing:** Training on small quantities of very high-quality math and code data was found to improve reasoning across the board for smaller models.

Post-training has two major components: a reward model and a language model. It uses supervised fine-tuning and direct preference optimization (DPO)[15]. An example of a post-training strategy is shown in Figure 4-1. First a preference dataset is constructed by prompting the model and generating outputs that are then annotated by humans in order of their preference. Human annotators can also edit the chosen output, and a ranking mechanism (edited ➤ chosen ➤ rejected) is used to rank preferences.

1. **Supervised fine-tuning (SFT):** During this stage, a reward model is trained on preferences. This reward model is then used at scale by prompting the pretrained language model and rejecting outputs based on reward to create a rejection-sampled dataset of LLM outputs. Additional synthetic data is generated for code and math by deploying various strategies:

    a. Training a code expert by branching the pretraining run and training it on high-quality code tokens. The intuition here is that continued pretraining on domain-specific data improves performance in that domain. This expert is then used to create high-quality generations that comprise a dataset.

b. Code interpreters and linting is used to improve the quality of synthetic generations. Self-correction via chain-of-thought reasoning is used to further improve the quality of the dataset.

The pretrained language model is then fine-tuned on high-quality datasets acquired via synthetic generation and rejection sampling with a learned reward model.

2. **Direct preference optimization (DPO):** While prior methods in language model instruction tuning used reinforcement learning on human feedback (RLHF)[16], DPO is increasingly popular now because it is simpler. By not learning an intermediate reward model, and it bypasses some of the instabilities in training actor-critic RL models. Chapter 7 covers RLHF in detail. DPO instead directly optimizes for a reward constrained by KL divergence by directly increasing the relative log likelihood of a preferred response over nonpreferred ones. The KL divergence measure in DPO balances the tradeoff between maximizing rewards by preference alignment with minimizing the divergence from the base pretrained model.

CHAPTER 4   FOUNDATION MODELS IN ROBOTICS

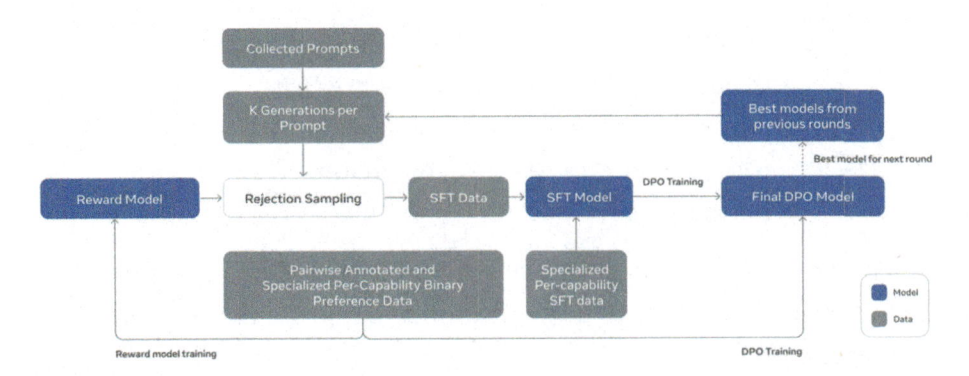

Figure 4-1.   LLM post-training strategy. Used with permission, source: https://arxiv.org/pdf/2407.21783 [10]

Llama3 is adapted for tool use by having a chat format where each message has a header specifying a source and a destination for a message. Each message also has a termination token specifying when to alternate between AI and human.

Scaling Laws for Language Models

Scaling laws for language models predict what performance or test loss one may achieve while scaling data, compute, or model parameter size. Since large model training runs are very expensive, understanding scaling behaviors helps forecast the capabilities of language models and make the right tradeoffs while investing into large model training runs. Note that almost all neural scaling laws are empirically determined and approximated via interpolation.

Kaplan et al.[17] came up with the first neural scaling laws for large language models by running lots of smaller models and interpolating their behavior. They considered three aspects of scale: the number of model parameters N, the dataset size D, and the amount of compute used C. They found that model loss has a power law relationship with each of the three factors when unconstrained by the others. This indicates that there is an efficient frontier that balances model size, dataset size, and compute to achieve the best possible performance with a given amount of resources. They also found that larger models are more sample-efficient than smaller models and that training to convergence is suboptimal. The results are shown in Figure 4-2.

$$L(N) = (N_c/N)^{\alpha_N} \; ; \quad \alpha_N \sim 0.076, \quad N_c \sim 8.8 \times 10^{13} \text{ (non-embedding parameters)}$$

$$L(D) = (D_c/D)^{\alpha_D} \; ; \quad \alpha_D \sim 0.095, \quad D_c \sim 5.4 \times 10^{13} \text{ (tokens)}$$

$$L(C_{\min}) = \left(C_c^{\min}/C_{\min}\right)^{\alpha_C^{\min}} \; ; \quad \alpha_C^{\min} \sim 0.050, \quad C_c^{\min} \sim 3.1 \times 10^8 \text{ (PF-days)}$$

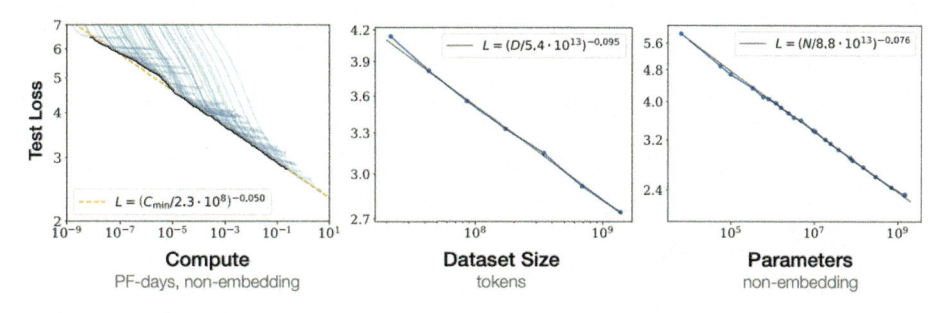

Figure 4-2. *Test loss as a function of compute, dataset size, and model size. Used with permission, source:* `https://arxiv.org/abs/2001.08361`*[17]*

Hoffman et al.[18], two years later, introduced Chinchilla optimal scaling laws. They tried to answer the question, given a fixed FLOPs[19] budget, what is the tradeoff between model size and dataset size?

They tried two approaches:

1. Fix the model size and train with different dataset sizes to plot the loss.

2. Fix the model size and train for different FLOP counts to plot the loss.

CHAPTER 4 FOUNDATION MODELS IN ROBOTICS

Based on the results of the two approaches, they fitted a parametric equation (Equation 4-1) for the data. The proposed fit is:

$$\hat{L}(N, D) \triangleq E + \frac{A}{N^\alpha} + \frac{B}{D^\beta}.$$

Equation 4-1

Then they estimated these parameters by minimizing the Huber loss between predicted log loss and observed log loss. A later study by Epoch AI that attempted to replicate the results from Hoffman et al.[20] came up with corrected versions of estimates for these parameters, as shown in Table 4-1.

Table 4-1. *Lost Function Estimates from Epoch AI Chinchilla Replication Attempt and Chinchilla's Original Estimates. Numbers in Brackets Are Standard Deviation. Used with Permission, Source* https://arxiv.org/pdf/2404.10102v1 *[20]*

Parameter	Our estimate	Hoffman et al's estimate
A	482.01 (124.58)	406.4
B	2085.43 (1293.23)	410.7
E	1.82 (0.03)	1.69
α	0.35 (0.02)	0.34
β	0.37 (0.02)	0.28
$a = \beta/(\alpha + \beta)$	0.512 (0.018)	0.454
Data points	240	> 400

Based on this, nearly optimal allocation of compute can be mapped to N and D separately, as follows:

CHAPTER 4   FOUNDATION MODELS IN ROBOTICS

$$N_{opt}(C) = G(C/6)^a \quad D_{opt}(C) = G^{-1}(C/6)^b$$

$$\text{where} \quad G = \left(\frac{\alpha A}{\beta B}\right)^{\frac{1}{\alpha+\beta}} \quad a = \frac{\beta}{\alpha+\beta} \quad b = \frac{\alpha}{\alpha+\beta}$$

Equation 4-2

The results from Chinchilla indicate that many of the frontier models trained at the time, such as GPT3, were undertrained on data, so the optimal model would be much smaller (15B from 175B) in parameter count. It also follows from the results that, for a fixed compute budget, scaling data is far more effective than scaling model parameters.

Muennighoff et al.[21] tried to predict scaling laws in data-constrained regimes and concluded that repeating data up to four epochs yields no difference in loss, and that to optimize performance while repeating data, training smaller models for longer is preferred. This is shown in Figure 4-3.

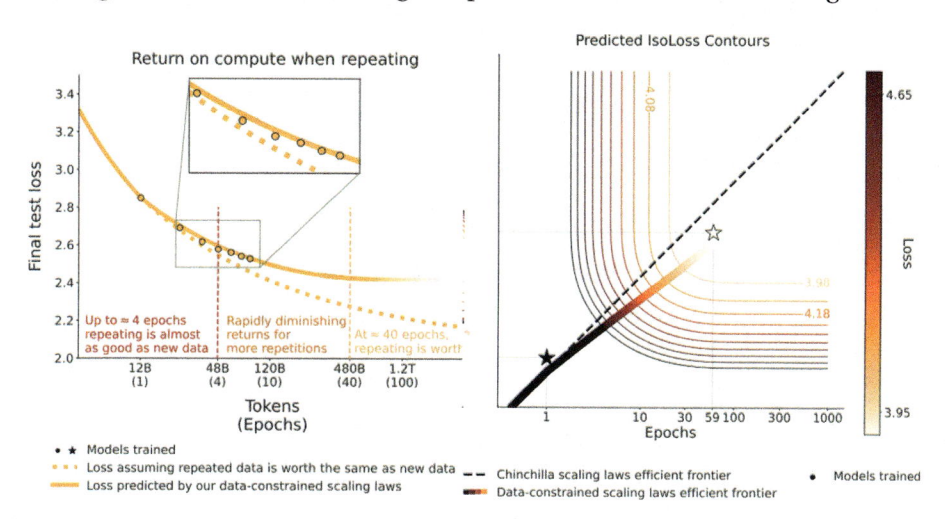

Figure 4-3. *Top: LLM loss trained on repeated data diminishes predictably. Bottom: How data-constrained scaling laws diverge from unconstrained optimal scaling: increasing distance of contours show diminishing gains from repeating data. Used with permission, source:* https://arxiv.org/pdf/2305.16264[21]

CHAPTER 4   FOUNDATION MODELS IN ROBOTICS

Recent research from Sorscher et al.[22] showed that the power law relationship between training loss and dataset token size could be broken by ranking the examples from hard to easy on a high-quality metric and then pruning the easy examples out of the dataset. They posit that such pruning can even lead to an exponential relationship.

Evaluating Language Models

Now that you understand how language models are trained and scaled, you'll learn how they are evaluated. Language model research predominantly uses benchmarks to evaluate performance. The following are a few widely used benchmarks, sorted by category and what they are testing for:

1. General Benchmarks

    - *MMLU[23] (Massive Multitask Language Understanding):* Evaluates understanding and reasoning across 57 academic disciplines with 14,000+ multiple-choice questions.

    - *MMLU-Pro[24]:* A more challenging version of MMLU with harder reasoning tasks.

    - *AGIEval[25]:* Assesses general intelligence of AI models across various tasks.

2. Math and Reasoning

    - *GSM8K[26]:* Tests grade-school math problem-solving with 8,000 step-by-step reasoning tasks.

    - *MATH[27]:* Evaluates high school and competition-level math problem-solving abilities.

    - *ARC-C [28] (AI2 Reasoning Challenge - Challenge Set):* Assesses complex reasoning and problem-solving skills.

3. Common Sense Understanding

    - *CommonSenseQA[29], OpenBookQA[30], PiQA[31], SiQA[32]:* Evaluates common sense reasoning and logical understanding in everyday scenarios.

    - *CommonSenseQA[29]:* A benchmark for evaluating common sense reasoning, testing models on multiple-choice questions.

    - *OpenBookQA[30]:* Tests open book question answering with a focus on science questions requiring additional knowledge beyond the text.

    - *PiQA[31]:* Evaluates physical common sense reasoning by requiring the choice of the most plausible completion for a given scenario.

    - *SocialIQA (SiQA)[32]:* Focuses on common sense reasoning about social situations and human interactions.

4. Code

    - *HumanEval[33], MBPP[34] (Multi-Task Benchmark for Programming Problems):* Assesses coding proficiency and problem-solving in programming contexts.

5. Reading Comprehension

    - *SQuAD[35] (Stanford Question Answering Dataset):* Tests reading comprehension and information extraction from Wikipedia articles.

    - *QuAC[36] (Question Answering in Context):* Evaluates contextual understanding and coherence in conversational question answering.

CHAPTER 4   FOUNDATION MODELS IN ROBOTICS

- *RACE[37] (Reading Comprehension from Examinations):* Assesses reading comprehension using middle and high school English exam questions.

6. Long Context Benchmarks

- *Needle-in-a-Haystack[38]:* Tests document comprehension with a focus on finding relevant information in long texts.

- *ZeroSCROLLS[39]:* Assesses the ability of models to perform various NLP tasks using minimal data annotations.

- *InfiniteBench[40]:* Focuses on long document understanding and reasoning tasks, including summarization and question answering.

7. Tool Use Benchmark

- *API-Bank[41], API-Bench[42]:* Evaluates the ability to interact with and utilize various external APIs effectively.

In addition to these, they're also evaluated on AP tests, GMAT, GRE, bar, and medical exams and various other specialized tests.

This section discussed how language models are trained, how they scale, and on what they are evaluated. The next section explains how language models are used in robotics. You learn that almost all robotics data can be modeled as tokens and sequences and processed like language for various downstream use cases.

CHAPTER 4   FOUNDATION MODELS IN ROBOTICS

Language as a Connective Tissue in Robotics

Language has long been used as a medium of communication among humans, so much so that a large part of the wealth of human knowledge and experience is encoded in language. Language is also fairly flexible and fluid, allowing for communication of a wide range of topics to a certain degree of precision. The rise of transformers[43] led to breakthroughs in language understanding, language generation, and reasoning with language. In recent years, language in robotics has become an interesting and promising new research area, allowing robots greater generalization with a language-predicated understanding of world context. Language models have provided a method for simple and extensible multimodal fusion of modalities, including vision and action.

Language-conditioned robotics is the idea of specifying robot goals in natural language and measuring success against correct achievement of those goals. Language conditioning can be used for planning ("how would you bring me a coke from the fridge"), control ("pick up the coke can"), mapping ("where is the coke can"), and navigation ("go to the coke can"). This section explains that almost all robotics data can be modeled as tokens and sequences and processed like language in a similar fashion for various downstream use cases.

Language for Planning

When humans interface with robots in natural language, a typical instruction could be "bring me a drink from the fridge," which consists of multiple steps such as "go to the fridge," "open the fridge door," "pick up the drink," "close the fridge," "go to the human," and "place the drink down." This requires a robot to perform embodied reasoning in an unseen

CHAPTER 4 FOUNDATION MODELS IN ROBOTICS

environment. If we ask a language model, say, "how do you clean up a spill," it may provide steps that are inaccessible to a robot. For example, it could say "find a vacuum cleaner" as one of the steps, but if a vacuum cleaner is not present in the scene, this would be an incorrect plan. A second mode of failure could be when the LLM suggests a plan that is feasible but that is not part of the robot's skillset, e.g., "use a vacuum cleaner," when a vacuum cleaner is present but the robot hasn't been trained to use one. One of the challenges of robot-human interaction is that robots have a finite set of skills that they're trained on and the vocabulary of task instructions conveyed to them needs to be limited by their skillset.

The property that refers to how feasible and successful a task could be in a particular scenario is formally referred to as *affordance*, usually expressed as a float value between 0.0 and 1.0, where 0.0 means highly impossible and 1.0 is possible and highly likely to succeed.

There are two classes of planning approaches:

1. **In-context learning:** Here a language model is shown examples of robot plans in context via prompting and asked to generate new plans. The language model is not specifically fine-tuned for robot learning. An example of an in-context learning algorithm is SayCan/Inner Monologue, explained in this next section.

2. **Fine-tuned/learned planning models:** Here, a general-purpose language/multimodal language model is specifically trained/fine-tuned for the purpose of robot planning. Roboplan and PaLM-E are examples of this.

CHAPTER 4   FOUNDATION MODELS IN ROBOTICS

Open Loop SayCan

SayCan[44] is an algorithm that integrates language for robotic planning. An overview of SayCan is shown in Figure 4-4.

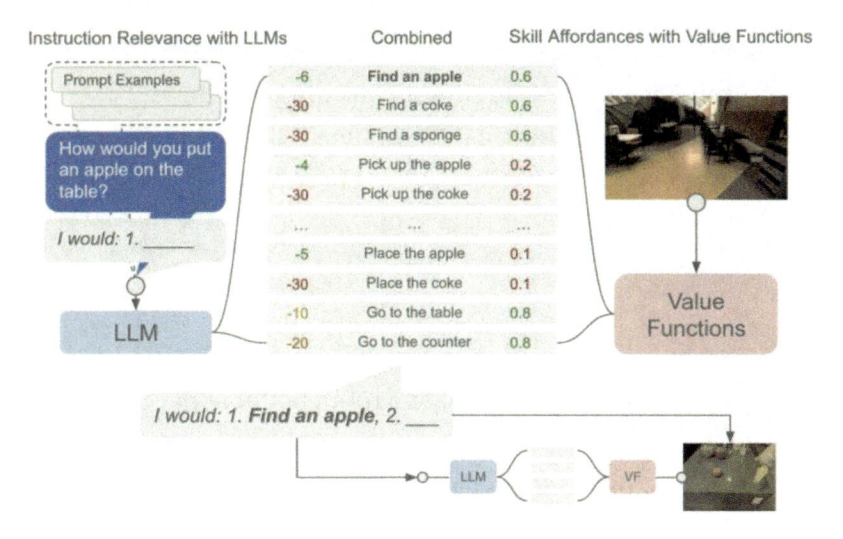

Figure 4-4. *SayCan combines probabilities from a LLM to determine which skill is useful for a given instruction with probabilities from a value function (VF) that assesses the feasibility of executing the skill. Used with permission, source:* `https://say-can.github.io/assets/palm_saycan.pdf`*[45]*

The main idea is as follows:

- A language model is used to break down high-level task instructions into skill primitives that a trained manipulation policy can then execute. The LLM models the probability that a high-level task instruction is achieved by compositionality of skill primitives.

- To ground the language model in the skill space of the robot, a second affordance model is independently learned. The affordance model learns the probability

153

of success of a skill primitive given the current state of the robot and environment. Value functions are a close approximation of such an affordance model and can be trained using reinforcement learning.

- The two probabilities are multiplied for each skill primitive and the one with the maximum value is picked for execution, as shown here.

$$p_\pi^{\text{LLM}} = p(\ell_\pi | i, \ell_{\pi_{n-1}}, ..., \ell_{\pi_0})$$ ▷ Evaluate scoring of LLM
$$p_\pi^{\text{affordance}} = p(c_\pi | s_n, \ell_\pi)$$ ▷ Evaluate affordance function
$$p_\pi^{\text{combined}} = p_\pi^{\text{affordance}} p_\pi^{\text{LLM}}$$

Another interesting observation from this work is that simply updating the underlying language model makes a robot better at planning. As shown in Figure 4-5, PaLM-SayCan has been shown to have increasing planning accuracy when scaling the language models.

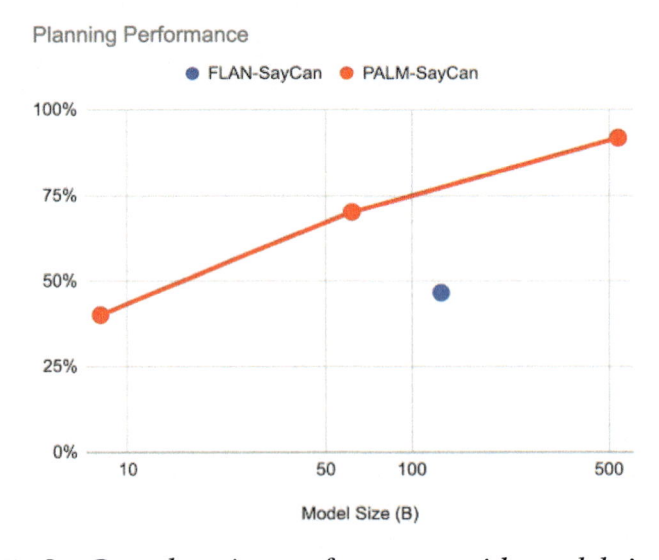

Figure 4-5. SayCan planning performance with model size. Used with permission, source: https://x.com/hausman_k/status/1559558929297727489 [96]

Closed Loop Planning

One of the problems with SayCan-style language predicated planning is that if any of the intermediate steps fails, the whole process execution fails. The algorithm is not interactively replanning or taking feedback from the environment. Inner Monologue[46] poses an improvement over SayCan by closing the feedback loop from the environment to the language model. It considers three types of feedback:

1. **Success detection:** A binary classification, in language form, of whether the task is successful or not at a given timestep.

2. **Passive scene description:** Feedback from the scene queried through external object detectors/scene descriptors. They're primarily of two types:

    a. *Object feedback:* Textual output of object locations from object detectors

    b. *Scene feedback:* Task progress scene description in text form

3. **Active scene description:** The language model interactively asks questions about the scene/task, which are answered by a visual question answering model.

Figure 4-6 shows Inner Monologue in different scenarios.

CHAPTER 4 FOUNDATION MODELS IN ROBOTICS

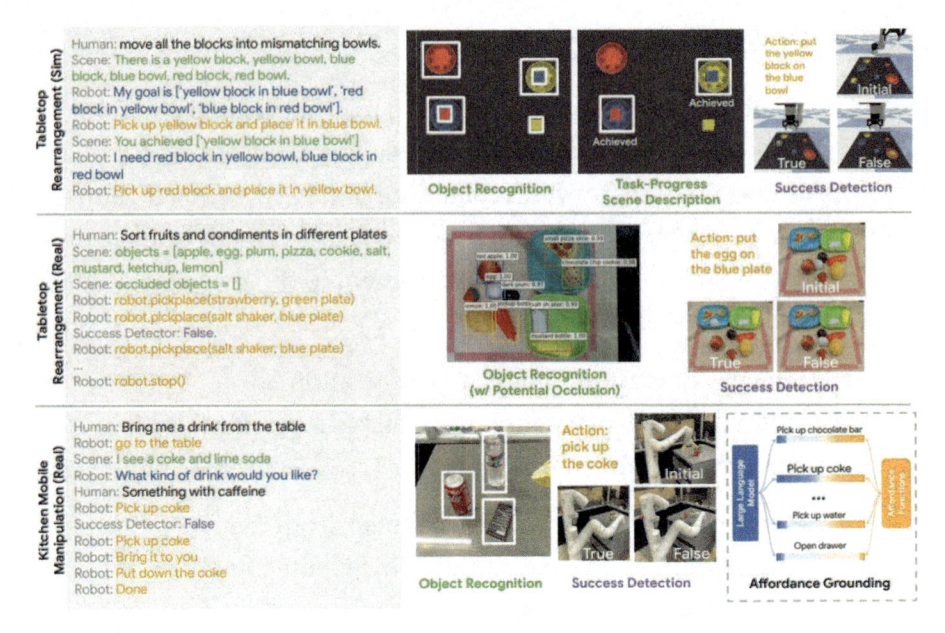

Figure 4-6. Example of Inner Monologue planning in robot manipulation tasks showing simulated and real-world scenarios for tabletop rearrangement and kitchen mobile manipulation. Used with permission, source: https://arxiv.org/pdf/2207.05608[46]

Inner Monologue reports a higher performance rate over SayCan and a few emergent capabilities, such as robot reacting to intermediary feedback from humans, replanning under infeasibility, and better interactive scene understanding demonstrated through question-answering.

Multimodal Planning

In addition to language feedback, multimodal models specifically fine-tuned for robotic planning have two advantages:

1. More precise than open loop planning in grounded, unseen environments

2. Ability to still remember larger context from Internet-scale knowledge

CHAPTER 4 FOUNDATION MODELS IN ROBOTICS

3. Visual feedback and states in the loop yield a model that can see and think simultaneously

One of the prominent works in multimodal planning is PaLM-e[47] (see Figure 4-7), also called Embodied PaLM.

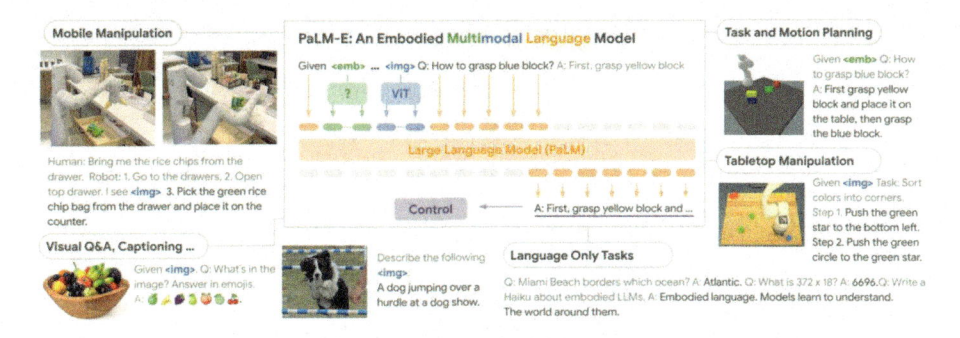

Figure 4-7. Overview of PaLM-E used for various tasks, such as mobile manipulation, visual Q&A, and task and motion planning. Used with permission, source: https://arxiv.org/pdf/2303.03378[47]

PaLM-e encodes continuous inputs into a decoder-only LLM (e.g., PaLM). To encode the image, it uses the ViT 22B[48] and feeds image tokens concatenated with language. The inputs to PaLM are multimodal sentences that interleave continuous inputs with text. An example of such a multimodal sentence from the paper is "Q: What happened between and ?" (represents an embedding of an image). PaLM-e then outputs text autoregressively. The outputs could be answers to questions, or a sequence of decisions/plans executed by a lower-level policy. One impressive result of the PaLM-e paper is that fine-tuning on robot text/plans doesn't significantly decrease performance in generic language skills, paving the way for one model for all high-level/low-level tasks in the future.

CHAPTER 4   FOUNDATION MODELS IN ROBOTICS

Planning via Multimodal Dialogue

A second approach to language-based planning is via visual question answering using multimodal models. RoboVQA[76] is an example of such a framework, using learned/fine-tuned models. This involves showing a model the last few seconds of a sensor feed and asking questions such as:

1. "The high-level goal is to stack the cups. What should the robot do next? So far we executed 1. Stack blue cup on red cup 2.".

2. "What just happened?"

3. In this context, generative affordance is a type of question that generates possible tasks, such as "What tasks can the robot do now?" and discriminative affordance is a type of question that classifies a proposed task into feasible or not. For example, "Is it possible to pick up a cup now?"

The output of the multimodal model is used for robot execution.

Challenges

Despite the rise of large language models and computation and planning via language, several problems remain to be solved with respect to LLM robotic planning:

1. There are fundamental failure modes when language models cannot see and do not have an understanding of physics. One way to bridge this gap is to use a VLM[49] in the loop or, as in PaLM-E, use a connected vision model. Both approaches are susceptible to fragility. In the former approach, the understanding of object relations is constrained

by the VLM's ability to describe, which can miss several things For example, if a VLM finds a phone and a person, a robot may plan to pick up the phone even if the phone is being held by the person, thus ignoring object relations. Object detection methods have their own failures that overall bring down planning accuracy.

2. A second failure mode is failure in understanding physics. For example, LLMs might not know how much an object weighs, whether it is safe to handle, how to approach the target object based on surrounding obstacles, and so on. Endowing language models with physics understanding[50] is an ongoing area of work.

3. A third failure mode is embodiment-specific planning. A specific robot may have certain strengths and weaknesses that need to be factored in with regards to its planning. This could include payload, number of arms, reachable height, blind spots of sensor suite, and so on. While affordance functions learned via RL can get some idea of expected reward, coverage over state action space may not be enough for thorough consideration of embodiment attributes.

4. Planning in an unseen environment is very hard for LLMs that suffer from hallucinations, faulty reasoning, and so on.

Language for Mapping

In addition to high-level planning, language can be used for mapping and localization. Earlier in this chapter, you learned that through affordance functions, which map the environment and potential interactions robots can have with objects, large language models can be used for robotic planning, such as in SayCan. One of the drawbacks of SayCan is that it is unable to reason over what a robot can do in a scene beyond what the robot can see. This is important because many indoor mobile manipulation tasks involve knowing where objects are even if they are not visible in the image.

NLMap-SayCan[51] aims to address this issue by proposing potential objects and generating plans to execute on actions even when those objects are not directly visible. As a recap, SayCan takes in an instruction through an LLM to generate and score possible actions from a predefined list, then uses value functions to select the most feasible action based on the current state of the environment. In contrast, NLMap[51] creates a map of the environment and uses the value functions to evaluate the feasibility of the actions. This ensures that the selected action is relevant to the instruction but also feasible within the context of the environment.

Figure 4-8 shows the main differences between SayCan and NLMap-SayCan. Overall, NLMap addresses SayCan's limitation in planning with a stronger contextual understanding through include maps.

CHAPTER 4   FOUNDATION MODELS IN ROBOTICS

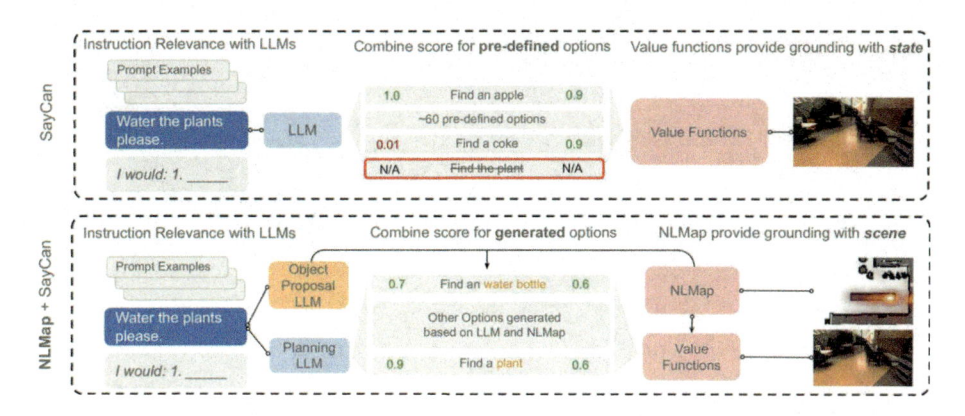

Figure 4-8. *Comparison of SayCan and NLMap-SayCan. NLMap-SayCan generates relevant objects dynamically using an LLM and queries the NLMap for object locations. This allows for more flexible and context-specific task planning. Used with permission, source: https://arxiv.org/pdf/2209.09874[51]*

NLMap starts with an instruction, such as "pick up the coke bottle," and by allowing the robot to explore, collects RGB-D images. Then ROIs are generated from these images to determine where the objects are. 3D information about the scene is learned through a VLM, which creates a detailed scene map. The LLM then proposes relevant objects based on the instructions, such as "coke" and "bottle," and the VLM checks the scene map to determine if and where these objects are, scoring how likely they are to be found in different places in the scene. Scene objects are then updated to match the most relevant ones and LLM plans how to execute the task. This process is shown in Figure 4-9.

CHAPTER 4 FOUNDATION MODELS IN ROBOTICS

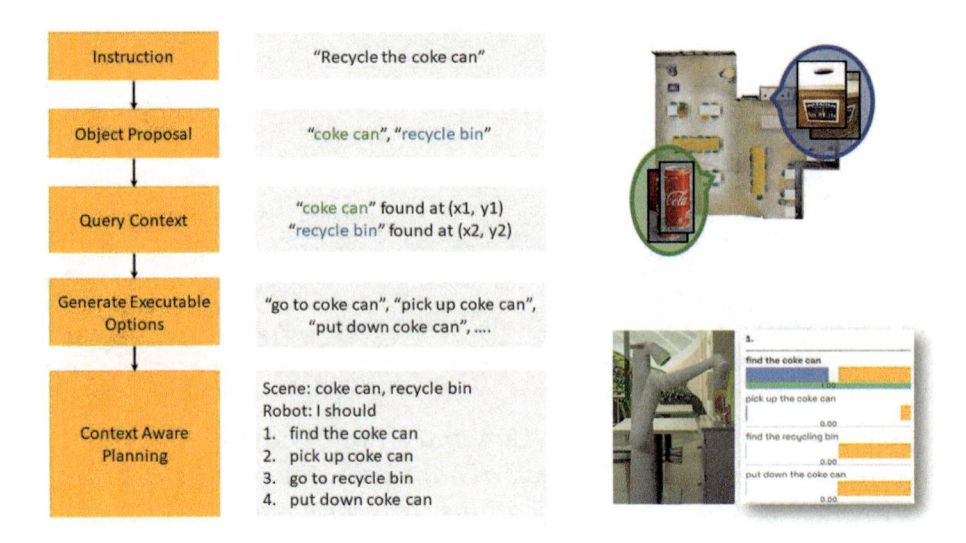

Figure 4-9. NLMap-SayCan generates relevant objects based on an instruction, then queries NLMap to filter the object list and find their locations. The system creates executable options and uses LLM-based planning to guide the robot through the task step-by-step. Used with permission, source: https://nlmap-saycan.github.io/[51]

This approach does have a few limitations worth mentioning that were highlighted in the paper: it struggles with understanding how objects relate to each other within the environment and handling changing scenes. It's also important to note that the accuracy of object proposals relies on the performance of the LLM. Although these limitations exist, NLMap-SayCan is still a leap forward in more robust robotic planning.

Language for Reward

Once planning is complete, the robot must execute the planned actions, which involves accurately doing robotic control and motion. LLM training data often lacks information about physical actions and embodiment, making it harder to use LLMs for direct robotic control and motion, as these scenarios fall outside their typical training distribution.

CHAPTER 4 FOUNDATION MODELS IN ROBOTICS

What if a robot could learn to execute the tasks more accurately by having an internal framework of what's "wrong" and what's "good"? This is exactly what reward functions can do for LLMs and control policies, by providing a robot with feedback to guide the robot's actions and improve its accuracy. "Language to Rewards for Robotic Skill Synthesis (L2R)"[52] is a paper that uses reward functions to determine controls for a robot to execute different tasks in its environment.

L2R works as follows:

- It takes in a user instruction, which in the case of Figure 4-10 is: "Make robot dog stand up on two feet."

- The bulk of the work is done by the Reward Translator, which breaks down the high-level command into more detailed and specific actions that the robot can execute. Within this component, the Motion Descriptor translates the user's instruction into a detailed motion description. For instance, it could say something like, "Both front feet should be lifted to 0.7 meters high."

- The motion description is passed to the Reward Coder, which converts the description into actionable reward functions for the robot's control system using a predefined API that the robot is set up to understand.

- These reward functions are sent to the Motion Controller, which is responsible for executing the actions using the reward functions to control the robot's movements and improve on them.

CHAPTER 4   FOUNDATION MODELS IN ROBOTICS

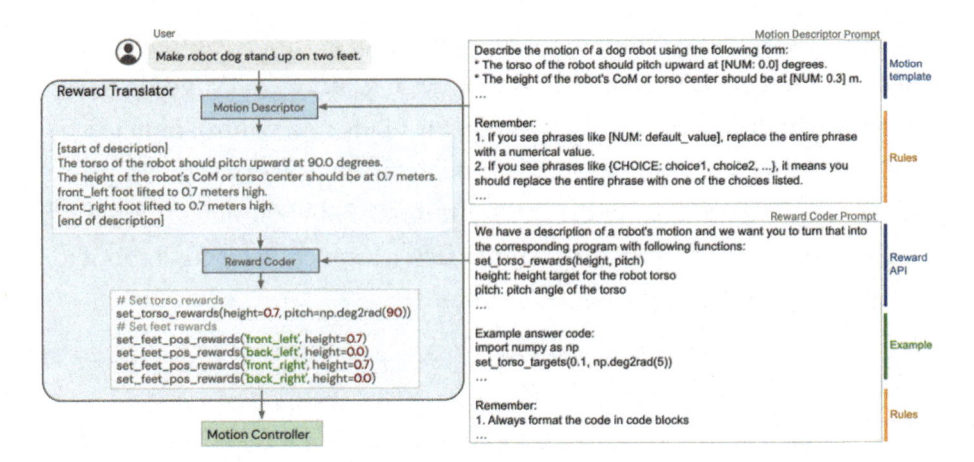

Figure 4-10. Depiction of L2R, with motion descriptor and reward coder. Used with permission, source: `https://arxiv.org/pdf/2306.08647`*[52]*

A key limitation to this approach is that it requires a predefined template that can be used for the robot to understand the motion descriptions. This can make it difficult to generalize across different robots and modalities.

Language for Robot Code

A key theme in methods like L2R is developing a "standardized" way for robots to understand the motion descriptions necessary to complete tasks. Given that robots can be programmed and controlled, can we leverage LLMs to improve their ability to execute these tasks more effectively by writing code?

Code as Policies[53] is a method that aims to do this by using LLMs to translate natural language commands into executable robot policy code. When a user provides an instruction, such as "put these fruits in a line," the LLM processes this input and generates corresponding code that includes a perception APIs to identify relevant objects in the environment

CHAPTER 4   FOUNDATION MODELS IN ROBOTICS

and control APIs to manipulate these objects. This code includes specific functions that identify these objects and execute the required actions by moving, arranging, or manipulating them. An example of code that could be generated by Code as Policies is shown in Figure 4-11. The interesting aspect of Code as Policies is that it relies solely on prompt engineering with a pretrained LLM without specific task training, meaning it could theoretically generalize to tasks involving objects and actions that fall within the distribution of the pretrained LLM. By generating and adjusting code from examples, robots can learn various tasks and follow new instructions in everyday language.

```
# stack the blocks in the empty bowl.
empty_bowl_name = parse_obj('empty bowl')
block_names = parse_obj('blocks')
obj_names = [empty_bowl_name] + block_names
stack_objs_in_order(obj_names=obj_names)
```

Figure 4-11. *Example of input instructions formatted as comments (green). The instructions are interpreted by an LLM, generating valid Python code (highlighted) to complete tasks such as stacking blocks in an empty bowl. Used with permission, source:* https://arxiv.org/pdf/2209.07753 [53]

To generate these policies, a language model is prompted with a few examples. An LLM then autoregressively and hierarchically creates them by combining well-known functions or invoking other programs to define functions that are not clearly defined already. As a framework, this is particularly interesting for robots because collecting large amounts of task-specific data is challenging, and using already trained large models allows for a more efficient solution. Visit [54] to learn more.

CHAPTER 4 FOUNDATION MODELS IN ROBOTICS

Recent research has shown that robot code generation can be improved in more ways as LLM research matured:

1. Robot code generation can be multimodal conditioned by using VLMs, which are good at code generation, and feeding them images from the robot's camera directly, rather than feeding natural language description of images into an LLM. This allowed reasoning to be more natively multimodal than in Code as Policies.

2. Feeding a VLM highly descriptive definitions of robot APIs and using long context models that are good at chain-of-thought reasoning improves code generation by large margins and can invoke unseen behaviors zero-shot without any training data.

3. Showing examples of correct code generation also improves the likelihood of being correct.

These and a few other tricks on how to prompt a robot can be further studied here[55]. An example is shown in Figure 4-12.

CHAPTER 4   FOUNDATION MODELS IN ROBOTICS

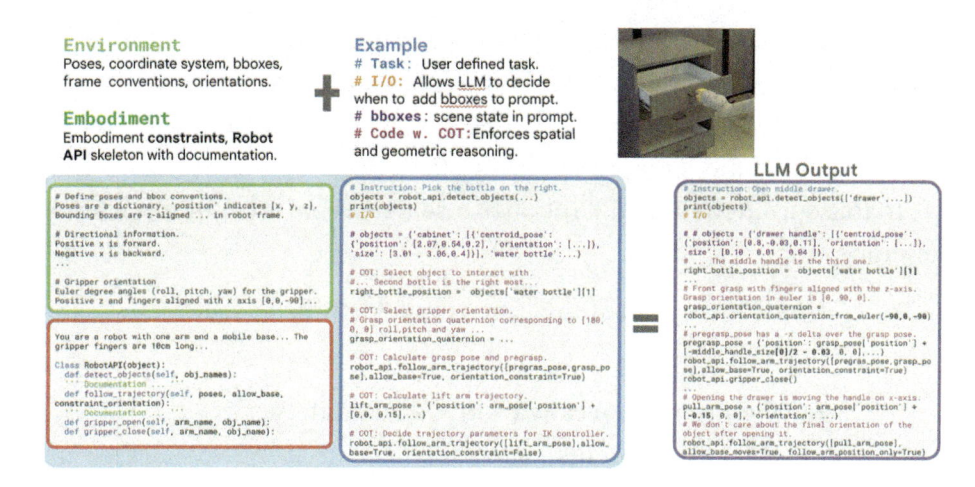

Figure 4-12. An example code gen input and output from "How to Prompt your Robot." Used with permission, source https://openreview.net/pdf?id=T8AiZj1QdN *[55]*

The success of code generation for zero-shot control signals that there's a lot more opportunity in extracting movements from the world knowledge contained in LLMs via code, and make it a learnable recipe for scaling robot learning. One potential opposition to this idea is the notion that, although some methods have very good zero-shot performance, they may not be enough to scale with learning resembling a method that has the potential to reach 100 percent success. Improving zero-shot methods via feedback is a common post-training strategy in LLM research, so it may quite be possible that Code as Policies can be framed into a continuously improving recipe, if it is also adapted to gather new skills, unknown to an LLM via demonstrations.

End-to-End Robot Control

An important concept in robotics is *generalization*: can a model perform a new task that it was not trained on? The importance of this concept emerges from the fact that generally intelligent robots have to be successful

CHAPTER 4 FOUNDATION MODELS IN ROBOTICS

across a wide range of deployment environments and tasks, but it would be hard to represent this diversity in training sets. So methods that can generalize to new situations, via learning better representations or reasoning about the semantics of tasks, stand to gain.

In this current moment of robotics research, there are multiple axes of generalization:

1. **Object generalization**: Can a model handle unseen objects?

2. **Environment generalization**: Can a model act in unseen environments?

3. **Motion generalization**: Can a model create new motions?

4. **Perspective generalization**: Can models transfer from third person to egocentric perspectives and vice versa? Does the pose of the camera matter?

5. **Embodiment generalization**: Can data on one type of robot body be used to improve skills on a different type of robot body?

Additionally, robot control models may be measured on symbolic understanding, reasoning, longer horizon planning, human recognition, physical safety, and so on.

End-to-end robot control encompasses the problem of learning low-level control in a way where gradients flow all the way from control outputs to robot inputs[56]. In conventional usage, this means learning actions directly from using camera images and other observations as input. How do we develop methods that can control robots end-to-end? Can we connect these models to the large trend of scaling and large models in AI? How do we measure and improve them on axes of generalization?

CHAPTER 4 FOUNDATION MODELS IN ROBOTICS

End-to-End Robot Control with Autoregressive Transformers

Until as recently as 2022, scaling large multitask reinforcement learning models was considered the key to solving robotics, but since then the wave has shifted toward large imitation learned models, mainly because imitation pretraining started to perform better than RL methods on multitask benchmarks. Robotics Transformer 1 (RT-1)[57], shown in Figure 4-13, was created in this era, and it was an early large-scale imitation learned multitask model. It is trained on a large corpus of real-world robot data covering a diverse range of tasks and is an early foundation model. RT-1 works as follows:

- A user instruction in text and a series of images (history) captured by the robot are the inputs to the RT-1 model. Text is encoded by a frozen text tokenizer and then images and text are jointly fused by a FiLM-efficient net[58][59]. Early fusion of vision and language tokens proved very important in extracting the right context from images.

- These visual language tokens are then compressed by a TokenLearner[60], which uses attention to focus on the most relevant parts of the images and text. TokenLearner compresses the total amount of tokens, and makes inference three times faster.

- These tokens are then processed by a transformer, which uses self-attention to understand the relationship between objects in the images and the commands in the text[43]. It then outputs actions: a single control parameter for every degree of freedom. In case of a cartesian end effector control, that becomes

169

a gripper 3D position, 3D rotation, base movements, and so on. Actions that are continuous values are tokenized by mapping them to a discrete uniform distribution, and then predicting the bucket number corresponding to the float action.

- The model is trained with cross-entropy loss, and the training data (action outputs and image/text inputs) is collected by humans teleoperating the robot.

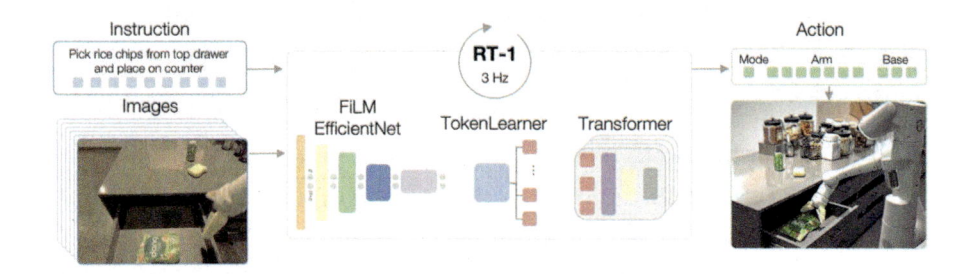

Figure 4-13. RT-1 processes images and natural language instructions to control a robot's actions. Used with permission, source: https://arxiv.org/pdf/2212.06817[57]

RT-1 performs closed-loop control and executes actions until it either produces a termination action or reaches a predetermined number of time steps. This work shows that given enough domain data, a transformer based model can fit to it to create a large multitask transformer for robot control. The work also showed that simulation data, when added in the right mix, can improve performance, as you saw with synthetic data in LLM domain. RT-1 also showed signs of transfer between robots that later emerged to be a wider area of research, as is covered later in this chapter.

One drawback of RT1 is that it was tested mainly on gross manipulation—pick and place—and this makes it harder to know if it can work for highly dexterous tasks. Dexterous tasks are manipulation tasks that require fine motor control and coordination similar to human hands

CHAPTER 4 FOUNDATION MODELS IN ROBOTICS

manipulating objects in their environment. Within robotics, this means dealing with objects of different texture, sizes, and shapes. A second drawback is that while it showed impressive generalization in domain, its ability to generalize outside of domain—with entirely new objects and with reasoning—was limited.

Another model, very similar to RT-1, that is widely used today is ACT (Action Chunking Transformer)[61]. ACT is an encoder-decoder model that introduces a concept called "action chunking." Action chunking refers to the method of modelling k steps into the future. At any given point in time, you get an observation but predict k actions, effectively reducing the horizon of the task by k fold. This outperforms single-step policies in situations where there are temporally correlated confounders, such as when there's a pause in demonstration and the next action is not just a function of state but also of timestep. As long as these confounders fall inside the length of the chunk, an action chunking policy can recover. In practice, chunk length corresponds to 1s of horizon into the future. That is, if your control is running at 30hz, a single observation, instead of predicting one action as in RT1, would predict 32 actions, or 1 second horizons. In practice, training on chunks also employs temporal ensembling, along with chunking. As shown in Figure 4-14 on the right, this means that instead of training on disjoint chunks, chunks for observations overlap, allowing for very dense modeling.

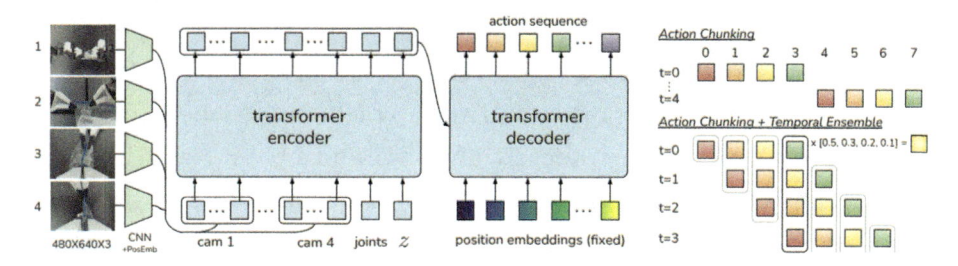

Figure 4-14. Action chunking encoder-decoder transformer architecture on top; chunking with temporal ensembling on bottom. Used with permission, source: `https://arxiv.org/pdf/2304.13705`*[61]*

171

CHAPTER 4 FOUNDATION MODELS IN ROBOTICS

RT-2[62] is a method built on top of RT-1 that uses the semantic information that pretrained vision-language models have to generate low-level robot actions.

RT-2 uses visual question answering (VQA) data, where there are images paired with questions and answers describing the context within the images, and robot action data, which pairs images and text instructions with robot actions that need to be executed to successfully complete that task. The model first uses a pretrained VLM trained on Internet-scale data, then co-trains it with robot action and Internet-scale data to get a vision-language-action (VLA) model for the robot control. This training recipe is depicted in Figure 4-15.

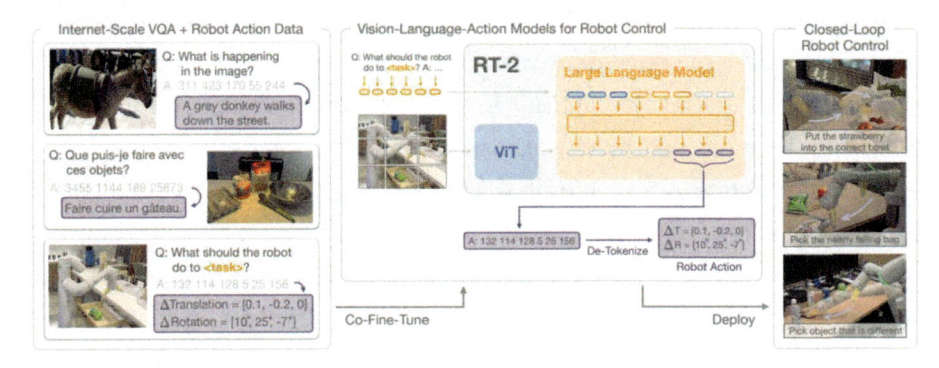

Figure 4-15. *RT-2 represents robot actions as text tokens, trained together with large-scale vision-language datasets. Used with permission, source: https://arxiv.org/pdf/2307.15818[62]*

Inference starts with a user query, such as "What should the robot do to pick up the apple?" This query and images of the scene where the user wants the task executed are processed by a ViT and a LLM. The ViT extracts visual features from the images and the LLM understands the language part of the query. These combined features generate a sequence of actions represented as specific translations and rotations for the robot to do.

The results of the paper denote that co-training on Internet and robot datasets jointly in this fashion allows robots to understand concepts only seen in the Internet and practically use them in real-world situations.

CHAPTER 4   FOUNDATION MODELS IN ROBOTICS

For example, the robots are able to identify celebrities ("move can to Taylor Swift"), do simple math ("move coke can to the sum of 1 + 2") and understand symbols ("move coke can to Google"). It even understands other relative concepts, such as "pick up the object with different color," "put strawberry into the correct bowl," and so on, that require one to reason among the options in the scene, when the question is asked in natural language. The work notices that Internet-scale pretraining leads to much better representations than previously possible. Some of these results are shown in Figure 4-16.

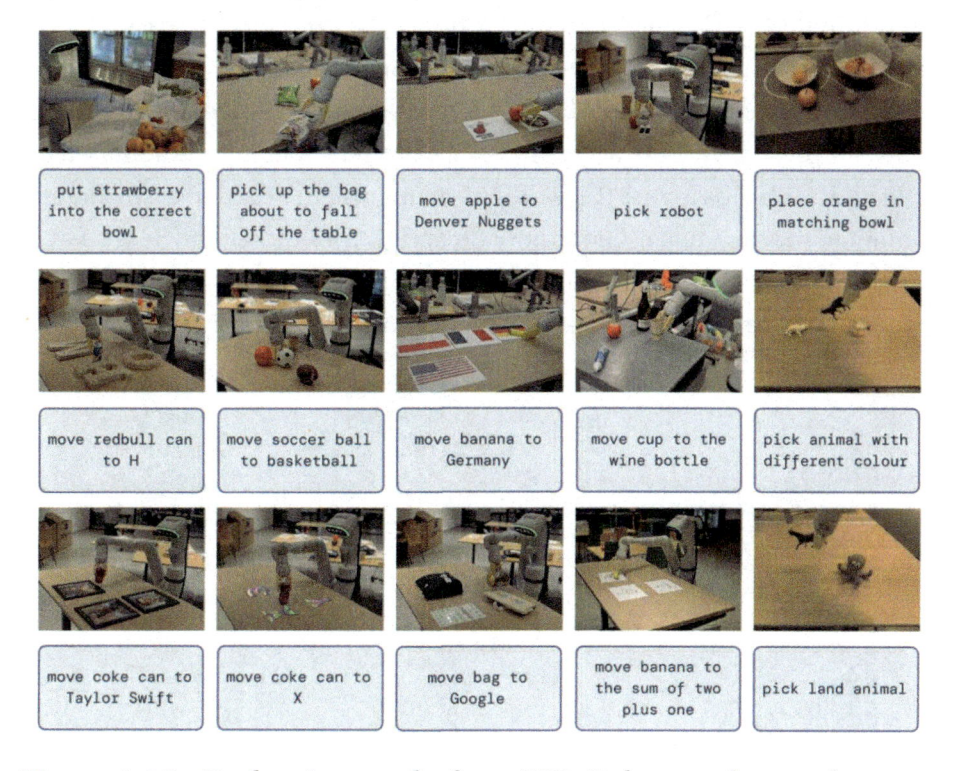

Figure 4-16. Evaluation results from RT2. Robots understand concepts from the Internet such as Google, letters, colors, Taylor Swift, and so on. Used with permission, source: https://arxiv.org/pdf/2307.15818[62]

CHAPTER 4 FOUNDATION MODELS IN ROBOTICS

A bonus of training on multimodal data is that the model, with a small amount of fine-tuning on robot plans with language and images, can also do high-level planning, as exhibited by the SayCan model. This indicates a future opportunity to unify high-level planning and low-level control into a single robot brain that's capable of both.

While RT1 and RT2 showed ways to model in-domain robot data, and model joint large-scale Internet and robot data respectively, each robot still had different action spaces, and it was unclear if learned representations transferred across robots. The Open X embodiment[63] effort initiated by Google DeepMind studied the problem of cross-embodied robot learning—can we build a singular brain to control many different robot embodiments? In order to study this problem, a huge dataset representing data across many different robot embodiments was collected, with labs across the world participating in this study. The dataset contained more than a million robot episodes from more than 34 research labs, representing 22+ different robot embodiments. Figure 4-17 shows details of the dataset. It is open source and can be accessed here[64].

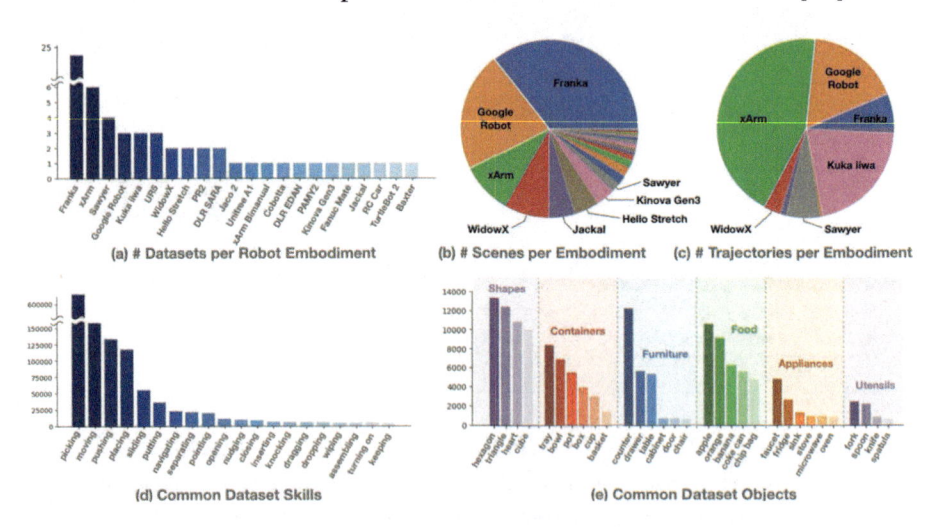

Figure 4-17. Distribution of data in the Open X embodiment dataset. Used with permission, source https://arxiv.org/pdf/2310.08864[63]

CHAPTER 4   FOUNDATION MODELS IN ROBOTICS

In order to train a model that can control multiple embodiments, two candidates were considered: RT1 and RT2. The versions of the model trained on cross-embodied data are respectively known as RT1-X[63] and RT2-X[63]. Only single arm robots were considered for this study and all datasets were transformed into cartesian position control parameters as learning targets, as in RT1 and RT2.

- Evaluations found that models trained on all robot data were able to beat individual models trained by each lab on their own data. In other words, generalist models performed better than specialists, as shown in Figure 4-18.

- The study also found that Internet co-trained data improved performance over training on robot data from scratch and that co-training with Internet data and robot data mixed in batches is better than fine-tuning on robot data followed by pretraining on Internet data. The former helped learn better joint representations between robot and Internet scale data, and to retain Internet concepts after several epochs of training.

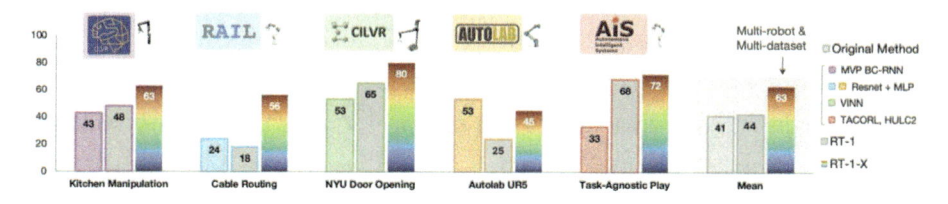

Figure 4-18. Evaluation of RTX models. Used with permission, source: https://arxiv.org/pdf/2310.08864[63]

175

RTX provides evidence to support the idea that a singular brain for many robots may be possible in the future, and that robots are not too different from each other. They may be only as different as say English and Chinese and many common synergies could be exploited. It also presents a way to collect data differently: if you can collect your data on a scalable cheaper embodiment and then transfer that to a more expensive embodiment that's harder to deploy, you improve the efficiency of operations and the ability to scale by a lot. This idea was further explored in work like CrossFormer[80], which expands cross-embodiment to navigation robots, bimanual robots, and even drones. Cross-embodied learning is now a key research area in robotics, with many upcoming developments.

End-to-End Robot Control with Diffusion Models

Recently, diffusion models have become very popular in the field of robot learning due to their ability to model multimodality well[65]. They also scale easily, and are better suited to deal with noisy datasets, especially when generating high-dimensional data that is extremely multimodal. For example, say your dataset has two trajectories for the same state: one with the robot going in clockwise circles and another with it going in counterclockwise circles. An autoregressive transformer is generally trained to predict the average of the two trajectories, since it uses a cross-entropy objective on the mean of the loss. This approach can yield nonsensical results in this situation. Diffusion, on the other hand, tends to fall back to one or more modes of the data and follow it consistently. This property can be extremely useful when there are distinct strategies, such as in manipulation. Another aspect where it becomes useful is when you need generated trajectories to be highly precise and have high fidelity. Denoising steps in diffusion allows correcting to get very fine trajectories needed for highly dexterous motions, when autoregressive models output an action, but cannot refine it to remove suboptimality.

This section explains how diffusion models work fundamentally and how diffusion can be applied in robotics.

Forward Diffusion Process

Diffusion models learn to generate data by reversing an iterative noising process. In the forward diffusion process, noise is added to data in small steps. In the reverse process, the model learns to remove this noise step by step.

Forward diffusion starts with an initial image that is sampled from a data distribution. At each time step t, noise is progressively added to the image until it resembles pure Gaussian noise. This process is described by a Markov chain, where each x(t) is a noisy version of x(t-1). Equation 4-3 shows how to generate the noisy image at time step t. The noisy image at some step is created by taking the original image, scaling it down (using square root of α), and then adding random noise. The random noise is added such that its variance is controlled by (1-α), making sure the added noise fits a normal distribution.

$$q(x_t|x_0) = \mathcal{N}(x_t; \sqrt{\bar{\alpha}_t}x_0, (1-\bar{\alpha}_t)I)$$

Equation 4-3

Reverse Diffusion Process

In the reverse process, you start with a pure noise sample. At every step, a trained neural network is used to predict the previous, less noisy image from the current noise sample. This prediction has two components: the expected value of the previous image and the uncertainty or variance of this prediction. The neural network takes the current noisy image, the time steps as inputs, and produces the expected value (mean) and variance for the previous step. This noise is then subtracted from the noisy image to get a clearer image. The process is repeated iteratively, allowing the network to gradually remove noise from the image step-by-step, shown in Equation 4-4.

$$p_\theta(x_{t-1}|x_t) = \mathcal{N}(x_{t-1}; \mu_\theta(x_t, t), \Sigma_\theta(x_t, t))$$

Equation 4-4

The goal of the training process is to make the reverse process as accurate as possible by minimizing the difference between the predicted distribution (reverse) and the true posterior distribution (calculated from the forward process), such that the output of the reverse process looks like the target output distribution seen during training. One way to do this is using KL divergence, which measures how one probability distribution differs from another. In the context of diffusion, the divergence measures the difference between the network's prediction of noise to be subtracted and the true noise added during the forward process. By reducing the KL divergence between these two, the neural network can be improved to make predictions more accurate and closer to the true noise that was added during the forward process, and thereby retrieving out of pure noise an image that resembles the training distribution.

An overview of this process is shown in Figure 4-19.

CHAPTER 4   FOUNDATION MODELS IN ROBOTICS

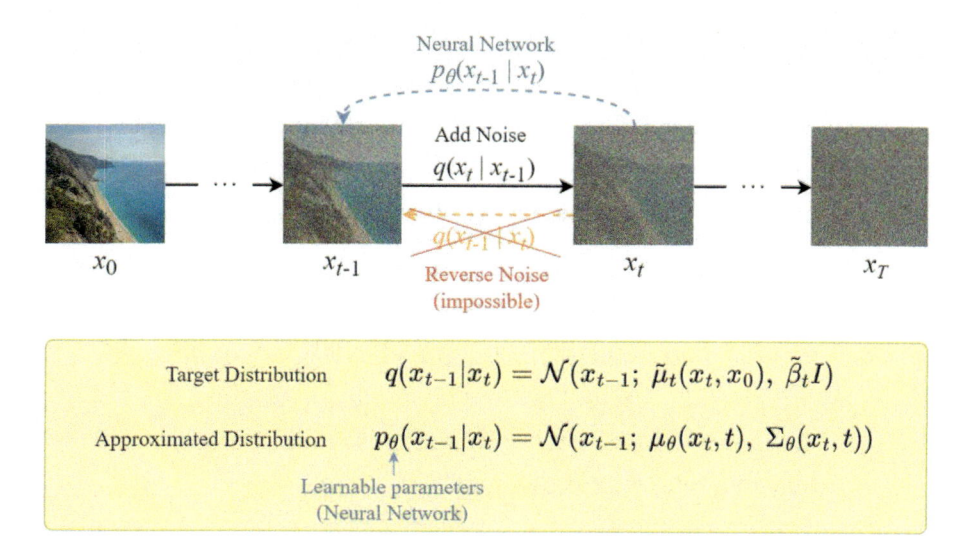

Figure 4-19. Illustration of the diffusion model's forward and reverse processes. In the forward process, noise is added step-by-step to an image until it becomes random noise. The reverse process, approximated by a neural network, predicts the denoised image at each step. Used with permission, source: `https://roysubhradip.hashnode.dev/a-beginners-guide-to-diffusion-models-understanding-the-basics-and-beyond`*[97]*

Putting this together, the core steps for diffusion training are:

- Load some dataset, preprocess images (augmentations, normalization), and use a data loader for loading data in batches.

- As described in Chapter 2, a U-Net model[66] can be used for image generation. Custom U-Net models can be created by specifying input/output channels and structuring downsampling and upsampling blocks.

CHAPTER 4   FOUNDATION MODELS IN ROBOTICS

- A noise scheduler can be used to define how (intensity and frequency) noise is added to the image progressively over multiple timesteps. This is important to ensure that noise is added in a controlled and gradual manner, which can help stabilize training.

- For each epoch, add noise to the current timestep. The model predicts the noise that was added and the loss is calculated as the Mean Squared Error between the predicted and actual noise. The model parameters are updated using backprop and these metrics (loss, learning rate, etc.) are logged.

On the inference side:

- You load the trained U-Net model and noise scheduler.

- Starting with a batch of noisy images (random noise), the model is used to predict the noise in the current image.

- The predicted noise is removed from the noisy image to get a less noisy image.

- This process is repeated for all timesteps to gradually remove noise from the image.

We recommend checking out HuggingFace docs as a way to get started implementing these steps[67].

DDPMs (Denoising Diffusion Probabilistic Models)

DDPM[72] was one of the first papers that showed how diffusion models could be used to generate images. The two main algorithms in the paper are for training and sampling, as shown in Figure 4-20.

CHAPTER 4   FOUNDATION MODELS IN ROBOTICS

The training stage aims to teach the model to predict noise added to images at various timesteps. These steps are done per epoch:

- Sample a batch of images from some data distribution (dataset of images). Sample a timestep t uniformly for each image in the batch. This will determine when in the diffusion process we add noise to the image. Each timestep is modeled individually, compared to modeling the entire diffusion process with one function. This improves training stability and speed, as the model only needs to predict noise for timestep t, as t is sampled uniformly.

- Generate random noise from a standard normal distribution. Each image up to timestep t in the batch is noised using an equation that combines the original image with sampled noise.

- The model predicts the noise given the noisy image and timestep t. Mean Squared Error (MSE) loss is computed between the predicted noise and the actual noise added at timestep t.

The goal of the sampling stage is to generate a new image by iteratively denoising a noisy image. The main steps for sampling are:

- Start by sampling a noisy image from a standard normal distribution. The image is currently at the final timestep T.

- Iteratively from t = T to 1, if t > 1, then sample noise from a standard normal distribution which will be used in the next timestep. If t = 1, then there's no noise needed since you are at the final denoising step. This noise will be used in the last stochastic part for computing the denoised image at each timestep.

CHAPTER 4   FOUNDATION MODELS IN ROBOTICS

- The goal is to estimate the denoised image at the previous timestep (t-1) given the current noisy image and the model's prediction of the noise. The trained model predicts the noise in the current image at t and then the prediction is used to remove the noise and move it to t-1.

- Computing the denoised image has two main parts to it. The first part is the deterministic part where the predicted noise is subtracted from the current noise. The stochastic part adds a controlled amount of noise back into the image using a scaling factor. This helps ensure there is variation and the model is not overly deterministic.

Algorithm 1 Training
1: **repeat**
2: $x_0 \sim q(x_0)$
3: $t \sim \text{Uniform}(\{1, \ldots, T\})$
4: $\epsilon \sim \mathcal{N}(0, I)$
5: Take gradient descent step on
   $\nabla_\theta \left\| \epsilon - \epsilon_\theta(\sqrt{\bar{\alpha}_t} x_0 + \sqrt{1-\bar{\alpha}_t}\epsilon, t) \right\|^2$
6: **until** converged

Algorithm 2 Sampling
1: $x_T \sim \mathcal{N}(0, I)$
2: **for** $t = T, \ldots, 1$ **do**
3: $z \sim \mathcal{N}(0, I)$ if $t > 1$, else $z = 0$
4: $x_{t-1} = \frac{1}{\sqrt{\alpha_t}} \left(x_t - \frac{1-\alpha_t}{\sqrt{1-\bar{\alpha}_t}} \epsilon_\theta(x_t, t) \right) + \sigma_t z$
5: **end for**
6: **return** x_0

Figure 4-20. DDPM training and sampling algorithms. Used with permission, source: https://arxiv.org/abs/2006.11239[72]

For an example of how DDPM is implemented, we recommend this tutorial[73].

The main issue with the original DDPM was its poor log-likelihood score. This meant that while it could generate high-quality images, it did not accurately fit the data distribution of real images. The improved DDPM[89] introduced some variations to address this:

CHAPTER 4   FOUNDATION MODELS IN ROBOTICS

- **Variance prediction:** Instead of using a fixed variance, the improved DDPM learns the variance of the noise distribution, which helps improve log-likelihood and model stability. This is done by interpolating variance between an upper bound and a lower bound.

- **Learning rate scheduling:** The original linear scheduling of noise was replaced with a cosine interpolation, which led to better control over noise addition and removal.

- **Increased timesteps:** The number of timesteps in the diffusion process was increased from 1,000 to 4,000, which improved sample quality. However, this did increase the time and compute required for training and sampling.

DDIM (Denoising Diffusion Implicit Models)

In the DDPM approach, the model is predicting noise that was added to a clean image to produce a noisy image. This is done gradually and can be over many steps. Instead of predicting noise added to the image at each step like DDPMs do, DDIMs[74] predict the noise so that when the noise and clean image are mixed, they make xt. This allows the model to get closer to the final clean image in fewer steps.

The process is controlled by σ, which determines how much noise is added at each step. If it is 0, then the process is deterministic. The model then predicts what the clean image looks like from the noisy image. This prediction is given by Equation 4-5, which subtracts the noise from the noisy image and then scales it:

$$f_\theta^t(x_t) := \left(x_t - \sqrt{1-\alpha_t}\epsilon_\theta(x_t, t)\right)/\sqrt{\alpha_t}$$

Equation 4-5

To generate the image at step t-1 from the image at step t, an equation (Equation 4-6) is used, which combines the predicted clean image and some noise to step back from xt to xt-1. The predicted clean image is scaled to fit the previous timestep. Some noise is added back and scaled back to fit the previous timestep. A bit of noise, which is controlled by σt, is also added to ensure there is some variability and the process isn't fully deterministic.

$$x_{t-1} = \sqrt{\alpha_{t-1}} f_\theta^t(x_t) + \sqrt{1 - \alpha_{t-1} - \sigma_t^2} \epsilon_\theta(x_t, t) + \sigma_t \epsilon$$

Equation 4-6

There is also a case of Equation 4-6 where σt is set to 0 so no noise is added and the process is more deterministic. The model and training objectives used are the same as DDPMs, but the idea is that the technique allows for fewer and larger steps to be taken.

Stable Diffusion

The diffusion process described so far works directly in the raw image pixel space. This can be time intensive for very large images (lots of pixels to denoise) and can suffer from modeling fine details in images incorrectly. Stable diffusion[68] is a popular method that tackles both these problems by compressing the image into a latent space before applying the diffusion model. Stable diffusion took the world by storm with its very remarkable and realistic AI generated output.

During the stable diffusion training phase, you take an image and pass it through a Variational Autoencoder (VAE)[69], which compresses it into a latent vector. The encoder helps reduce the image into a lower-dimensional representation while preserving important features—think of this as a semantic compression. This vector is then used as the starting point of the diffusion process and noise is added to the latent vector at each timestep to make it nosier. The resultant noisy vector becomes the

CHAPTER 4   FOUNDATION MODELS IN ROBOTICS

starting point of the reverse diffusion process, where a U-Net is used to predict and remove noise at each step. This U-Net has text-conditioning, which means that it also uses text embeddings from a user's input query (during inference) or captioning (in training datasets) to guide the image generation. The denoised latent vector is passed through a decoder, which reconstructs the original image—think of this as decompression. Overall, the goal of this approach is to make sure a high-quality image is output that matches the user's input at inference time by reconstructing this from the conditioned latent vector. Because the diffusion process operates on the latent encoding instead of the original image, it's fast enough to be used in products. You can see these steps in Figure 4-21.

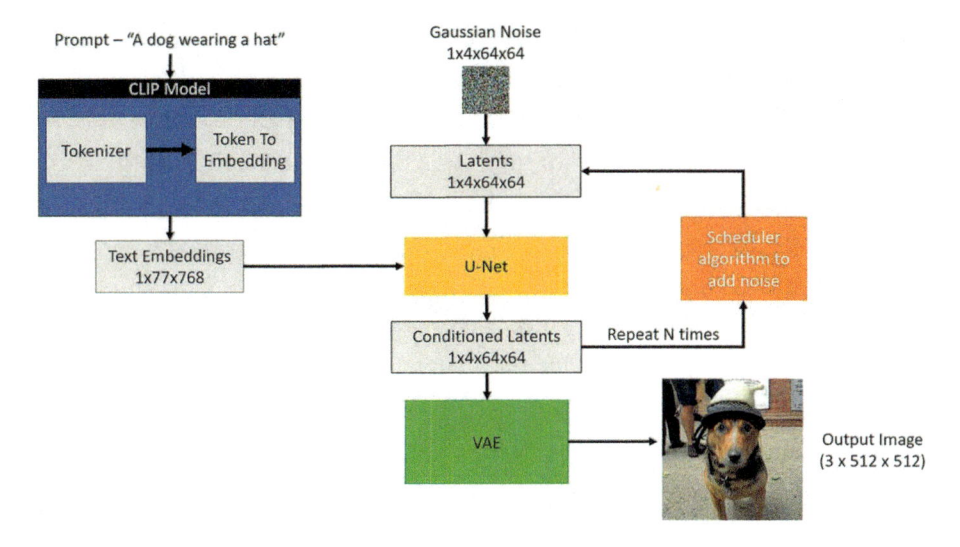

Figure 4-21. *Stable diffusion generates an image by iteratively denoising a latent image representation, conditioned on text embeddings produced by a CLIP model. U-Net refines the noisy latents over multiple iterations, and the final latent representation is decoded by a VAE into the output image. Used with permission, source:* https://towardsdatascience.com/stable-diffusion-using-hugging-face-501d8dbdd8 *[98]*

CHAPTER 4 FOUNDATION MODELS IN ROBOTICS

Conditioned Generation

You have seen in stable diffusion how a user input query or caption is used to guide the generated image via conditioning. This section digs deeper into conditioning for diffusion.

There are two variants of conditioning popularly used: classifier guidance[70] and classifier-free[71]. The main difference between the two is that in the case of classifier guidance, a separate classifier needs to be trained, whereas classifier-free guidance does not need an external classifier network.

Classifier-Guided Diffusion

The goal in classifier-guided diffusion is to generate images that lie within a certain class or have certain attributes. This is accomplished by training a separate classifier network on the labels of the same dataset that the diffusion network is trained on. Then, the classifier network is used to guide the image-generation process by using the Bayes rule to combine the probability of the image and the class (see Equation 4-7):

$$p(x_t|c) \propto p(c|x_t)p(x_t)$$

Equation 4-7

The image generation is updated based on the combination of the gradients of the log probabilities of the image and the class (see Equation 4-8).

$$\nabla_{x_t} \log p(x_t|c) = \nabla_{x_t} \log p(x_t) + \nabla_{x_t} \log p(c|x_t)$$

Equation 4-8

Combining the signal's sub-times requires a linear parameter called guidance scale, γ. A higher guidance means images pay more attention to class and are less diverse, and a lower guidance means that images are less class specific but are more diverse.

Classifier-Free Guidance

The main idea in classifier-free guidance is to train the diffusion model to understand class labels directly without needing a separate classifier, that is, your classifier is the model itself. The model is trained to generate images with and without specific class information, by dropping in labels 20-30 percent of the time during training. During denoising, two predictions are made—one with the label as conditioning and one without. The probabilities of the conditional generation and unconditional generation are combined while applying a weight to balance the importance of class guidance, γ. Once this combined gradient equation is expanded and simplified, you get Equation 4-9:

$$\nabla_{x_t}[(1-w)\log p(x_t) + w\log p(x_t|c)]$$

Equation 4-9

Where $\gamma = w/(1-w)$. As in classifier-guided generation, higher γ implies more attention to class labels at the cost of diversity.

Classifier-free guidance is simpler while providing more control and fidelity over generation. It also avoids any adversarial generation prompted by a separate classifier network.

Text-Conditioned Guidance

Text-conditioned guidance is used in many generative AI products that generate images/videos that a user is specifically asking for. To this end, a text prompt that describes the output image is converted into embeddings with a text encoder. The text embeddings are integrated into the U-Net model using cross-attention (see Equation 4-10):

$$\epsilon_\theta(x_t, t, c)$$

Equation 4-10

Equation 4-11 represents the model's prediction at step t, using the image and the text embeddings. During image generation, the model uses the text embeddings to guide each step:

$$\nabla_{x_t} \log p(x_t|c)$$

Equation 4-11

The gradient will adjust the image based on the text description, just like the guidance discussed previously. Overall, a text description is used to guide the model in generating images that match the description.

Action Diffusion for Robot Control

Diffusion Policy[75] is a method that, instead of directly deciding the robot's actions, uses conditional denoising diffusion to generate the best actions to take. Diffusion policy can sample from a very high dimensional output space. The iterative refinement in diffusion also allows for more flexible and adaptable actions. Some other common policies used for actions include:

- **Explicit policy:** This directly maps visual inputs to robot actions and can have different action representations, like regression, mixture of Gaussians, and so on. One limitation of this policy is that directly outputting actions can be less flexible and harder to train for more complex behaviors because there is a direct learned mapping from input to output.

- **Implicit policy:** These methods use an energy function to implicitly define the policy, where the policy learns an energy function conditioned on both actions and observations. One limitation of this method is that

CHAPTER 4 FOUNDATION MODELS IN ROBOTICS

optimization during inference can be computationally intensive when it comes to solving the minimum energy configuration.

Instead of directly outputting actions, in diffusion policy, the policy infers the action-score gradient based on visual observations. The process starts with a sequence of images as input, which are encoded to extract relevant features. The diffusion policy takes these observations and generates a sequence of actions over some time horizon. This can be seen as initial noisy actions given as a rough guess, which is refined over K iterations. As shown in Figure 4-22, the model refines noise into actions using a gradient field in diffusion policy.

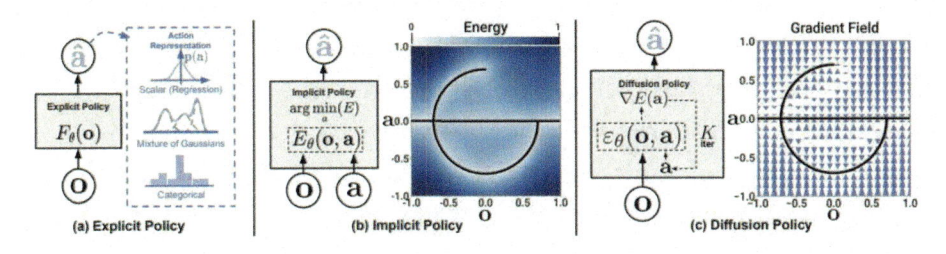

Figure 4-22. The difference between explicit, implicit and diffusion policy. Used with permission, source: https://arxiv.org/pdf/2303.04137[75]

As shown in Figure 4-23, the general steps are that the robot gets a sequence of images from the environment, which are encoded. The policy uses the encoded observations to generate a sequence of actions, which are refined iteratively. Based on new observations, these actions are refined to guide the robot's movement. The specific policy implementation uses two approaches: a) CNN-based, which uses convolutional layers and FiLM[58] to condition each layer with some observation features and refine actions and b) transformer-based, which uses cross-attention to integrate observations and actions while leveraging strengths of transformers to handle sequential data. For more details, we recommend checking out the paper[75].

CHAPTER 4 FOUNDATION MODELS IN ROBOTICS

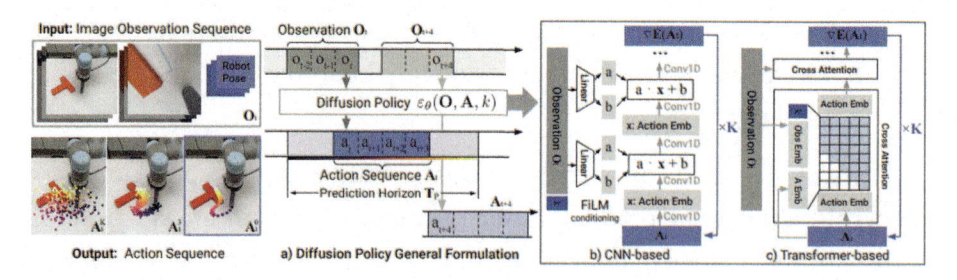

Figure 4-23. Overview of diffusion policy pipeline outputting the action sequence for the robot to perform. Used with permission, source: https://arxiv.org/pdf/2303.04137[75]

Since this work, there have been other variants of diffusion policies for robot control, such as Aloha Unleashed[94], where a transformer decoder iteratively refines actions from noise, while taking in an embedding of observations and diffusion timestep as input, via cross-attention. Diffusion models provide an alternative to autoregressive models for control of robots, and they can model high-fidelity actions, due to outputting float actions directly instead of outputting tokens. In practice, they can also model multimodality and noisy datasets much better, as discussed earlier in this chapter.

Combining VLMs and Diffusion Models

Recently, we have seen architectures that combine VLAs and diffusion models as a system 1 / system 2 model and trained jointly. Pi0[99], Groot N1[100] and Gemini Robotics[101] are some of these variants. Combining a large VLM with diffusion allows exploiting the generality and capacity of a VLA while retaining the smooth behaviours of diffusion. In Groot N1, the input to the diffusion model is the output of one of the final layers of the VLM. During inference time, the models may be split, with the large model living in cloud and the smaller model on robot as in [101].

Learning from Video Demonstrations

A key paradigm in robot learning is learning from video demonstrations (LfV)[90]. The biggest challenge to learning from video demonstrations is that they do not have action labels, which makes it harder to convert video data into actionable robotic data. A second challenge is that most _Internet-scale video data has a distribution shift relative to robot deployment, but in instances this may be an advantage as the diversity of video data allows for generalization to wide variety of scenes the robot hasn't seen before. A third challenge, or opportunity, is that well curated, open source video datasets with language annotations that can be used for robotics are still relatively limited or small in size.

The first challenge may be overcome in the following ways:

1. **Inferring motion from video:** Adding proxy action labels to video data, via using optimal flow. Such as by annotating hand positions, or inferring hand positions via monocular/stereo tracking.

2. **High-level language as actions:** Annotations using captioning models, language description of motions, can be derived. While this type of information is useful for learning semantics, language lacks the fidelity required for precise control.

3. **Learning latent actions from video:** Usually an auto-encoder compresses and decompresses the video, and then uses the learned compressions as a latent variable.

CHAPTER 4   FOUNDATION MODELS IN ROBOTICS

World Modeling Using Video Data

A world model is an internal representation of the world: the physics of the world, how objects interact and their dynamics, and what a robot may possess to anticipate outcomes for its own or another dynamic agents' actions. One way to look at it is that it is an internal simulation.

A common way of creating a world model is by learning good latent representations of observations. Videos provide a very rich opportunity to create extensive world models. Usually, an encoder network encodes video into a latent representation, and then a dynamics model predicts the next latent state and the rewards, using current state and embedded actions as input. A reconstruction loss is typically used to train them. In a video, the dataset presents future states, thus providing a useful training signal. A world model thus learned can then be fine-tuned to learn general action dynamics based on specific robot embodiments.

Once you create this world model, you can use it for a variety of downstream robotic applications.

Reasoning from Video

Video language planning[95] is an example of this method. It reasons about the world and creates plans in multimodal video and text, akin to PaLM-e, discussed earlier, but with a temporal component. Internet-scale video data with language annotations holds a ton of contextual knowledge, presenting an opportunity for acquiring generalization. Robots can learn low-level actions from their current "robot dataset" while gaining high-level task understanding from the video dataset through visual cues and steps needed to complete tasks.

Actions from Video by Fine-Tuning World Models

Structured World Models for Intentionality, or SWIM[78], is a work that helps robots learn manipulation tasks by first creating a world model from human videos and then fine-tuning it on robot data. SWIM trains both a world model (which predicts how the environment changes when actions are taken) and an affordance model (which learns where and how to interact with objects, i.e. grasping) using human video data. This affordance model is used to guide the behavior of a robot collecting data in the real world, and the data from the robot is used to fine-tune the world model, thus embodying it in an unsupervised manner.

One can argue that today's foundation models already have an implicit world model from being trained on the Internet and a wide variety of data. Therefore, fine-tuning them on actions is an expression of transferring their world model understanding to the concept of actions and movement. Results from RT2 discussed earlier in this chapter provide evidence to some of these claims.

Action Models by Conditioning on Video Representations

Video-conditioned policy learning uses the idea that a robot can observe a human (or robot) performing a task and infer actions the needed to complete the task in its own environment. This is useful, as it allows non-experts to specify tasks for robots in an intuitive way without needing complex reward function design or explicit goal definitions. For example, you can just show a robot how to do a task, then that video is the input to policy. Vid2Robot is an example of this type of method[77]. In addition to directly prompting with the video, users can also extract high-level task representations from video via large-scale pretraining, and then use it for downstream policy learning, thus conditioning a policy on embeddings extracted from videos. For example, Time-Contrastive Networks (TCN)[81] extracts visual features like hand-object interactions and spatial relations from multi-view videos and Contrastive Video Representation Learning (CVRL)[82] learns spatiotemporal representations from video data.

CHAPTER 4   FOUNDATION MODELS IN ROBOTICS

Action Models from Thinking in Video

An interesting approach toward action prediction we want to highlight is Policy-as-Video[79]. It models the world in images. It introduces Unified Predictive Decision Process (UPDP) as an alternative to MDP that uses images for representing environments instead of numerical states. The video generator predicts trajectories of images showing how the robot should progress from its initial state to achieve the goal described by the text. Once the video is generated, an inverse dynamics model (IDM) extracts the corresponding actions (e.g., motor commands) from the predicted frames. Diffusion models are used to generate the video sequences in UPDP and they progressively denoise the images toward the end goal. A key feature of UPDP is that it separates the planning process (video generation) from the execution (action extraction), which can make this method more flexible to different robots and environments.

AI Safety for Robotics

As we trust AI models to perform actions in the physical world autonomously, we need to be aware of the security risks that stem from these applications.

A practical study of security risks show that physical, network, and software attacks are probable scenarios[86]. Physical attacks on robots, like tampering with hardware or feeding them false sensor data (sensor spoofing) can directly alter robot behavior, causing them to crash or fail. Attacks on the perception layer (sensors such as IMU, GPS, LiDAR, camera, etc.) can disrupt the robot's ability to understand its environment and this can further impact its navigation and control abilities. On the hardware side, hacking actuators and control systems can directly lead to dangerous behavior. Network attacks, like Denial-of-Service (DoS), can overload the robot's network and cause slowdown or system failures. Software-level attacks, such as injecting malicious code, can alter a robot's actions and lead to harmful actions.

To protect robots from various security threats, some defense strategies include:

1. Anomaly detection systems that monitor sensor data for irregularities and then trigger failsafe modes when issues are detected.

2. Filtering and sensor fusion can mitigate spoofing and jamming by cross-referencing multiple data sources (e.g., GPS, LiDAR) to verify if data is consistent.

3. Data sanitization and validation can help ensure that inputs are checked for malicious content before processing.

4. Cryptographic integrity checks and secure firmware updates protect against data tampering and unauthorized code execution.

5. Failsafe mechanisms and secure update protocols can be useful for robots to revert to safe states and receive only verified updates.

For more details on attacks and potential defense mechanisms, we recommend reading [87].

While these threats are more practical, there is also a longer range and philosophical resistance to autonomy. As AI becomes more intelligent and more integrated into everyday life, their situational awareness could improve, making them able to deceive humans, persuade us to execute harmful behavior, spread misinformation, and so on. There are several organizations working on addressing longer-term existential threats to humanity from AI. This is in sharp contrast to folks working on AI ethics, who focus on nearer-term risks from AI such as deep fakes, bias in AI outputs, privacy concerns, and so on.

Anthropic's core AI safety principles[88] is an industry-defining standard for designing safer AI systems. These are some of the ways we might use these principles to improve AI-driven robotics systems:

- **Mechanistic interpretability** focuses on understanding the internal decision-making processes of AI models (their interpretability). In robotics, this means being able to detect if a robot might misinterpret a situation before potentially making any harmful actions. An area of AI safety research is reverse-engineering models and building tools for model interpretability so engineers can detect whether the robot may have unintended behaviors, like moving into a dangerous area or misidentifying an object. A recent advance in interpretability is the study from OpenAI that showed that a smaller model can be explained with a larger model[91]. In the study, GPT-4 generates explanations for the role of neurons in GPT-2, and then the neuron is simulated to score the explanation, providing a way for scalable interpretability. There remains the question though whether all neurons could be interpretable, if they do not have clean and human understandable semantic explanations for their role.

- **Red teaming** is a technique where AI models are stress-tested with difficult or adversarial conditions prior to their deployment to identify and correct vulnerabilities before applying them in the real world. Stress-testing AI models in an environment that is controlled can help identify and fix any failure points. In the context of robotics, for example, robots could encounter scenarios where their sensors are fed misleading information, which could lead to dangerous actions, such as crashing or navigating incorrectly.

CHAPTER 4   FOUNDATION MODELS IN ROBOTICS

By deliberately placing robots in these controlled adversarial conditions, engineers can detect and mitigate failure points when under attack from physical or cyber threats, like sensor spoofing.

- **Alignment** refers to the problem of enabling AI systems to follow human values and adhere to human preferences. As discussed earlier in this chapter, the post-training phase is mainly used to improve the alignment of models, along with their instruction following abilities. In many situations, especially in robots being deployed in real-time, it is not practical for humans to provide constant, real-time supervision. Developing techniques that allow robotic systems to self-monitor and correct their own behaviors without human supervision will be important. Constitutional AI[92] and RLAIF[93] are two works that advance this. RLAIF provides a way to scale human feedback by training a model from human preferences that can provide feedback without the bottleneck of human involvement and act as a critic during training large networks. Constitutional AI introduces a constitution or value system, which is used by language models to reason about their responses and edit them to align them to the constitution during post-training.

- Robots must be able to operate in various environments without reprogramming. Building generalization into model training, by training on varying conditions and environments and testing on these, can be critical. If a robot was trained in a specific factory, it may have unintended behavior without reprogramming in a new environment. Ensuring and testing for generalization can be helpful not only for performance but also for safety in robotics.

CHAPTER 4   FOUNDATION MODELS IN ROBOTICS

For more aspects of practical safety during deployment, refer to Chapter 11.

Summary

In summary, this chapter covered the following points:

- How robotic planning, control, and mapping can be framed as a language problem and various ways of solving it with LLMs using open loop and closed loop techniques.

- Large foundation models are pretrained on a large corpus of data and later refined using supervised fine-tuning (SFT) or direct preference optimization (DPO). Scaling laws help guide their development by predicting performance improvements as models grow in size, data, or compute. Benchmarks like MMLU and HumanEval evaluate language models' performance across tasks.

- Models like SayCan break down high-level instructions for a robot to execute and PaLM-e incorporates multimodal inputs (text, vision) for planning. These models help robots understand tasks in context, generate actions, and adjust.

- Methods like NLMap-SayCan allow robots to build a map of their surroundings and plan based on unseen objects. Code as Policies allows LLMs to translate natural language instructions into executable code for robots using pretrained models and examples to guide task execution.

CHAPTER 4   FOUNDATION MODELS IN ROBOTICS

- Models like RT-1, RT-2, and RT-X use transformers to directly control robot actions from visual and textual inputs, allowing multitask and real-world control across different robots and tasks.

- Diffusion-based policies offer flexibility in generating high-dimensional action sequences through iterative refinement.

- Robots can learn tasks by observing video demonstrations, extracting high-level task representations, and applying those tasks to their own actions, bridging the gap between human demonstrations and robot tasks.

- AI-driven robots face practical security risks, such as physical tampering, network attacks, and software vulnerabilities, while also posing long-term risks related to autonomy and generalization. Robust safety mechanisms, anomaly detection, and interpretability are crucial for safe robotic operations.

The next chapter discusses how simulation can be used in robotics, common simulators, their tradeoffs and benefits, and methods like domain randomization and domain adaptation, which are used to address the Sim2Real gap. The next chapter also discusses common learning methods using simulators for RL and IL.

References

[1]   Achiam, Josh, et al. "Gpt-4 technical report." *arXiv preprint arXiv*:2303.08774 (2023).

[2]   Team, Gemini, et al. "Gemini: A family of highly capable multimodal models." *arXiv preprint arXiv*:2312.11805 (2023).

[3] Touvron, Hugo, et al. "Llama: Open and efficient foundation language models." *arXiv preprint arXiv*:2302.13971 (2023).

[4] Devlin, Jacob. "Bert: Pretraining of deep bidirectional transformers for language understanding." *arXiv preprint arXiv*:1810.04805 (2018).

[5] Brown, Tom B. "Language models are few-shot learners." *arXiv preprint ArXiv*:2005.14165 (2020).

[6] Radford, Alec, et al. "Language models are unsupervised multitask learners." OpenAI blog 1.8 (2019): 9.

[7] Radford, Alec, et al. "Improving language understanding by generative pre-training." (2018).

[8] Chowdhery, Aakanksha, et al. "Palm: Scaling language modeling with pathways." *Journal of Machine Learning Research* 24.240 (2023): 1-113.

[9] Touvron, Hugo, et al. "Llama: Open and efficient foundation language models." *arXiv preprint arXiv*:2302.13971 (2023).

[10] Dubey, Abhimanyu, et al. "The llama 3 herd of models." *arXiv preprint arXiv*:2407.21783 (2024).

[11] Ainslie, Joshua, et al. "GQA: Training generalized multi-query transformer models from multi-head checkpoints." *arXiv preprint arXiv*:2305.13245 (2023).

[12] Lages, João. "Transformers KV Caching Explained - João Lages." *Medium, Medium*, 8 Oct. 2023, medium.com/@joaolages/kv-caching-explained-276520203249.

[13] Radford, Alec, et al. "Learning transferable visual models from natural language supervision." International Conference on Machine Learning. PMLR, 2021.

[14] https://opencompass.readthedocs.io/en/latest/advanced_guides/needleinahaystack_eval.html

[15] Rafailov, Rafael, et al. "Direct preference optimization: Your language model is secretly a reward model." *Advances in Neural Information Processing Systems* 36 (2024).

[16] Ouyang, Long, et al. "Training language models to follow instructions with human feedback." *Advances in Neural Information Processing Systems* 35 (2022): 27730-27744.

[17] Kaplan, Jared, et al. "Scaling laws for neural language models." *arXiv preprint arXiv*:2001.08361 (2020).

[18] Hoffmann, Jordan, et al. "Training compute-optimal large language models." *arXiv preprint arXiv*:2203.15556 (2022).

[19] Bahdanau, Dzmitry. "The Flops Calculus of Language Model Training." *Medium*, 9 Jan. 2022, medium.com/@dzmitrybahdanau/the-flops-calculus-of-language-model-training-3b19c1f025e4.

[20] Besiroglu, Tamay, et al. "Chinchilla Scaling: A replication attempt." *arXiv preprint arXiv*:2404.10102 (2024).

[21] Muennighoff, Niklas, et al. "Scaling data-constrained language models." *Advances in Neural Information Processing Systems* 36 (2024).

[22] Sorscher, Ben, et al. "Beyond neural scaling laws: beating power law scaling via data pruning." *Advances in Neural Information Processing Systems* 35 (2022): 19523-19536.

[23] Hendrycks, Dan, et al. "Measuring massive multitask language understanding." *arXiv preprint arXiv*:2009.03300 (2020).

[24] Wang, Yubo, et al. "MMLU-Pro: A more robust and challenging multi-task language understanding benchmark." *arXiv preprint arXiv*:2406.01574 (2024).

[25] Zhong, Wanjun, et al. "Agieval: A human-centric benchmark for evaluating foundation models." *arXiv preprint arXiv*:2304.06364 (2023).

[26] Cobbe, Karl, et al. "Training verifiers to solve math word problems." *arXiv preprint arXiv*:2110.14168 (2021).

[27] Hendrycks, Dan, et al. "Measuring mathematical problem solving with the math dataset." *arXiv preprint arXiv*:2103.03874 (2021).

[28] Clark, Peter, et al. "Think you have solved question answering? try Arc, the AI2 reasoning challenge." *arXiv preprint arXiv*:1803.05457 (2018).

[29] Talmor, Alon, et al. "Common sense QA: A question answering challenge targeting commonsense knowledge." *arXiv preprint arXiv*:1811.00937 (2018).

[30] Mihaylov, Todor, et al. "Can a suit of armor conduct electricity? A new dataset for open book question answering." *arXiv preprint arXiv*:1809.02789 (2018).

[31] Bisk, Yonatan, et al. "PiQA: Reasoning about physical common sense in natural language." Proceedings of the AAAI Conference on Artificial Intelligence. Vol. 34. No. 05. 2020.

[32] Sap, Maarten, et al. "SocialIQA: Common sense reasoning about social interactions." *arXiv preprint arXiv*:1904.09728 (2019).

[33] Chen, Mark, et al. "Evaluating large language models trained on code." *arXiv preprint arXiv*:2107.03374 (2021).

[34] Austin, Jacob, et al. "Program synthesis with large language models." *arXiv preprint arXiv:*2108.07732 (2021).

[35] Rajpurkar, Pranav, et al. "Squad: 100,000+ questions for machine comprehension of text." *arXiv preprint arXiv*:1606.05250 (2016).

[36] Choi, Eunsol, et al. "QuAC: Question answering in context." *arXiv preprint arXiv*:1808.07036 (2018).

[37] Lai, Guokun, et al. "Race: Large-scale reading comprehension dataset from examinations." *arXiv preprint arXiv*:1704.04683 (2017).

[38] Li, Mo, et al. "NeedleBench: Can LLMs Do Retrieval and Reasoning in 1 Million Context Window?." *arXiv preprint arXiv*:2407.11963 (2024).

[39] Shaham, Uri, et al. "ZeroSCROLLS: A zero-shot benchmark for long text understanding." *arXiv preprint arXiv:*2305.14196 (2023).

[40] Zhang, Xinrong, et al. "∞ Bench: Extending Long Context Evaluation Beyond 100K Tokens." arXiv preprint arXiv:2402.13718 (2024).

[41] Li, Minghao, et al. "API-bank: A comprehensive benchmark for tool-augmented LLMs." *arXiv preprint arXiv:*2304.08244 (2023).

CHAPTER 4 FOUNDATION MODELS IN ROBOTICS

[42] Peng, Yun, et al. "Revisiting, benchmarking and exploring API recommendation: How far are we?." IEEE Transactions on Software Engineering 49.4 (2022): 1876-1897.

[43] Vaswani, Ashish. "Attention is all you need." *arXiv preprint arXiv*:1706.03762 (2017).

[44] Ahn, Michael, et al. "Do as I can, not As I say: Grounding language in robotic affordances." *arXiv preprint arXiv*:2204.01691 (2022).

[45] `https://sites.research.google/palm-saycan`

[46] Huang, Wenlong, et al. "Inner monologue: Embodied reasoning through planning with language models." *arXiv preprint arXiv*:2207.05608 (2022).

[47] Driess, Danny, et al. "Palm-E: An embodied multimodal language model." *arXiv preprint arXiv*:2303.03378 (2023).

[48] Dehghani, Mostafa, et al. "Scaling vision transformers to 22 billion parameters." International Conference on Machine Learning. PMLR, 2023.

[49] Noyan, Merve and Edward Beeching. "Vision Language Models Explained." *Hugging Face – The AI Community Building the Future,* 11 Apr. 2024, `huggingface.co/blog/vlms`.

[50] Gao, Jensen, et al. "Physically grounded vision-language models for robotic manipulation." 2024 IEEE International Conference on Robotics and Automation (ICRA). IEEE, 2024.

[51] Chen, Boyuan, et al. "Open-vocabulary queryable scene representations for real world planning." 2023 IEEE International Conference on Robotics and Automation (ICRA). IEEE, 2023.

[52] Yu, Wenhao, et al. "Language to rewards for robotic skill synthesis." *arXiv preprint arXiv*:2306.08647 (2023).

[53] Liang, Jacky, et al. "Code as Policies: Language model programs for embodied control." 2023 IEEE International Conference on Robotics and Automation (ICRA). IEEE, 2023.

[54] https://code-as-policies.github.io

[55] Arenas, Montserrat Gonzalez, et al. "How to prompt your robot: A promptbook for manipulation skills with Code as Policies." 2024 IEEE International Conference on Robotics and Automation (ICRA). IEEE, 2024.

[56] Levine, Sergey, et al. "End-to-end training of deep visuomotor policies." *Journal of Machine Learning Research* 17.39 (2016): 1-40.

[57] Brohan, Anthony, et al. "Rt-1: Robotics transformer for real-world control at scale." *arXiv preprint arXiv*:2212.06817 (2022).

[58] Perez, Ethan, et al. "Film: Visual reasoning with a general conditioning layer." Proceedings of the AAAI conference on artificial intelligence. Vol. 32. No. 1. 2018.

[59] Tan, Mingxing, and Quoc Le. "EfficientNet: Rethinking model scaling for convolutional neural networks." International Conference on Machine Learning. PMLR, 2019.

[60] Ryoo, Michael S., et al. "TokenLearner: What can 8 learned tokens do for images and videos?." *arXiv preprint arXiv*:2106.11297 (2021).

[61] Zhao, Tony Z., et al. "Learning fine-grained bimanual manipulation with low-cost hardware." *arXiv preprint arXiv*:2304.13705 (2023).

[62] Brohan, Anthony, et al. "Rt-2: Vision-language-action models transfer web knowledge to robotic control." *arXiv preprint arXiv*:2307.15818 (2023).

[63] Padalkar, Abhishek, et al. "Open x-Embodiment: Robotic learning datasets and rt-x models." *arXiv preprint arXiv*:2310.08864 (2023).

[64] https://docs.google.com/spreadsheets/d/1rP BD77tk6OAEIGZrGSODwyyzs5FgCU9Uz3h-3_t2A9g/ edit?gid=0#gid=0

[65] https://github.com/mbreuss/diffusion-literature-for-robotics

[66] Ronneberger, Olaf, Philipp Fischer, and Thomas Brox. "U-net: Convolutional networks for biomedical image segmentation." Medical image computing and computer-assisted intervention–MICCAI 2015: 18th international conference, Munich, Germany, October 5-9, 2015, proceedings, part III 18. Springer International Publishing, 2015.

[67] https://huggingface.co/docs/diffusers/en/tutorials/basic_training

[68] Esser, Patrick, et al. "Scaling rectified flow transformers for high-resolution image synthesis." Forty-first International Conference on Machine Learning. 2024.

[69] Kingma, D. P. "Auto-Encoding Variational Bayes." *arXiv preprint arXiv*:1312.6114 (2013).

[70] Dhariwal, Prafulla, and Alexander Nichol. "Diffusion models beat GANs on image synthesis." *Advances in Neural Information Processing Systems* 34 (2021): 8780-8794.

[71] Ho, Jonathan, and Tim Salimans. "Classifier-free diffusion guidance." *arXiv preprint arXiv*:2207.12598 (2022).

[72] Ho, Jonathan, Ajay Jain, and Pieter Abbeel. "Denoising diffusion probabilistic models." *Advances in Neural Information Processing Systems* 33 (2020): 6840-6851.

[73] Nain, Aakash Kumar. "Keras Documentation: Denoising Diffusion Probabilistic Model." *Keras*, 7 Dec. 2022, keras.io/examples/generative/ddpm/.

[74] Song, Jiaming, Chenlin Meng, and Stefano Ermon. "Denoising diffusion implicit models." *arXiv preprint arXiv*:2010.02502 (2020).

[75] Chi, Cheng, et al. "Diffusion policy: Visuomotor policy learning via action diffusion." *arXiv preprint arXiv*:2303.04137 (2023).

[76] Sermanet, Pierre, et al. "RobovQA: Multimodal long-horizon reasoning for robotics." 2024 IEEE International Conference on Robotics and Automation (ICRA). IEEE, 2024.

[77] Jain, V., Attarian, M., Joshi, N.J., Wahid, A., Driess, D., Vuong, Q., Sanketi, P.R., Sermanet, P., Welker, S., Chan, C. and Gilitschenski, I., 2024. Vid2robot: End-to-end video-conditioned policy learning with cross-attention transformers. *arXiv preprint arXiv:2403.12943*.

[78] Mendonca, Russell, Shikhar Bahl, and Deepak Pathak. "Structured world models from human videos." *arXiv preprint arXiv*:2308.10901 (2023).

[79] Du, Yilun, et al. "Learning universal policies via text-guided video generation." *Advances in Neural Information Processing Systems* 36 (2024).

[80] Doshi, Ria, Homer Walke, Oier Mees, Sudeep Dasari, and Sergey Levine. "Scaling cross-embodied learning: One policy for manipulation, navigation, locomotion and aviation." *arXiv preprint arXiv:2408.11812* (2024).

[81] Sermanet, Pierre, et al. "Time-contrastive networks: Self-supervised learning from video." 2018 IEEE International Conference on Robotics and Automation (ICRA). IEEE, 2018.

[82] Qian, Rui, et al. "Spatiotemporal contrastive video representation learning." Proceedings of the IEEE/CVF Conference on Computer Vision and Pattern Recognition. 2021.

[83] Jiang, Yunfan, et al. "Vima: General robot manipulation with multimodal prompts." *arXiv preprint arXiv*:2210.03094 2.3 (2022): 6.

[84] Mees, Oier, Lukas Hermann, and Wolfram Burgard. "What matters in language conditioned robotic imitation learning over unstructured data." *IEEE Robotics and Automation Letters* 7.4 (2022): 11205-11212.

[85] Do, Thanh-Toan, Anh Nguyen, and Ian Reid. "AffordanceNet: An end-to-end deep learning approach for object affordance detection." 2018 IEEE International Conference on Robotics and Automation (ICRA). IEEE, 2018.

[86] Botta, Alessio, et al. "Cyber security of robots: A comprehensive survey." *Intelligent Systems with Applications* 18 (2023): 200237.

[87] Neupane, Subash, et al. "Security Considerations in AI-Robotics: A Survey of Current Methods, Challenges, and Opportunities." *IEEE Access* (2024).

[88] https://www.anthropic.com/news/core-views-on-ai-safety

[89] Nichol, Alexander Quinn, and Prafulla Dhariwal. "Improved denoising diffusion probabilistic models." International Conference on Machine Learning. PMLR, 2021.

[90] McCarthy, R., Tan, D.C., Schmidt, D., Acero, F., Herr, N., Du, Y., Thuruthel, T.G. and Li, Z., 2024. Towards Generalist Robot Learning from Internet Video: A Survey. *arXiv preprint arXiv:2404.19664*.

[91] https://openaipublic.blob.core.windows.net/neuron-explainer/paper/index.html

[92] Bai, Y., Kadavath, S., Kundu, S., Askell, A., Kernion, J., Jones, A., Chen, A., Goldie, A., Mirhoseini, A., McKinnon, C. and Chen, C., 2022. "Constitutional AI: Harmlessness from AI feedback." *arXiv preprint arXiv:2212.08073*.

[93] Lee, Harrison, Samrat Phatale, Hassan Mansoor, Thomas Mesnard, Johan Ferret, Kellie Lu, Colton Bishop et al. "RLAIF: Scaling reinforcement learning from human feedback with AI feedback." *arXiv preprint arXiv:2309.00267* (2023).

[94] Zhao, T.Z., Tompson, J., Driess, D., Florence, P., Ghasemipour, K., Finn, C. and Wahid, A., 2024. "Aloha unleashed: A simple recipe for robot dexterity." *arXiv preprint arXiv:2410.13126.*

[95] Du, Yilun, Mengjiao Yang, Pete Florence, Fei Xia, Ayzaan Wahid, Brian Ichter, Pierre Sermanet, et al. "Video language planning." *arXiv preprint arXiv:2310.10625* (2023).

[96] Hausman, Karol. "X.com." X (Formerly Twitter), 2024, x.com/hausman_k/status/1559558929297727489.

[97] Roy, Subhradip. "A Beginner's Guide to Diffusion Models: Understanding the Basics and Beyond." Subhradip Roy's Blog, 4 Mar. 2023, roysubhradip.hashnode.dev/a-beginners-guide-to-diffusion-models-understanding-the-basics-and-beyond.

[98] Agrawal, Aayush. "Stable Diffusion Using 🤗 Hugging Face." *Medium*, 9 Nov. 2022, towardsdatascience.com/stable-diffusion-using-hugging-face-501d8dbdd8.

[99] Black, Kevin, Noah Brown, Danny Driess, Adnan Esmail, Michael Equi, Chelsea Finn, Niccolo Fusai et al. "π_0: A Vision-Language-Action Flow Model for General Robot Control." arXiv preprint arXiv:2410.24164 (2024).

[100] Bjorck, Johan, Fernando Castañeda, Nikita Cherniadev, Xingye Da, Runyu Ding, Linxi Fan, Yu Fang et al. "GR00T N1: An Open Foundation Model for Generalist Humanoid Robots." arXiv preprint arXiv:2503.14734 (2025).

[101] https://storage.googleapis.com/deepmind-media/gemini-robotics/gemini_robotics_report.pdf.

CHAPTER 5

Simulation

Simulation for Robots

Obtaining real-world data to train robot models can be costly and time-consuming. Many models require trillions of trajectories to learn all scenarios in which a robot can operate and every potential combination of actions it can perform. This could take decades to collect, a ton of capital to deploy, and is harder to scale due to having hardware in the loop.

To address this issue, several researchers and companies have implemented simulation technologies that aim to create synthetic data that closely resembles the real world. One example of such an engine is NVIDIA's Isaac Sim[1], which generates synthetic data for lifelike graphics.

Including simulation data in robotics training pipelines has various benefits. Simulators provide a controlled environment in which robots can learn and execute their actions without the dangers involved with real-world testing. Simulation data covering diverse scenes allows models to generalize more effectively to new scenarios. See Figure 5-1. However, it is essential that simulated data adequately represents real-world dynamics and bridges the "simulated-to-real" (Sim2Real) gap, which is the gap that explains that models trained in simulation may not perform as well in the real world.

CHAPTER 5   SIMULATION

Figure 5-1. *Simulated robotic grasping environment in PyBullet. Used with permission, source:* https://towardsdatascience.com/sample-efficient-robot-training-on-pybullet-simulation-with-sac-algorithm-71d5d1d4587f *[56]*

Considerations for Simulation in Robotics

Before you learn how these simulation engines work and can be used for robotics, it's important to consider the benefits and limitations of using simulations in your applications.

Benefits of simulation include:

- **Scaling faster with diversity**: A key advantage of using simulation is that you can use generative AI to produce highly diverse scenes in simulation, which can then be used to train models. This is cost effective and can scale better than setting up those scenes in the real world and collecting them with robots.

- **Faster iteration cycles**: Simulated environments allow for easier development and testing of novel algorithms, leading to faster engineering iterations. In comparison, these iteration cycles on a physical robot in the real world may take incredibly long or may be incredibly expensive.

- **Safety critical feature testing**: Many safety features that are critical in dangerous situations are harder to test in the real world. These are better tested thoroughly in simulation, where you can easily and safely simulate dangerous agent behaviors, crashes, and so on, without causing actual danger. This is especially useful for safety critical applications like self-driving or surgical robots.

- **Reproducibility and comparison for experimentation**: It is often challenging to reproduce a certain result in the real world since real-world results and data can be noisy or constantly changing. Simulated environments are designed to be controllable and reproducible and thus reliable, which allows them to also be used to compare algorithms in identical experiments. Simulated environments can also be used to create clean data to ablate for highly specific experiments.

CHAPTER 5   SIMULATION

A few limitations of using simulators include these:

- **World modeling**: Learning only in toy environments can lead to toy intelligence. Ultimately, the models may fail to learn real-world dynamics, and this knowledge may fail to transfer to real physical robots.

- **Sim2Real gap**: A gap still exists where the performance of a robot trained in simulation may not exactly translate to the real world. This is due to physics, dynamics, and external factors that are not fully captured in a virtual environment but are present in physical environments.

Although there are many benefits to simulated data, the main limitation is that there can be discrepancies between the real world and the simulated world. A combination of simulation and real-world data for training and careful domain adaptation/randomization strategies are frequently needed to close this gap. Later in this chapter, we expand on these strategies.

Components of a Robot Simulator

This section discusses the commonly used robot simulators—how they work and their benefits and tradeoffs.

A rigid body physics engine simulates a variety of interactions, such as collisions and contact forces. A rigid body is defined as a solid item that does not alter its form as it interacts with the environment. It has mass, location, velocity, volume, and shape. This allows it to rotate around its center of mass, as shown in Figure 5-2, which represents the average position of the object's mass. When a rigid body rotates, it has three angular properties: angular velocity (how fast it spins), torque (the force

CHAPTER 5 SIMULATION

that causes it to rotate), and moment of inertia (the resistance to altering rotation). When you apply force to a spot on a rigid body, torque is created, causing it to rotate. Torque calculation is simple in two dimensions and requires quaternions and a 3x3 matrix in three dimensions. Integrals can help find the center of mass and the moment of inertia because the object is continuous (not made of separate particles).

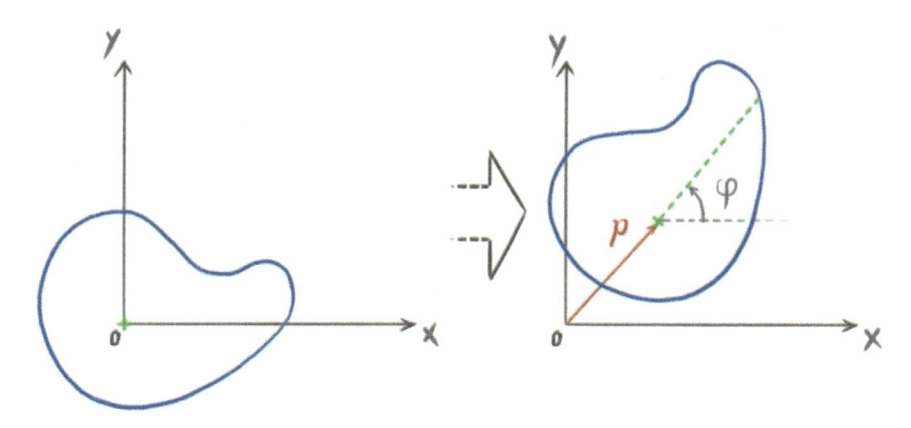

Figure 5-2. *The position and rotation of a rigid body at any given time is defined as an offset from the initial state. Used with permission, source:* `https://www.toptal.com/game/video-game-physics-part-i-an-introduction-to-rigid-body-dynamics`*[25]*

Soft body simulation, on the other hand, deals with objects that may deform when forces are applied, which means that the soft body changes shape. Often, more advanced simulation approaches are required to account for interior forces and deformations. Bending, stretching, and compressing are modeled using techniques such as spring-mass models or finite element approaches, making soft body simulations computationally costly but necessary for robotic simulators.

CHAPTER 5   SIMULATION

Most robot simulators frequently include the following features:

1. **Physics engine:** A physics engine is used to simulate physical dynamics of a robot, including rigid dynamics, soft body modeling, and fluid dynamics. Rigid body dynamics often focuses on collision detection and modeling, friction, and other interactive dynamics. Soft body modeling focuses on the behavior of deformable objects like tissue, cloth, or flexible materials. Fluid dynamics is used to simulate the behavior of liquids and gasses, which is especially important in situations involving aerodynamics (drones, planes, moving parts) and underwater robots.

    a. Simulating physics includes solving equations that model the motion of rigid bodies and other forces that may act on a robot body.

    b. For soft body simulation, finite element[2] or mass-spring methods[3] are often used for understanding internal forces and for modeling how deformations occur in materials.

    c. Fluid dynamics use various computational fluid dynamics (CFD)[4] methods, which solve Navier-Stokes equations[5] under the hood to simulate the flow of liquids and gasses.

2. **Modeling robots:** Simulators use standardized formats like URDF (Unified Robot Description Format)[6] to describe and load robot kinematics. This is usually done by parsing XML-based files describing joints (such as the weight, dimensions of a link, and its center of gravity) and transformations

CHAPTER 5   SIMULATION

between joints. They usually also include support for sensors such as cameras and LiDAR, and they map camera intrinsics and extrinsics (transformations of the camera with respect to the robot). Once you have the full specification of a robot and its sensor, you can construct images from the point of view of the sensors and simulate motion by applying forces or specifying positions for end effectors and then calculating the position of all joints using forward or inverse kinematics.

3. **Modeling environments:** Simulators usually have libraries that enable you to configure the scenes and environment. Adding objects, changing their locations, and adding terrains and surfaces, for example. In addition to objects and surfaces, they also allow you to simulate weather (e.g., sunshine), lighting (indoor/outdoor) and shadows, and other conditions, like humidity/temperature, and so on. In addition to static objects, simulators can also generate dynamic agents such as humans by using humanoid models to imitate realistic human movements. Once a user configures the environment and settings, the physics engine then models them, and a graphics library renders them onscreen.

4. **Interacting with users:** To interface with the user, simulation engines provide APIs that allow users to write code to control the environment and the robot and to run specific models/policies or specific high-level goals. In addition to APIs, simulators also have GUIs for users to interact with the simulator.

217

CHAPTER 5   SIMULATION

Most simulators have built-in trajectory planning algorithms, such as A*[7], Dijkstra's[8], or model predictive control (MPC)[54] for path planning. We cover model predictive control in detail later in this chapter.

5. **Visualization and rendering:** The scenes themselves are rendered using graphics libraries like OpenGL[9] so that robot movements and interactions can be seen onscreen in real time. Most simulators also integrate with middleware frameworks such as Robot Operating System (ROS)[10]. ROS is a distributed framework for publishing and subscribing to information, enabling sensors/joints/programs within a robot to operate at different frequencies (such as when a camera updates at 30fps, but an IMU updates at 100 Hz) and still communicate with each other. Many simulators allow you to publish information out of and in to the simulator as ROS messages or other commonly used communication protocols.

Now that you understand the key components of robot simulators, take a look at how they interact to mimic real-world object behavior[25][26]:

1. It begins with a mathematical model that represents the system's current state, such as the object's positions and speeds.

2. The engine uses equations to predict how these states will change over time depending on forces, object shapes, and movements.

3. The engine uses numerical methods to solve these equations. Runge-Kutta is a widely used technique that estimates varying positions and velocities step by step.

4. When objects collide, the engine identifies the collision and adjusts the objects' speeds and directions to reflect the impact correctly.

5. There are many performance and reliability optimization tasks involved, such as rendering objects onscreen, managing user inputs, and ensuring that everything runs in real time.

Now let's look at some popular simulators used by robotics researchers and developers and their specific properties.

The PyBullet Module

The Python module PyBullet[11] uses the Bullet physics engine to simulate physical interactions and robot motion. An important notable application of PyBullet was the QT-Opt system[12], an algorithm that we explore in detail in the learning section of this chapter, where Bullet helped mimic a grasping environment in order to train a robotic arm on a large corpus of grasping tasks. PyBullet allows you to import robot models in multiple formats:

1. SDFormat (Simulation Description Format)[13]

2. Unified Robotics Description Format (URDF)[6]

3. MJCF[14]

PyBullet supports contact, collision, friction, and rigid-body dynamics for existing robot models or the robots imported by users. PyBullet is commonly used to simulate various robotic tasks, like robotic

CHAPTER 5   SIMULATION

manipulation tasks. Pictures of some of the supported robots are shown in Figure 5-3. For advanced motion planning, you can integrate PyBullet with external libraries, such as the Open Motion Planning Library (OMPL)[15], which is widely used in robotics. There is a tutorial on getting started with PyBullet at the end of this chapter.

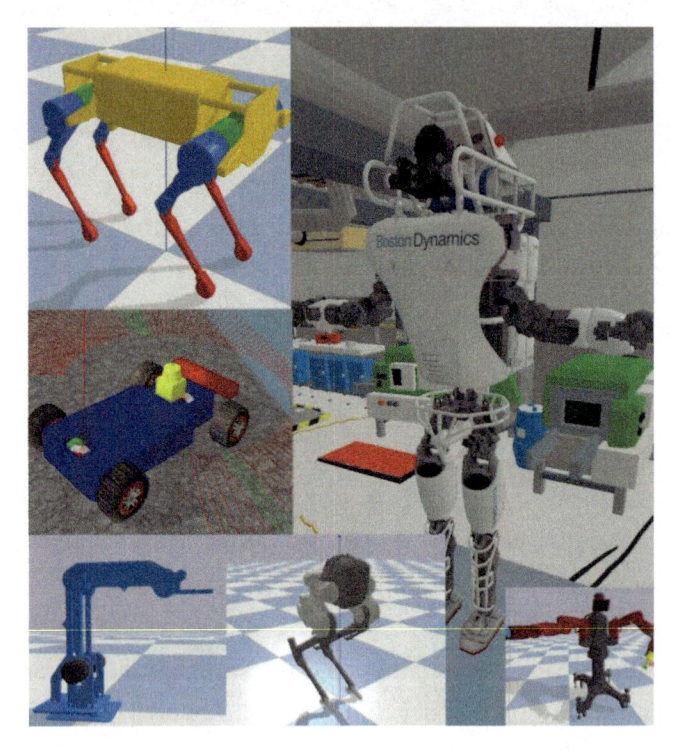

Figure 5-3.   A variety of robots supported by PyBullet simulation. These include legged robots like Boston Dynamics' Atlas, quadrupeds, wheeled robots, robotic arms, and manipulators. Used with permission, source: https://github.com/erwincoumans/pybullet_robots[57]

MuJoCo

MuJoCo (Multi-Joint Dynamics with Contact)[16] is a platform owned and open-sourced by DeepMind[17] that was developed for robotics. MuJoCo focuses on traditional robotics applications, including multiple link arms, grasping, and bipedal walking. It has garnered a lot of popularity because of its speed, precision in highly accurate physics, and user-friendly robotics design. It supports friction, contact, and rigid-body dynamics, and can accommodate more flexible bodies. As an example, the shadow hand[18] from Open AI, which was one of the first times a robotic hand could manipulate its environment with high dexterity, was built using the MuJoCo physics engine. We recommend going through MuJoCo's documentation[19] to get started with it.

Gazebo

Another popular robotics simulation platform is Gazebo[20]. It is often utilized for mobile manipulation, robotic grasping, and off-road mobility, as well as for more conventional robotics applications. We recommend going through Gazebo's documentation[21] to get started with it. Specifically, here is a tutorial[22] we recommend for using a simple Gazebo environment.

Gazebo's middleware interface with The Robot Operating System (ROS)[23] allows it to be used in combination with existing robotics stacks used in industrial robots. In contrast, PyBullet and MuJoCo provide more integration with DL and RL frameworks, as well as gym environments, making it more favorable for these use cases. The simulator you choose may be determined by a variety of criteria, including the features you require, speed, and application. A more comprehensive comparison of various simulators is shown in Figure 5-4.

CHAPTER 5 SIMULATION

Simulator	RGBD + LiDAR	Force Sensor	Linear + Cable Acutator	Multi-Body Import	Soft-Body Contacts	DEM Simulation	Fluid Mechanics	Headless Mode	ROS Support	HITL	Teleoperation	Realistic Rendering	Inverse Kinematics
Airsim	✓	✗	✗	✗	✗	✗	✗	✓	✓	✓	✓	✓, unreal	✗
CARLA	✓	✗	✗	✗	✗	✗	✗	✓	✓	✗	✓	✓, unreal	✗
CoppeliaSim	✓	✓	Linear only	✓	✗	✓	✗	✓	✓	✓	✓	✗	✓
Gazebo	✓	✓	Linear only	✓	✗	Through Fluidix	Through Fluidix	✓	✓	✓	✓	✗	✓
MuJoCo	✓	✓	✓	✓	✓	✓	Limited	✓	✗	HAPTIX only	HAPTIX only	✗	✗
PyBullet	✓	✓	Linear only	✓	✓	✓	✗	✓	✗	✗	✓	✗	✓
SOFA	✗	✗	✓	✓	✓	✓	✓	✓	✓	✓	✓	✓, Unity	✓
UWSim	RGBD only	✓	✗	✓	✗	✗	✗	✓	✓	✓	✓	✓, custom	✗
Chrono	✓	✓	✓	✗	✓	✓	✓	✓	✗	✗	✓	✓, offline	✓
Webots	✓	✓	linear	✓	✗	✗	Limited	✓	✓	✗	✓	✗	✗

Figure 5-4. *Comparison of robotics simulation platforms based on various features, such as sensor support (RGBD, LiDAR, force sensors), actuator types, multi-body import, and soft-body contacts. This table shows the strengths and limitations of each simulator, including PyBullet, MuJoCo, Gazebo, and others. Used with permission, source* https://ieeexplore.ieee.org/stamp/stamp.jsp?tp=&arnumber=9386154 *[58]*

Overall, Gazebo offers extensive ROS integration and the ability to simulate different environments. PyBullet integrates well with machine learning frameworks and MuJoCo outperforms simulation speed and accuracy[24].

Concepts in Sim2Real

Sim2Real refers to strategies that utilize simulation to learn policies to act in the real world. They may or may not also utilize real-world data. In this part, we look at some of the most frequent strategies utilized in Sim2Real. Many of these principles may be discussed in the context of reinforcement learning (RL), but they may also be applied to more general tasks, such as object identification, real-world control, and so on. In the context of RL specifically, the goal is to train an RL agent in a simulation environment and deploy/test it in the real world using techniques like domain adaptation or domain randomization.

Domain Adaptation

The primary principle behind domain adaptation (DA)[27][28][29] is to use data from one area (source) to improve a model's performance in another area (target) when there is less data. To accomplish this, you want to make the data from both sources appear more similar. To understand what source domain and target domain mean, consider the Markov decision process (MDP)[30]. An MDP is a model of decision-making in which an agent makes decisions in a succession of steps. At each step, the agent performs an action that changes the current scenario (state) to a new scenario (next state) with varying probability. The agent is rewarded for the actions it takes. RL seeks to identify the most effective technique (policy) that maximizes the total rewards over time. MDP is represented with states (S), actions (A), transitions (P) that happen with a certain probability, and rewards (R):

$$D \equiv (\mathcal{S}, \mathcal{A}, \mathcal{P}, \mathcal{R})$$

Equation 5-1

A Markov Decision Process is Markovian, which means that the state is fully observable and the information of how it got to the state is not relevant in deciding any aspect of the future.

The source domain is the environment to which you have complete access (in this case, the simulator), and the target domain is the actual physical world. The source domain in RL tasks and the target domain are designed as custom MDPs. Their states can be very different, while actions, transitions, and rewards have similarity, as you want actions from simulation to translate to the real world.

CHAPTER 5   SIMULATION

There are several common domain adaptation techniques[27][29], which ultimately intend to make source data look like target data:

- One method uses statistical techniques to assess and correct disparities between features from several sources. This can be done by aligning the mean and variances of features or minimizing the distance between distributions of the domains in a high-dimensional space.

- Another method trains a model to recognize which source the features come from before adjusting them to be more alike. Here, you can use a domain classifier to distinguish between features of each domain, which can be trained adversarially to make the classification difficult.

- Finally, a technique will identify shared characteristics by learning to reproduce the original data using these shared features, ensuring that essential information is captured from both sources. An encoder-decoder structure can be used, where the encoder maps the data from both domains to a latent space and the decoder is used to reconstruct the input data based on this shared representation.

We recommend [29] as a source to learn more.

Domain Randomization

Domain randomization[31] is a strategy for increasing model robustness by exposing the model to a wide range of simulated situations. The purpose of randomizing the simulation is to make the model responsive

CHAPTER 5   SIMULATION

to real-world fluctuations and situations that may arise. In this way, the model learns to deal with a wide range of circumstances, making it more successful when applied to real-world data.

In Figure 5-5, the diagram depicts domain randomization by demonstrating how simulated data is randomized to replicate the distribution of real-world data. It shows two tasks—Rope shaping and Assembling—trained in both standard and randomized simulation environments. Domain randomization adds visual and physical variability to help models generalize better to real-world robot testing. In contrast, domain adaptation aligns features from the source (simulation) and target (real) domains to create a shared feature space for reinforcement learning.

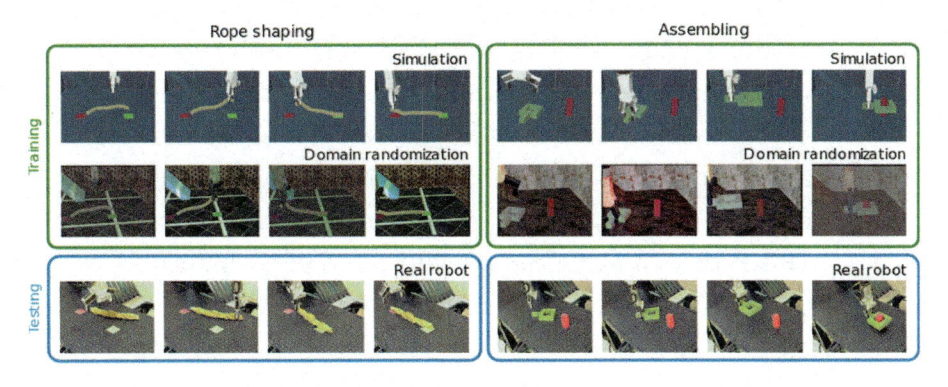

Figure 5-5. *Example of domain randomization: Visuomotor manipulation policies are trained in simulation (top row) with domain randomization using varied textures, lighting, object colors, and camera settings (middle row). These policies transfer directly to real-world tasks (bottom row). Used with permission, source: https://arxiv.org/pdf/2307.15320[59]*

Domain randomization has three primary training approaches[28]:

- **Static randomization,** which introduces random changes to the training environment at the beginning and maintains these variations throughout the training process. This is often done through varying light conditions, textures, and placement. These changes are then set throughout the training process of the model. This strategy is straightforward and quick, but it may not be as useful in real-world situations because it may not capture the full range of variations encountered by a robot.

- **Adaptive randomization** modifies these random modifications during training based on the model's learning progress. For example, the environment might have smaller variations at the start of the training but as the model improves, more complex changes might be introduced to ensure that the model is learning. This approach might result in improved performance but requires substantial real-world data to guide the adjustments.

- **Adversarial randomization** adds adversarial disturbances during training by using a different model to generate challenging scenarios, thus making the training environment more difficult for the "main" model. This helps the model grow more robust, but it needs to be carefully calibrated to prevent making the tasks overly difficult and worsen learning.

We recommend [28] as a source where we gathered ideas for this section.

Another way of exposing the model to different scenarios during training is through Uniform Domain Randomization (UDR)[32], which is the process of randomly modifying the training environment's properties at every step. These variations can include aspects in lightning, textures, and placement, so that the model does not overfit on a specific environment[32]. This implies that the model is continually exposed to new scenarios, making it more adaptive and powerful in the face of variations between training and real-world contexts. However, uniform sampling assumes that all changes in the training environment are equally important, which is incorrect because certain scenarios are more difficult to learn than others and require more focus during the training process.

Guided Domain Randomization

GDR (Guided Domain Randomization)[32] was proposed to solve some of these issues with a more directed sampling technique. This section covers two GDR methods: Active Domain Randomization and Automatic Domain Randomization.

Active Domain Randomization

Instead of simply randomizing environment settings uniformly, Active Domain Randomization (ADR), which was presented in Mehta et al., 2020[33], focuses on identifying and training the model in the most difficult settings it encounters. This technique involves constructing diverse and complex simulated situations, evaluating which ones are the most difficult for the model, and then focusing training on these difficult scenarios.

A simulator is specifically used to generate a variety of simulated environments with varied parameters. The model's policy is then tested in these simulated environments. A discriminator assesses the level of difficulty of these environments by comparing them to a reference

CHAPTER 5   SIMULATION

environment; it provides a reward accordingly. The reference environment could be chosen based on typical conditions that the model is expected to encounter. The Stein Variational Policy Gradient (SVPG)[34] then uses this reward to update the model's parameters by directing the model to spend more time training in difficult scenarios. This is achieved by iteratively adjusting the policy's parameters to maximize the expected reward. The main steps of how active domain randomization works are shown in Figure 5-6.

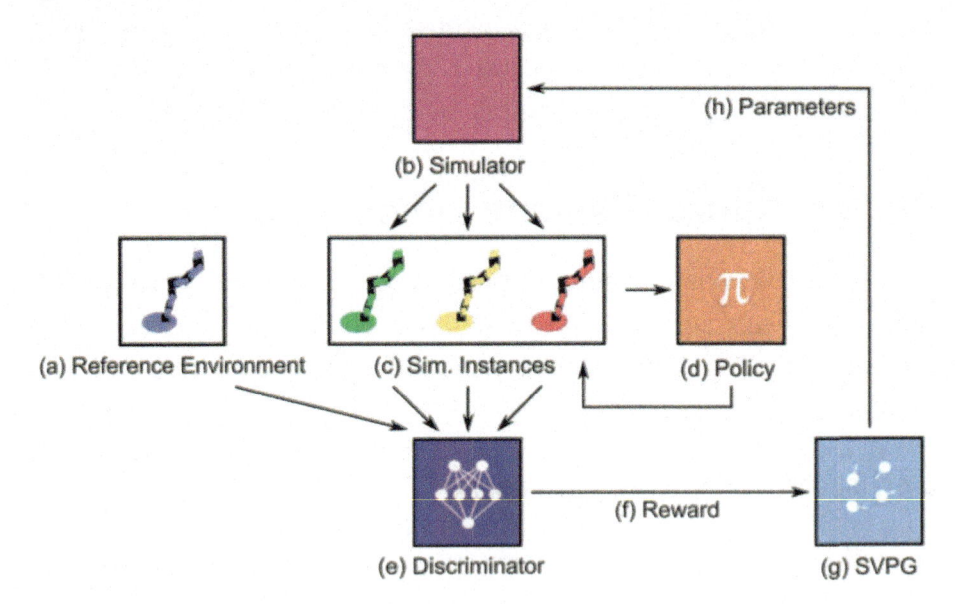

Figure 5-6. ADR framework uses a simulator to generate randomized environments for training an agent policy. A discriminator distinguishes between reference and randomized environments, generating a reward signal used to train SVPG particles. These particles explore difficult environment parameters to improve the policy's robustness. Used with permission, source: https://arxiv.org/abs/1904.04762[33]

CHAPTER 5   SIMULATION

The focus of active domain randomization is to dynamically adapt the training process based on difficulty. On the other hand, Automatic Domain Randomization provides a more systematic approach to domain randomization.

Automatic Domain Randomization

The relatively straightforward GDR technique Automatic Domain Randomization was introduced by Open AI in 2018[35], and it was successful in helping a real robotic hand solve a Rubik's Cube[36].

Automatic Domain Randomization begins by training a policy in an environment with default parameters. It goes through several training episodes in which it randomly decides whether to put lower or higher constraints on parameters to create variability. The policy interacts with the environment, and its performance is evaluated. If the policy performs consistently well, the parameters' limits are gradually increased, making the environment more difficult. If performance decreases, the boundaries are tightened to make the environment easier. This iterative process teaches the policy how to handle a wide range of scenarios based on direct feedback.

For more details on the algorithmic design, we recommend reading[35].

CHAPTER 5   SIMULATION

Closing the Sim2Real Gap for RL

RL-CycleGAN[37] is an important method that aims to close the Sim2Real gap for RL and was tested for vision-based robotic grasping tasks. At a high level, RL-CycleGAN combines RL with CycleGAN[38] (which we explain in more detail later) to transfer knowledge from simulated to real-world environments. The main components of RL-CycleGAN are:

- **Sim2Real generator**: This component converts images from the simulator to look closer to real-world images. The generator takes an image from the simulator and translates it into an image that has the texture, lightning, and noise characteristics of real-world images.

- **Real2Sim generator**: To ensure that the transformation between two domains (simulated and real images) is consistent, the real images are converted to simulated ones and back again. This means that when an image is converted from simulated to real and back to simulated, it should closely resemble the original simulated image.

The RL learns by interacting with the environment and getting some sort of reward as feedback. For example, a robot might get a positive reward for successfully picking up an object and a negative reward for dropping it. Overall, the agent's goal is to maximize its cumulative reward by learning the best actions to take in different states. In the case of CycleGAN, the RL agent is trained using images converted by the Sim2Real generator (see Figure 5-7). Ideally, these images look as real as possible, so that the agent learns to effectively perform tasks outside the simulator. Once the RL agent is well-trained in the simulated environment with Sim2Real images, it is deployed and tested in the real world.

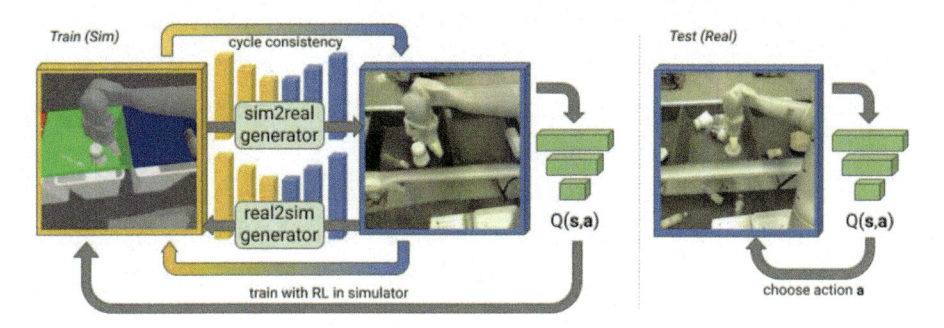

Figure 5-7. RL-CycleGAN trains a CycleGAN to map simulator images to realistic images and vice versa. The RL model is transferred to a real robot for testing and task execution. Used with permission, source: https://arxiv.org/pdf/2006.09001[37]

CycleGAN

To fully understand how CycleGAN[38] can be used for RL-based robotic grasping tasks, it can be helpful to understand the architecture and key losses. CycleGAN itself is a type of Generative Adversarial Network (GAN)[39] used to learn mapping between two image domains (simulated and real images) without paired examples. It is made up of two generators—the Sim2Real generator and the Real2Sim generator. The Sim2Real generator maps images from the simulated domain to the real domain. The Real2Sim generator maps images from the real domain back to the simulated domain. There is a discriminator for each image domain type: one discriminator will distinguish between real images from the real-world domain and fake images generated by the Sim2Real generator from simulated images, while the other discriminator distinguishes simulated images from the simulated domain and fake images generated by the Real2Sim generator from the real images.

Three main losses are used in CycleGAN—adversarial loss, cycle consistency loss, and final total loss.

CHAPTER 5   SIMULATION

The *adversarial loss* is used to make the generated images look as realistic as possible. There are two discriminators: one for real images and one for simulated images. Each generator (Sim2real and Real2Sim) tries to fool its corresponding discriminator by generating images that look real. The discriminators, in turn, try to correctly identify real and fake images. The generators aim to minimize the adversarial loss by producing highly realistic images, while the discriminators aim to maximize it by accurately identifying fake images. The first term in each loss function measures how well the discriminator identifies real images and the second term measures how well it identifies fake images produced by the generators.

$$L_{GAN}(G, D_Y, X, Y) = \mathbb{E}_{y \sim p_{data}(y)}[\log D_Y(y)] + \mathbb{E}_{x \sim p_{data}(x)}[\log(1 - D_Y(G(x)))]$$

$$L_{GAN}(F, D_X, Y, X) = \mathbb{E}_{x \sim p_{data}(x)}[\log D_X(x)] + \mathbb{E}_{y \sim p_{data}(y)}[\log(1 - D_X(F(y)))]$$

Equation 5-2

The *cycle consistency loss* ensures that the transformation between simulated and real images is consistent and reversible. The cycle consistency loss function ensures that when an image is transformed from one domain to another and back again, it remains similar to the original. It does this by penalizing the model if the twice-transformed image differs significantly from the original. Overall, this encourages the model to preserve important features during the transformations.

$$L_{cyc}(G, F) = \mathbb{E}_{x \sim p_{data}(x)}[\|F(G(x)) - x\|_1] + \mathbb{E}_{y \sim p_{data}(y)}[\|G(F(y)) - y\|_1]$$

Equation 5-3

The *final total loss* function combines the adversarial losses for both generators and discriminators with the cycle consistency loss. This ensures that the generated images are realistic and the transformation between domains are consistent.

CHAPTER 5 SIMULATION

$$L(G, F, D_X, D_Y) = L_{GAN}(G, D_Y, X, Y) + L_{GAN}(F, D_X, Y, X) + \lambda L_{cyc}(G, F)$$

Equation 5-4

RL-CycleGAN

The main components of RL-CycleGAN include:

- GAN, which transforms images between the simulated and real domains. This includes the Sim2Real GAN, which converts simulated images to look realistic, and the Real2Sim GAN, which converts real images back into simulated.

- Cycle consistency ensures that when an image is transformed from simulated to real and back to simulated (or vice versa), it retains the essential features.

- Two Q-networks are trained: one on simulated images to help the robot interact with objects in the simulated environment and one on real images to help the robot interact with objects in the real world.

- There is an RL scene consistency component that ensures that robot actions are consistent across simulated and real environments. For example, if the robot learns how to grasp an object in simulation, it should be able to apply the same grasping technique in the real world. As well, this ensures that q-values for similar scenes (whether in real or sim) are similar. This is done because the RL-scene consistency loss penalizes differences in q-values for corresponding scenes across different transformations (e.g., sim-to-real-to-sim, real-to-sim-to-real).

CHAPTER 5   SIMULATION

Figure 5-8 shows how these components fit together.

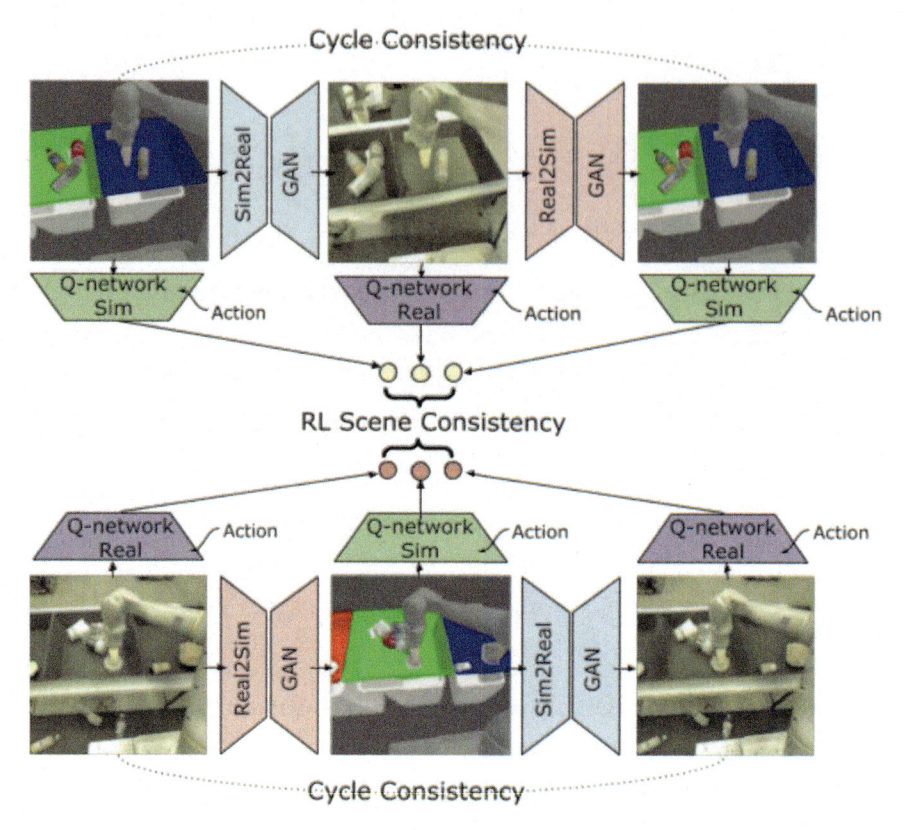

Figure 5-8. RL-CycleGAN combines a CycleGAN with RL scene consistency. Used with permission, source: https://arxiv.org/pdf/2006.09001[37]

The training of RL-CycleGAN combines several objectives:

- **Realistic Image Generation (L_GAN)**: Ensures that the images generated by the GANs look realistic.

- **Cycle Consistency (L_cycle)**: Ensures that the image transformed from simulated to real and back to simulated (or real to simulated and back to real) remains consistent with the original image.

234

- **RL-Scene Consistency (L_RL-scene)**: Ensures that similar scenes have similar q-values.

- **Traditional RL Loss (L_RL)**: Trains the q-networks using standard RL loss; the model learns the q-values based on the actions and rewards.

The distinction between style and semantics is important in this context. Style refers to visual aspects such as lighting and textures, which should ideally not impact the robot's performance. For example, a change in the lightning of the environment should not change how the robot grasps an object. Semantics, on the other hand, refers to features like positions and what objects are. These impact how a robot performs its tasks. For example, knowing the exact position and identity of an object is crucial for a robot to grasp an object. RL-CycleGAN focuses on preserving semantics while transforming images between simulated and real environments. This is important, as semantics for the task should be maintained while visuals aspects (style) can change during the transformation.

Learning from Simulation

Now that you understand how domain randomization can improve training models from simulation and translate them more effectively to the real world, this section explores a few key examples of how simulation has been used in robotics.

Specifically, robots can be trained to handle complex and dynamic environments using reinforcement learning, but the high cost of data acquisition is a significant limitation. Many methods have been developed to address this by combining real-world and simulation data for model development and training.

CHAPTER 5 SIMULATION

Simulation for Bootstrapping RL

QT-Opt[12] trains a Q-function with tens of thousands of real-world and simulated grasps. In this case, a Q-function is the expected future rewards for taking a given action in a given state. This approach can generalize grasping to 96 percent grip success on unseen objects, representing a substantial improvement over existing approaches.

QT-Opt starts with a large collection of offline data consisting of 580,000 recorded grasps, which includes state-action-reward tuples. This data is stored in replay buffers, where it is supplemented with data from online real robot interactions. The Bellman Updater samples transitions from these replay buffers to generate training examples and refines the model's value estimates using the Bellman equation[40]. This equation updates the Q-values, which represent the expected rewards for each state-action pair, based on both immediate and future rewards. The training workers then use these examples to update the Q-function parameters. To select the best grasping action, QT-Opt uses the Cross Entropy Method[41], which identifies the action that maximizes the model's estimated value (Q). Finally, the robot uses the trained model to perform grasping tasks in the real world. This process is shown in Figure 5-9.

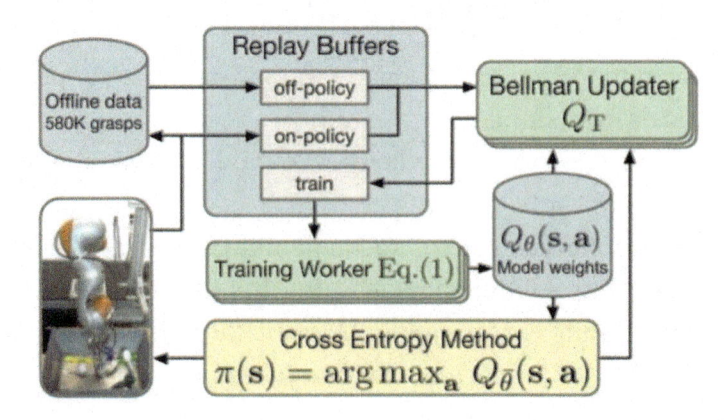

Figure 5-9. *Pipeline for QT-Opt. Used with permission, source:* `https://arxiv.org/pdf/1806.10293[12]`

CHAPTER 5 SIMULATION

Alongside using real data, an important component of this model is training it on a diverse set of objects using simulation data from the Bullet Physics simulator. The same model, training method, and control were used to train in real and simulation. This ensures that the learned policy can generalize and the robot can execute grasping in varying environments. The simulation setup is shown in Figure 5-10.

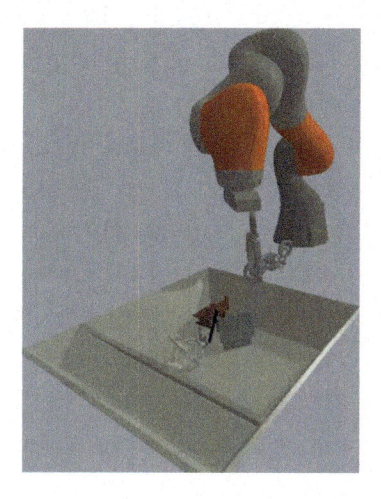

Figure 5-10. *QT-Opt robot manipulation set up in simulation. Used with permission, source:* https://arxiv.org/pdf/1806.10293 [12]

Through simulation, they were able to quickly run a large-scale experiment with up to 1,000 virtual robots operating simultaneously. These experiments used real-world policies learned from real-world data, finding real-world learning more challenging with more data needed and longer training time for the performance to be as effective as simulation. Overall, this experiment highlighted the importance of using both simulation and real-world data for testing and training robot models.

237

CHAPTER 5   SIMULATION

Foundation Agents in Simulation

Over the past few years, a key direction in machine learning has been developing foundation models trained on a large corpus of data that can be generalized to many tasks. Foundation models are frequently used in simulations to allow robots to accomplish a wider range of activities and better understand their environment. Large language models can generate action plans, but many robots still need help acquiring and updating knowledge over long periods.

In order to drive exploration and generalize to a wider range of abilities in a Minecraft simulation environment, Voyager[42] tries to address this issue by using LLMs for agents like robots. This is an example of how learning in simulation may be helpful for exploring a wide range of tasks and contexts and evaluate tasks at a larger scale. Voyager has a loop where, based on the player's inventory, what's located nearby, health, hunger, and the environment (biome and time of day), the agent in simulation suggests the next best action. But its exploration, learning, and memory are managed via the three following components:

- **Automatic curriculum**: New tasks for the agent to complete are generated by GPT-4[43], by analyzing state information. The model's reasoning explains why the suggested task is beneficial, and from this list, a specific task is queued for the player to execute. Since the curriculum or tasks an agent improves on is determined by an LLM, it is programmatically generated. Examples of tasks generated by the agent are shown in Figure 5-11.

CHAPTER 5 SIMULATION

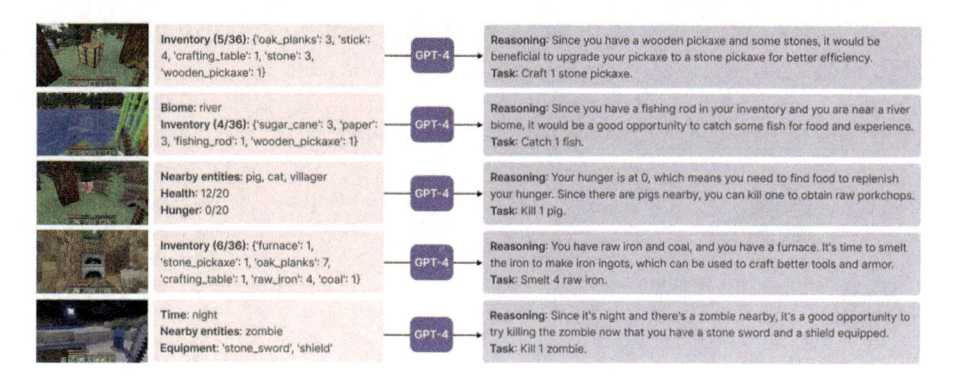

Figure 5-11. *Example of tasks that automatic curriculum has proposed. Used with permission, source:* `https://arxiv.org/pdf/2305.16291[42]`

- **Skill library**: In order to accumulate and build on the skills gathered in simulation, agents have a skill library. When an agent performs a task, by generating code, this task and the program generated are stored in a dictionary called the skill library. The keys of the dictionary are embeddings of the skill description and the value corresponding is the program itself. New skills are added to the dictionary when the corresponding tasks are encountered, post their execution. Older skills are retrieved via querying and updated using any improvements or feedback from the last round when they were invoked.

- **Iterative prompting mechanism**: There is an iterative process for refining actions by executing them, reporting errors, and refining the actions using feedback from the language model and errors from the simulator.

CHAPTER 5   SIMULATION

An overview of how all of these ideas fit together is shown in Figure 5-12.

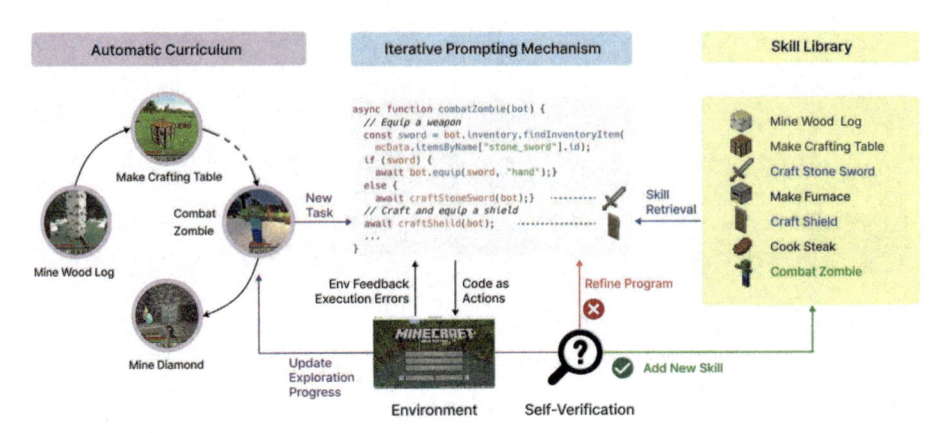

Figure 5-12.   Automatic curriculum, iterative prompting mechanism, and skill library in Voyager. Used with permission, source: `https://arxiv.org/pdf/2305.16291`*[42]*

Voyager learns and improves at playing Minecraft better than previous methods. It also adapts to new tasks in a new Minecraft world using its library of learned skills. Figure 5-13 shows a comparison of other methods with Voyager.

CHAPTER 5   SIMULATION

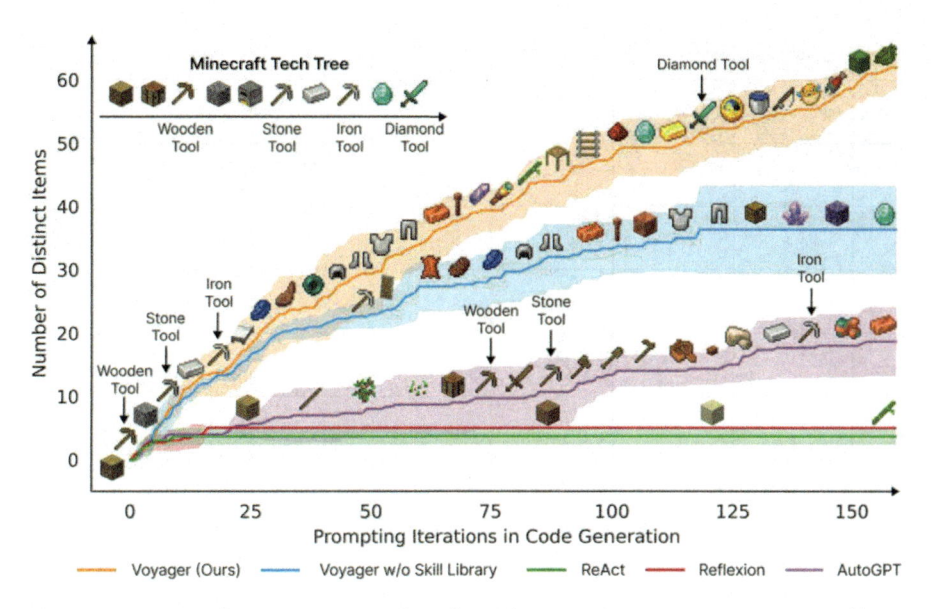

Figure 5-13. *When compared to baselines, Voyager is continually finding new Minecraft items and skills through self-exploration. Used with permission, source:* `https://arxiv.org/pdf/2305.16291`*[42]*

Alongside Voyager, the team at NVIDIA also developed MineDojo[44], which is a platform built on Minecraft that can be used to train AI agents using a wide range of tasks in a simulated environment (see Figure 5-14). The MineDojo platform:

- Provides a standardized way to define tasks, world settings, and agent behaviors, all within Minecraft, making it easier to develop and test different models.

- Includes a large benchmark with thousands of different Minecraft tasks.

- Curates a large-scale multimodal knowledge base from tutorial videos, live streams, and so on, which can be used to teach the agent tasks and strategies for accomplishing a goal.

241

CHAPTER 5   SIMULATION

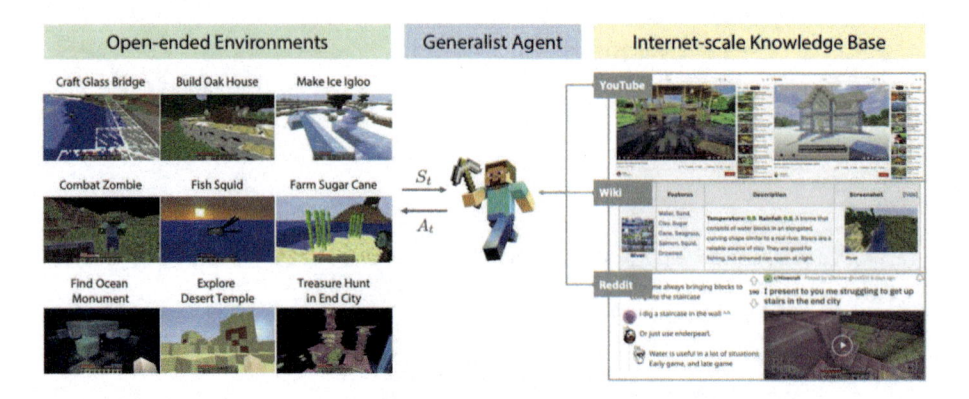

Figure 5-14. MineDojo is a framework for developing generalist agents that learn from open-ended tasks and Internet-scale data. Used with permission, source: `https://arxiv.org/pdf/2206.08853`*[44]*

MineCLIP[45] is a novel learning method that uses pretrained video models to learn a reward function that is used to guide agents in Minecraft. MineCLIP connects large language models with visual understanding of the Minecraft world, allowing agents to:

- **Understand natural language instructions:** Interpret human-like commands and goals (e.g., "build a house").

- **Learn from videos:** Gain skills and strategies by watching Minecraft gameplay videos.

- **Act autonomously:** Perform complex tasks in the game based on language instructions and learned knowledge.

In essence, MineCLIP acts as a bridge between language and action in Minecraft. It does this by building a contrastive video-language model pretrained on MineDojo's YouTube videos. This model processes video frames and their descriptions to learn correlations. It takes a sequence

CHAPTER 5 SIMULATION

of the last 16 RGB frames, processes them to extract features, and aggregates them into a video feature. The textual goal is encoded into a text feature and then the video and text features are compared to compute a correlation score, indicating how well the video frames the agent observes match the goal, thus learning whether the agent is actually successful in accomplishing the goal or not. For example, if the task is "do x task to get y food" the correlation score will reflect how well the video frames align with this task description.

This score is used as a reward for training the RL agent, helping it prioritize actions aligned with goals. MineCLIP is designed for multi-task RL, using the correlation score to guide the agent on various tasks without manually engineering reward functions (see Figure 5-15).

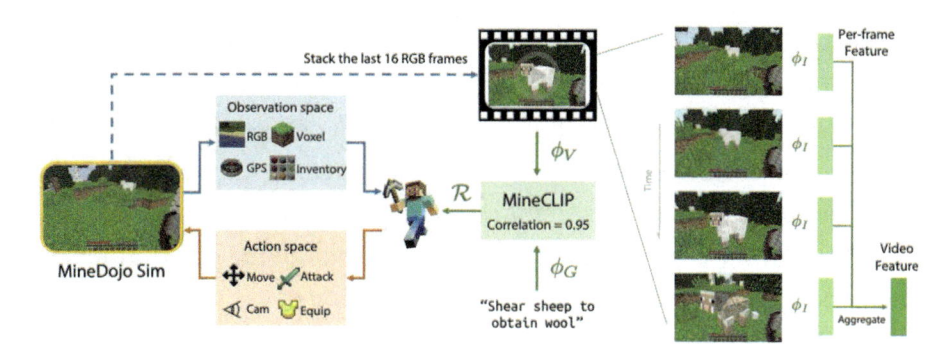

Figure 5-15. Pipeline for MineCLIP uses contrastive learning on video-language data. Used with permission, source: `https://arxiv.org/pdf/2206.08853`*[44]*

Although the focus for this specific project was in Minecraft, similar simulation environments can be used to show how robotics can benefit from simulation. These environments provide a safe space for robots to practice, refine, and store their skills, which enables better learning and adaptation to new tasks and environments.

CHAPTER 5   SIMULATION

Simulation for Reward Design

Teaching robots to perform complex tasks often requires hand-engineering reward functions to guide the robot's learning process. However, designing these reward functions can be time-consuming and might not effectively capture the complexity of tasks like dexterous manipulation. Using simulation data can be useful as a way to generate, evaluate, and refine reward functions in a more controlled and safe environment.

For example, Eureka[46], developed by NVIDIA, uses IsaacGym[1] to create diverse and complex robotic manipulation tasks. These tasks range from simple object handling to bimanual manipulations like rotating a cup by 180 degrees or performing pen spinning tricks with a Shadow Hand[47].

The core idea behind Eureka is using LLMs such as GPT-4 to generate and improve reward functions using an iterative process within simulation. See Figure 5-16.

In the Eureka framework, the robot is given a detailed description of the task to perform: details on the environment, goals, and any constraints the robot needs to abide by. An initial prompt is passed into the LLM, which includes the task description and details about the environment that guide the generation of initial reward functions.

Given a specific number of iteration, the following are done:

- **Sample reward code**: The LLM generates multiple reward functions based on the initial prompt. These reward functions are different "strategies" that provide the robot with feedback on how well it is performing the task.

- **Evaluate reward functions in simulation**: Each reward function is tested in a simulated environment where the robot attempts to complete the task using the reward function and then performance is measured by using a fitness function that quantifies the effectiveness of each reward function in teaching the robot

- **Reward reflection**: The best performing reward functions are selected based on analyzing how rotation, distance, and angular velocity penalties contributed to the robot's performance. Values of these components and the task fitness function are monitored during the training process. The initial prompts are then refined to help the LLM generate more optimized reward functions using feedback from the components and adjustments.

- **Update reward**: If the best reward function from the current iteration performs better than the previous best, then Eureka updates its reference reward.

Algorithm 1 EUREKA

1: **Require**: Task description l, environment code M, coding LLM LLM, fitness function F, initial prompt prompt
2: **Hyperparameters**: search iteration N, iteration batch size K
3: **for** N iterations **do**
4:     // Sample K reward code from LLM
5:     $R_1, ..., R_k \sim \text{LLM}(l, M, \text{prompt})$
6:     // Evaluate reward candidates
7:     $s_1 = F(R_1), ..., s_K = F(R_K)$
8:     // Reward reflection
9:     prompt := prompt : Reflection(R_{best}^n, s_{best}^n), where $best = \arg\max_k s_1, ..., s_K$
10:     // Update Eureka reward
11:     $R_{\text{Eureka}}, s_{\text{Eureka}} = (R_{best}^n, s_{best}^n),$ if $s_{best}^n > s_{\text{Eureka}}$
12: **Output**: R_{Eureka}

Figure 5-16. Eureka algorithm. Used with permission, source: https://arxiv.org/pdf/2310.12931[46]

CHAPTER 5   SIMULATION

An evolutionary search technique is used to sample multiple reward functions from the LLM in each iteration, increasing the chances of finding at least one good reward function by increasing the number of samples. These functions are refined by mutating the best performers and using random restarts to avoid local optima and find better solutions.

The reward function itself consists of a few main components:

- Rotation reward measures how closely the object's orientation matches the goal orientation.

- Distance reward focuses on the proximity between the robot's fingertips and the object.

- An angular velocity penalty is applied, which discourages any sudden or rapid movements and helps create more controlled actions.

- The total reward is a weighted sum of these components based on how much they influence the robot in completing its task successfully.

This continuous improvement would not be possible without a simulated environment where it is easy to control the robot and its environment and test different reward functions. Using this technique, Eureka found that the generated reward functions improved over human designed reward functions and in some cases even generated novel reward formulations.

Simulation for World Modelling

Traditional methods for generating large-scale datasets for robot learning often rely on[48]:

CHAPTER 5   SIMULATION

- **Demonstrations:** Require significant human involvement to teach the robot tasks, thus can be very time-consuming and labor-intensive.

- **Autonomous data collection:** Can be engineering-intensive for bootstrapping skills and challenging to scale even if it is autonomous.

Since both methods are challenging to implement at a large scale, this can limit the diversity and volume of data available for training robot models. Simulation, as discussed, is one way to solve this problem because it allows robots to learn in a virtual environment. Another solution is through advanced world modeling, which can help generate realistic and diverse data.

World modeling is important for robots to understand their environment, but it's also a good way to gather data that can be used for robot learning. Consider the example of ROSIE (Scaling Robot Learning with Semantically Imagined Experience)[48], work from Google DeepMind, in which a text-guided diffusion model is used for world modeling. See Figure 5-17.

ROSIE augments demonstration data to improve the variation in robot learning datasets and increase the adaptability of imitation learned policies.

- Using an open vocabulary segmentation model, ROSIE identifies and localizes different regions of interest in the image to identify where augmentations should be applied.

- The identified regions are edited using text-guided image-editing. For example, it might add new objects or alter existing ones based on text prompts that are variations of the initial task. Using the text prompts, the image editor creates new, varied versions of the original

247

CHAPTER 5 SIMULATION

task. This augmented data includes different objects, backgrounds, and so on. Each image in the episode is processed to include these new elements, which as a result increases the diversity of the dataset.

- The augmented data is then used to train an RT-1[49] manipulation policy, which is a multitask imitation policy covered in Chapter 4. The idea is that, by incorporating a wide range of scenarios, the policy becomes more robust to new and unseen tasks.

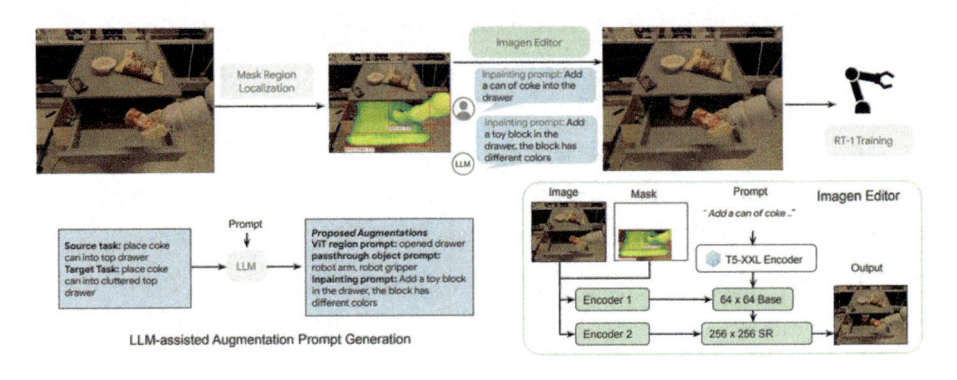

Figure 5-17. Architecture for ROSIE. Used with permission, source: https://diffusion-rosie.github.io/ [48]

Other methods, such as InstructPix2Pix[50], have tried using similar approaches. However, ROSIE has found that their generated augmentations end up being more physically realistic and consistent within the context of the original task. This is done by ensuring there are no broader, global changes to the image that could change the context of the original image. See Figure 5-18.

CHAPTER 5   SIMULATION

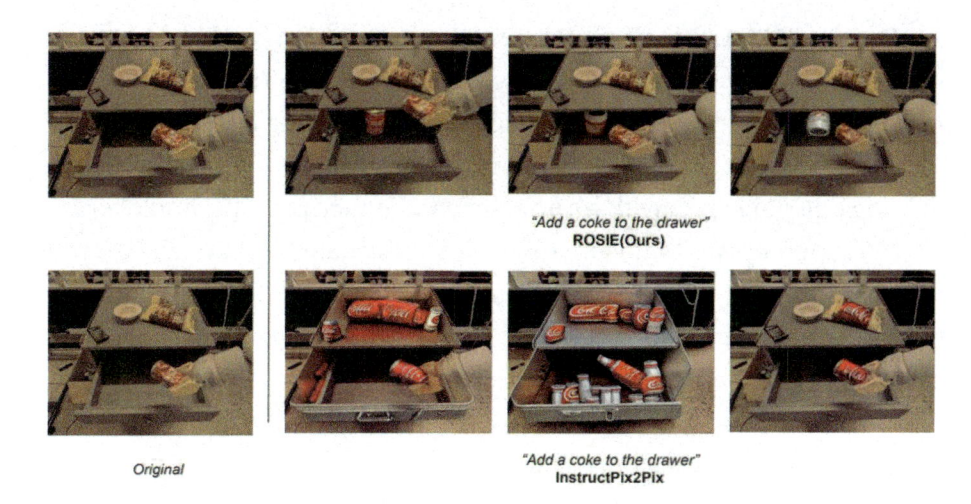

Figure 5-18. Augmentation process used in ROSE. Used with permission, source: https://diffusion-rosie.github.io/[48]

From an infrastructure perspective, NVIDIA has released a lot of new tools for simulation and world modeling that may be of interest to robotics developers interested in building realistic and scalable robot learning policies. At GTC 2024, they announced Project GR00T, a foundation model for humanoid robots and Jetson Thor, a new computer specifically designed for humanoid robots[51]. Isaac Lab from NVIDIA supports running thousands of parallel simulations to support RL and large-scale data generation, model training, and distributed workflows[51]. These tools collectively advance the capabilities of robots to be able to perform more complex tasks and improve some of the data-scarcity issues that exist in robotics through large-scale simulation support.

Simulation for Imitation Learning

Simulation environments can provide a means to scale imitation learning approaches and iterate on scalable research ideas. VIMA: General Robot Manipulation with Multimodal Prompts is a work from NVIDIA that pushes along this direction, with specific focus on multimodal task conditioning[52]. While VIMA is a generic method that can be executed in real-world settings, it is an impressive example of using simulation for end-to-end learning, ground concepts, and iterating fast at scale. VIMA constructs a simulation benchmark with thousands of generated tabletop grasping tasks with a mix of text and image prompts and over 600,000 expert episodes to help robots learn through imitation. VIMA uses multimodal prompting—that is prompting with a combination of images and texts—with a transformer-based architecture to execute an extensive range of robotic manipulation tasks.

VIMA demonstrates some interesting ways to specify tasks for robot learning, via multimodal prompting:

- **Goal conditioning for orientation**: In a task such as rearranging objects, it is harder to specify that goal via text alone, and needs multimodal specification. One way to do that is to give an image and prompt the robot to rearrange the scene to match the expected final configuration. Let's say you have a table with various fruits (apples, bananas, oranges) that are scattered around. The prompt includes an image of a neatly arranged fruit platter, where each fruit is arranged in a specific order. The corresponding text "Rearrange the fruits to match this layout" is provided.

- **One-shot learning**: The goal here is for the robot to learn to perform a task when given a single expert demonstration only. For example, if you want a robot to learn to stack books in a specific order, you can provide a sequence of images that show a person stacking different books in each frame alongside the text "Stack books in this order."

CHAPTER 5   SIMULATION

- **Concept grounding**: Checks if a robot can understand symbolic concepts and perform tasks that manipulate them. Let's say you introduce new objects called "x" and "y" to the robot. The prompt includes images showing "This is x" next to a picture of a cylindrical object and "This is y" next to a spherical object. The task might be to "Place x next to y."

- **Visual constraint**: Here you specify constraints through a mix of images and text. For example, if you want the robot to clean up a play area, you might provide it with the prompt "Move all the toys into the box without knocking over the tower of blocks," but part of the prompt is visual tokens.

Figure 5-19 shows examples of these use cases and how VIMA is prompted multimodally for each.

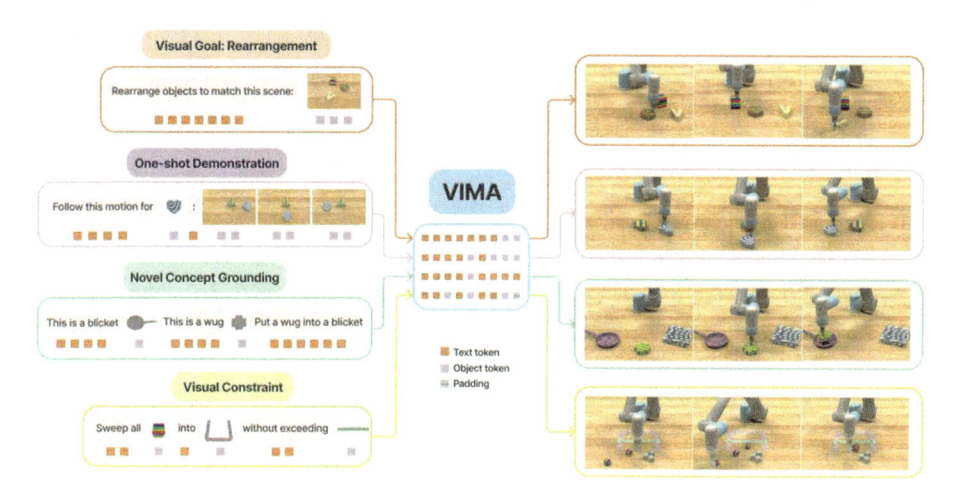

Figure 5-19. VIMA processes multimodal prompts combining language and visual inputs to specify robot manipulation tasks. Used with permission, source: https://arxiv.org/pdf/2210.03094 [52]

251

CHAPTER 5   SIMULATION

The architecture of VIMA is shown in Figure 5-20. They use a trained T5[55] tokenizer to tokenize the multimodal prompts and use self- and cross-attention to interleave prompt tokens and tokenized interaction history. This context is then used by a causal decoder to output actions, and the entire system is trained via behavioral cloning on a dataset of expert demonstrations.

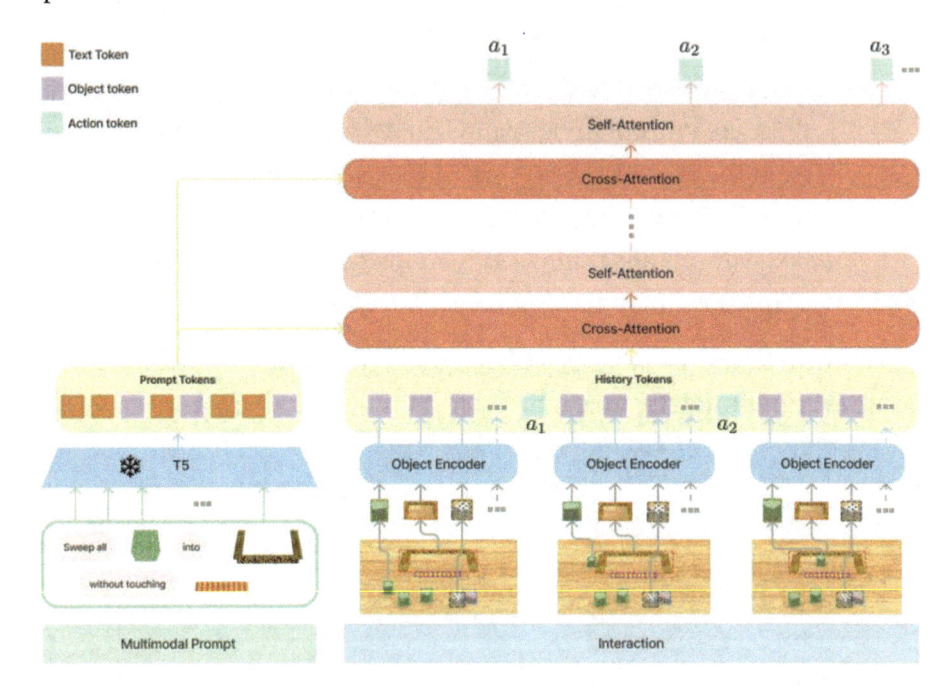

Figure 5-20. *Architecture of VIMA. Used with permission, source: https://arxiv.org/pdf/2210.03094[52]*

CHAPTER 5 SIMULATION

Trained VIMA variants were tested for various levels of generalization:

1. Scene generalization: By placing the objects in different orders/orientations in the scene.

2. Seen objects, but novel combinations: Manipulating two different objects in a combination that was not seen in the training set, but in which each object was seen in a different context while training. This checks for transfer of concept grounding from the training stage to the inference stage.

3. Novel objects: New objects are introduced to tasks not seen during training to check for generalization.

4. Novel tasks: Entirely unseen tasks with unseen objects.

Figure 5-21 shows the generalization evaluation framework used in VIMA. Each of these levels differs progressively from the training data, making the task more challenging and acting as a strong evaluation framework for generalization. While this study was done entirely in simulation, one can transfer all these concepts to real-world domains, providing a stark example of how simulation is used to advance robot learning research.

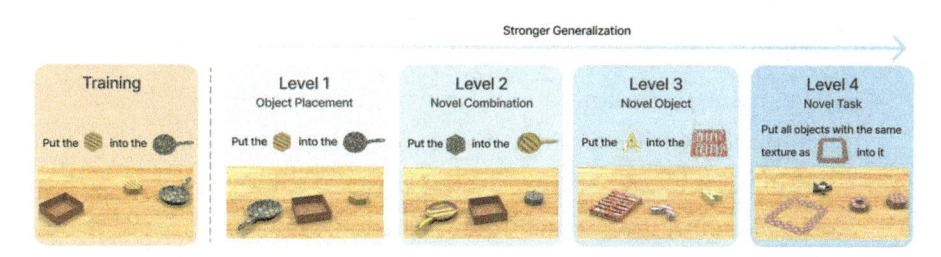

Figure 5-21. *Four levels of evaluation for VIMA to measure zero-shot generalizability. Used with permission, source:* `https://arxiv.org/pdf/2210.03094`*[52]*

253

CHAPTER 5   SIMULATION

Summary

During this chapter, you learned that:

- Simulations offer a cost-effective and scalable solution to generate diverse synthetic data, which can help robots learn safely and efficiently. However, limitations like the Sim2Real gap require combining simulation with real-world data.

- Robot simulators, like PyBullet, MuJoCo, and Gazebo, include core features such as physics engines for rigid and soft body dynamics, robot modeling through URDF, environment simulation, and user interaction via APIs and GUIs for control and visualization.

- Techniques like domain adaptation and domain randomization are used to bridge the gap between simulation and the real world. Approaches such as guided domain randomization focus on training models with progressively more challenging scenarios.

- Methods like RL-CycleGAN address the Sim2Real gap by translating simulated images into realistic ones while preserving semantics. This can help improve the ability of RL agents to generalize to real-world tasks.

- Learning from simulation using both simulated and real-world data improves task exploration and skill acquisition. This was shown by QT-Opt for robotic manipulation, and projects like Voyager and MineDojo, with LLMs and foundation models. Simulators also help in reward design, as seen with Eureka. Additionally, world modeling methods like ROSIE augment training data to improve

CHAPTER 5   SIMULATION

policy adaptability, while VIMA uses multimodal prompting and transformers to train robots on diverse manipulation tasks.

The next chapter explores the use of deep learning methods for mapping and localization in robotics, sensor setups, geometric and semantic mapping techniques, and various localization methods.

Tutorials
PyBullet Tutorial

This tutorial walks you through foundational concepts in PyBullet, including loading a URDF, running a simulation, and controlling joints.

To start, you need to create a URDF file of your robot's physical structure. If you haven't created a URDF file before, you can use an example one from online or generate one using SolidWorks, which has a URDF export plugin.

To install PyBullet in your system, you can copy and paste the following into your terminal (Mac and Linux):

```
sudo pip3 install pybullet
```

For Windows, you can use the following command in your terminal:

```
pip install pybullet
```

Create a new Python file where your URDF is stored. Name it pybullet_simulation.py and open it in some code editor. The first pieces of code to write in your script are `import` statements. PyBullet gives you access to its simulation API and the time module allows you to add delays to control the speed of the simulation.

```
import pybullet as p
import pybullet_data
import time
```

CHAPTER 5   SIMULATION

You can establish a connection to the PyBullet simulation GUI using this line, which allows you to visualize and interact with the simulation:

```
p.connect(p.GUI)
```

You can run the simulation with an empty world by adding this loop:

```
while True:
    p.stepSimulation()
    time.sleep(1./240)
```

This environment should be completely empty. You can add a flow by loading a preexisting URDF of a plane available in PyBullet's data path:

```
p.setAdditionalSearchPath(pybullet_data.getDataPath())
floor = p.loadURDF("plane.urdf")
```

You can then load the robot URDF and set gravity in your environment. Make sure to replace [robot_file_name].urdf with the actual name of your URDF file.

```
robot = p.loadURDF("my_robot.urdf")
p.setGravity(0, 0, -9.81)
```

Before controlling the robot joints, it can be useful to inspect them to understand the joints well. The getNumJoints() and getJointInfo() functions can be used for this:

```
for i in range(p.getNumJoints(robot)):
    joint_info = p.getJointInfo(robot, i)
    print(f"Joint {i}: {joint_info}")
```

To move some joints, you can use the setJointMotorControl2() function. This function lets you specify a target position, velocity, and other control parameters.

This code will control the first joint, which usually represents the base joint of your robot:

```
target_position = 1.57   # 90 degrees
p.setJointMotorControl2(bodyUniqueId=robot,
                        jointIndex=0,   # can be changed
                        as needed
                        controlMode=p.POSITION_CONTROL,
                        targetPosition=target_position)
```

You can explore different control modes:

- POSITION_CONTROL: Moves the joint to a specific angle.
- VELOCITY_CONTROL: Controls the speed of the joint.
- TORQUE_CONTROL: Directly controls the torque applied to the joint.
- PD_CONTROL: Uses a PD controller for the joint.

Finally, you can add the simulation loop to the end of your script to run everything together:

```
while True:
    p.stepSimulation()
    time.sleep(1./240)
```

Once you have this basic setup working, you can start experimenting with more the advanced features of PyBullet:

- **Collision detection**: Add objects to the environment and check for collisions.
- **Sensors**: Simulate sensors like cameras or LiDAR.
- **Reinforcement learning (RL)**: Train your robot using RL algorithms in PyBullet.

For debugging and fixing any errors, we recommend looking through PyBullet's official documentation[53].

CHAPTER 5   SIMULATION

References

[1] https://developer.nvidia.com/isaac/sim

[2] https://www.simscale.com/blog/what-is-finite-element-method/

[3] https://restackor.com/physics/response/spring-mass-damper

[4] https://www.simscale.com/docs/simwiki/cfd-computational-fluid-dynamics/what-is-cfd-computational-fluid-dynamics/

[5] https://www.grc.nasa.gov/www/k-12/airplane/nseqs.html

[6] https://wiki.ros.org/urdf

[7] https://www.geeksforgeeks.org/a-search-algorithm/

[8] https://www.geeksforgeeks.org/introduction-to-dijkstras-shortest-path-algorithm/

[9] https://www.opengl.org/

[10] https://www.ros.org/

[11] https://pybullet.org/wordpress/

[12] Kalashnikov, Dmitry, et al. "Scalable deep reinforcement learning for vision-based robotic manipulation." Conference on Robot Learning. PMLR, 2018.

[13] http://sdformat.org/

[14] https://mujoco.readthedocs.io/en/latest/modeling.html

CHAPTER 5   SIMULATION

[15]   https://ompl.kavrakilab.org/

[16]   https://gymnasium.farama.org/main/environments/mujoco/

[17]   https://github.com/google-deepmind/mujoco

[18]   https://openai.com/index/learning-dexterity/

[19]   https://mujoco.readthedocs.io/en/latest/modeling.html

[20]   https://gazebosim.org/home

[21]   https://classic.gazebosim.org/tutorials

[22]   https://classes.cs.uchicago.edu/archive/2021/winter/20600-1/gazebo_simulator.html

[23]   https://www.ros.org/

[24]   Körber, Marian, et al. "Comparing popular simulation environments in the scope of robotics and reinforcement learning." *arXiv preprint arXiv*:2103.04616 (2021).

[25]   Souto, Nilson. "Video Game Physics Tutorial - Part I: An Introduction to Rigid Body Dynamics." Toptal Engineering Blog, www.toptal.com/game/video-game-physics-part-i-an-introduction-to-rigid-body-dynamics.

[26]   Souto, Nilson. "Video Game Physics Tutorial - Part III: Constrained Rigid Body Simulation." Toptal Engineering Blog, www.toptal.com/game/video-game-physics-part-iii-constrained-rigid-body-simulation.

[27]   Tanveer, Muhammad Hassan, et al. "An In-Depth Analysis of Domain Adaptation in Computer and Robotic Vision." *Applied Sciences* 13.23 (2023): 12823.

[28] Muratore, Fabio, et al. "Robot learning from randomized simulations: A review." *Frontiers in Robotics and AI* 9 (2022): 799893.

[29] Zhao, Wenshuai, Jorge Peña Queralta, and Tomi Westerlund. "Sim-to-real transfer in deep reinforcement learning for robotics: a survey." 2020 IEEE Symposium Series on Computational Intelligence (SSCI). IEEE, 2020.

[30] `https://builtin.com/machine-learning/markov-decision-process`

[31] Tobin, Josh, et al. "Domain randomization for transferring deep neural networks from simulation to the real world." 2017 IEEE/RSJ International Conference on intelligent Robots and Systems (IROS). IEEE, 2017.

[32] Weng, Lilian. "Domain Randomization for Sim2Real Transfer." lilianweng.github.io, 5 May 2019, `https://lilianweng.github.io/posts/2019-05-05-domain-randomization/`.

[33] Mehta, Bhairav, et al. "Active domain randomization." Conference on Robot Learning. PMLR, 2020.

[34] Liu, Yang, et al. "Stein variational policy gradient." *arXiv preprint arXiv*:1704.02399 (2017).

[35] Akkaya, Ilge, et al. "Solving Rubik's Cube with a robot hand." *arXiv preprint arXiv*:1910.07113 (2019).

[36] `https://openai.com/index/solving-rubiks-cube/`

[37] Rao, Kanishka, et al. "Rl-cyclegan: Reinforcement learning aware simulation-to-real." Proceedings of the IEEE/CVF Conference on Computer Vision and Pattern Recognition. 2020.

[38] Zhu, Jun-Yan, et al. "Unpaired image-to-image translation using cycle-consistent adversarial networks." Proceedings of the IEEE International Conference on Computer Vision. 2017.

[39] https://developers.google.com/machine-learning/gan/gan_structure

[40] https://huggingface.co/learn/deep-rl-course/en/unit2/bellman-equation

[41] https://web.mit.edu/6.454/www/www_fall_2003/gew/CEtutorial.pdf

[42] Wang, Guanzhi, et al. "Voyager: An open-ended embodied agent with large language models." *arXiv preprint arXiv*:2305.16291 (2023).

[43] Achiam, Josh, et al. "GPT-4 technical report." *arXiv preprint arXiv*:2303.08774 (2023).

[44] Fan, Linxi, et al. "MineDojo: Building open-ended embodied agents with Internet-scale knowledge." *Advances in Neural Information Processing Systems* 35 (2022): 18343-18362.

[45] https://github.com/MineDojo/MineCLIP

[46] Ma, Yecheng Jason, et al. "Eureka: Human-level reward design via coding large language models." *arXiv preprint arXiv*:2310.12931 (2023).

[47] https://www.shadowrobot.com/dexterous-hand-series/

[48] Yu, Tianhe, et al. "Scaling robot learning with semantically imagined experience." *arXiv preprint arXiv*:2302.11550 (2023).

[49] Brohan, Anthony, et al. "Rt-1: Robotics transformer for real-world control at scale." *arXiv preprint arXiv*:2212.06817 (2022).

[50] Brooks, Tim, Aleksander Holynski, and Alexei A. Efros. "Instructpix2pix: Learning to follow image editing instructions." Proceedings of the IEEE/CVF Conference on Computer Vision and Pattern Recognition. 2023.

[51] `https://nvidianews.nvidia.com/news/foundation-model-isaac-robotics-platform`

[52] Jiang, Yunfan, et al. "Vima: General robot manipulation with multimodal prompts." *arXiv preprint arXiv*:2210.03094 2.3 (2022): 6.

[53] `https://pybullet.org/wordpress/index.php/forum-2/`

[54] `https://www.do-mpc.com/en/latest/theory_mpc.html`

[55] Raffel, Colin, Noam Shazeer, Adam Roberts, Katherine Lee, Sharan Narang, Michael Matena, Yanqi Zhou, Wei Li, and Peter J. Liu. "Exploring the limits of transfer learning with a unified text-to-text transformer." *Journal of Machine Learning Research* 21, no. 140 (2020): 1-67.

[56] Yazici, Baris. "How to build self-learning grasping robot." *Medium*, 20 Dec. 2020, `towardsdatascience.com/sample-efficient-robot-training-on-pybullet-simulation-with-sac-algorithm-71d5d1d4587f`.

[57] Coumans, Erwin. "GitHub - Erwincoumans/Pybullet_robots." GitHub, 2017, `github.com/erwincoumans/pybullet_robots`.

[58] Collins, Jack, et al. "A review of physics simulators for robotic applications." *IEEE Access* 9 (2021): 51416-51431.

[59] Garcia, Ricardo, Robin Strudel, Shizhe Chen, Etienne Arlaud, Ivan Laptev, and Cordelia Schmid. Robust Visual Sim-to-Real Transfer for Robotic Manipulation. International Conference on Intelligent Robots and Systems (IROS), 2023.

CHAPTER 6

Mapping, Localization, and Navigation

Humans use multimodal sensory perception to identify objects and move in complicated spaces. This allows us to navigate, perform complex tasks, and interact with objects in our environment. Similarly, using varying sensors and multimodal sensory perception, robots (such as self-driving cars, drones, and home robots) can sense their surroundings and make decisions by estimating their current state and environment. These capabilities are enabled through localization and mapping, which are the focus of this chapter.

Localization helps a robot understand where it stands in a current scene; it involves determining the "states" of a robot (location, orientation, velocity, etc.) relative to other objects. *Mapping* involves capturing a model of its environment through geometry, appearance, and different 2D or 3D space semantics. Mapping is a precursor to localization, as localization of a robot is done with respect to its map. *Navigation* is the act of moving to a desired location, and it encompasses mapping, localization and reasoning. Before you dive into methods for each of these, it's essential to understand why and when to use deep learning for mapping and localization tasks.

CHAPTER 6   MAPPING, LOCALIZATION, AND NAVIGATION

Why Use Deep Learning?

Numerous classical methods and algorithms have been developed over the years to address the challenges in localization and mapping. Some of these methods include odometry estimation—such as visual odometry[1], LIDAR odometry[2], SLAM[3], and structure from motion (SfM)[4]. Choosing between deep learning-based and traditional techniques, like Kalman filters[5], involves considering tradeoffs. Kalman filters are effective for visual odometry, sensor fusion localization, or SLAM. Take for example, Comma AI's open-source implementation, RedNose[6] that is used for real world self-driving. While this chapter focuses on deep learning methods, the choice of methods used depends on the application requirements, as each approach has advantages and disadvantages.

Traditional Methods

Traditional methods excel at handling noisy or inconsistent data through well-established mathematical techniques. They can be more interpretable, which is essential for applications where understanding the decision-making process is crucial. These methods typically require less labelled data for training, making them suitable for scenarios with limited data collection.

However, traditional methods often require manual feature engineering, which can be time-consuming and might only capture some relevant information in complex sensor data. They may also need significant modifications to adapt to new or changing environments.

Deep Learning Methods

Deep learning methods can achieve high accuracy when trained on large and diverse datasets, as they can learn complex relationships in data. Relevant features can be learned directly from complex and high-dimensional sensor input data, reducing the need for manual feature engineering. If trained on representative data, these methods are adaptable to different environments and scenarios, making them suitable for applications with frequently changing environments. Additionally, they capture intricate spatial relationships within data, useful for understanding complex maps or 3D reconstructions.

On the downside, for deep learning models to learn effectively, a lot of the labeled data can be needed for training and annotating data can be time-consuming and expensive. These methods also demand significant computational resources for training and running. Furthermore, while they can excel in environments they've been trained in, they might only generalize to new or unseen scenarios when trained on a large corpus of diverse data.

A Hybrid Approach

More practically, a hybrid strategy is often deployed in the real world, which incorporates the benefits of both deep learning and conventional methods. For instance, deep learning models can extract features and learn complicated patterns, while conventional techniques can be used for sensor fusion or noise reduction. Currently, tasks such as object detection and semantic segmentation are done via a neural network, but functions like odometry and 3D reconstruction are often done using more "traditional" methods. See Figure 6-1.

CHAPTER 6 MAPPING, LOCALIZATION, AND NAVIGATION

However, we are starting to see a change, as methods like Dust3r[7] use neural networks to reconstruct geometry. Dust3r works with the following main steps:

- Takes multiple images from different viewpoints and passes these through a ViT[8] to get feature maps.

- These feature maps are then fed into transformers[9] that share information between the different viewpoints to produce pointmaps and confidence scores for each image. The pointmaps represent the 3D coordinates of points in the scene, and each has an assigned confidence score.

- The pointmaps from different viewpoints are aligned to a common coordinate frame, allowing for the reconstruction of a 3D scene model.

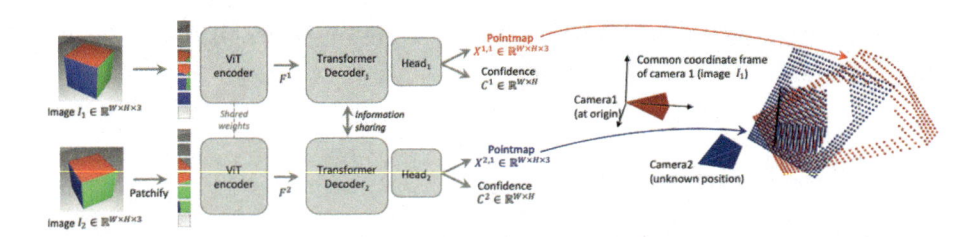

Figure 6-1. *The network uses a shared ViT encoder, transformer decoders, and regression heads to generate pointmaps and confidence maps from two scene views. Used with permission, source:* `https://arxiv.org/pdf/2312.14132`*[7]*

Similarly, neural network-based methods are becoming more promising for monocular depth estimation. Recent developments in video models are a strong example of this, as they can be conditioned on an input image and could allow people to perform depth estimation from video sequences. A 2023 survey[10] highlights that while NN monocular depth estimation is not fully mature, progress is being made: "In our view, the majority of existing MDE algorithms suffer from a limitation in training diversity, specifically in terms of various areas and image types. For instance, when the training dataset lacks a sufficient number of images featuring the sky, it can lead to challenges in accurately estimating depth in those particular regions." As training datasets become more diverse and detailed, the performance of neural network-based depth estimation methods could improve significantly.

The decision of where and when to use deep learning for mapping and localization tasks is ultimately influenced by variables such as data availability, computational resources, and application needs.

Typical Mobile Robot Setup

While there is a lot of variation in robot sensor configurations and embodiments, most general-purpose systems (such as autonomous cars or humanoid robots) require specific sensors to be effective at mapping and localization. Having an RGB camera and an IMU sensor is essential, and most systems will likely have multiple RGB cameras (possibly in a stereo camera setup for depth estimation). A LiDAR depth sensor can provide highly accurate 3D maps, but some companies (e.g., Tesla) prefer vision-based systems due to LiDAR's cost and complexity. Finally, GPS sensors are commonly used for robots that may travel considerable distances (such as autonomous cars). Determining the proper sensor configuration and pairing it with the right estimation techniques is a complex problem without a "one size fits all" answer and often involves some experimentation.

CHAPTER 6 MAPPING, LOCALIZATION, AND NAVIGATION

Mapping

In robotics, *mapping* refers to a robot's ability to create a reliable, detailed model of its environment, which helps it be aware of its surroundings. Deep learning methods can be used to perceive the environment and create a scene map. This scene map can be helpful in planning and navigation.

Geometric Mapping

Geometric mapping represents a scene's shape and structural details. In geometric mapping, voxel representation is a scene representation that divides the 3D environment into a grid of small cubes (voxels) that can tell you if the space in the scene at that point is occupied or free. Depth representation involves creating an image from depth sensors where each pixel has information on the distance to the nearest object. These representations provide an understanding of the scene's spatial layout and spatial geometry.

Depth Representation

Depth maps are crucial for understanding the geometry and structure of a scene. Many deep learning methods, such as supervised learning methods, use CNNs with RGB images and depth maps to predict depth. However, these methods require a large amount of depth data, which might not be available in all cases. To address this, semi-supervised methods and self-supervised methods that rely on unlabeled data can be used, but they face challenges in generalization.

CHAPTER 6   MAPPING, LOCALIZATION, AND NAVIGATION

The main steps in the depth representation pipeline are as follows:

- Collect RGB images using a monocular camera or datasets such as NYU-v2[11] or KITTI[12], which provide RGB images with ground truth depth maps.

- To extract features from the images, use a CNN, such as pretrained models like ResNet[13] or DenseNet[14].

- Various deep-learning models can be used to predict depth values for each image pixel. In supervised learning methods, a CNN would be trained on RGB images and corresponding depth maps to directly output depth information. In self-supervised methods, networks would be trained using unlabeled images by getting additional information from stereo images or visual odometry. Semi-supervised methods often combine labelled and unlabeled data for training.

- Often, an additional module refines the depth map. This can be done by aligning the initial depth estimates with expected geometric constraints or integrating global context through self-attention[15].

We don't go into too much detail about these methods, but an overview of specific deep learning-based depth estimation models is shown in Figure 6-2, and you can learn more here[16].

Chapter 6 Mapping, Localization, and Navigation

Method	Architecture	Category
EMDEOM [32]	FC	Supervised
ACAN [46]	Encoder-Decoder	
DenseDepth [47]	Encoder-Decoder	
DORN [18]	CNN	
VNL [48]	Encoder-Decoder	
BTS [49]	Encoder-Decoder	
DeepV2D [50]	CNN	
LISM [51]	Encoder-Decoder	Self-supervised
monoResMatch [38]	CNN	
PackNet-SfM [52]	CNN	
VOMonodepth [53]	Auto-Decoder	
monodepth2 [42]	CNN	
GASDA [54]	CNN	Semi-supervised

Figure 6-2. Methods for depth estimation. Used with permission, source: https://www.ncbi.nlm.nih.gov/pmc/articles/PMC7219073/ *[16]*

Voxel Representation

A 3D space's volume element is known as a *voxel*, and a natural way to represent 3D geometry is with a voxel-based formulation. The pipeline for voxel representation often looks as follows:

- Sensors such as LiDAR, RGB-D, or stereo cameras collect raw 3D data.

- This raw 3D data can be converted into a *point cloud*, which is a collection of data points in which each point represents a point on the surface of an object.

CHAPTER 6　MAPPING, LOCALIZATION, AND NAVIGATION

- In some cases, this point cloud can also be converted into a *3D mesh,* which is a group of vertices, edges, and faces representing the shape of objects.

- This 3D data can undergo voxelization, where, based on the points from the point cloud, each voxel is identified as either occupied by an object or free.

- A voxel grid can be used as an input to a CNN to extract features that represent the spatial structure and occupancy of the environment.

Tesla previously popularized an Occupancy Network[17], a 3D voxel representation of the world from their eight camera streams[18]. Their system uses eight cameras, which feed into models designed for feature extraction (see Figure 6-3), where an attention module generates an occupancy feature volume, which is combined with previous volumes to create a 4D occupancy grid. Finally, deconvolutions restore the original size, resulting in the occupancy volume, a 3D bird's-eye view, and the occupancy flow, which shows the movement of each pixel from one frame to the next.

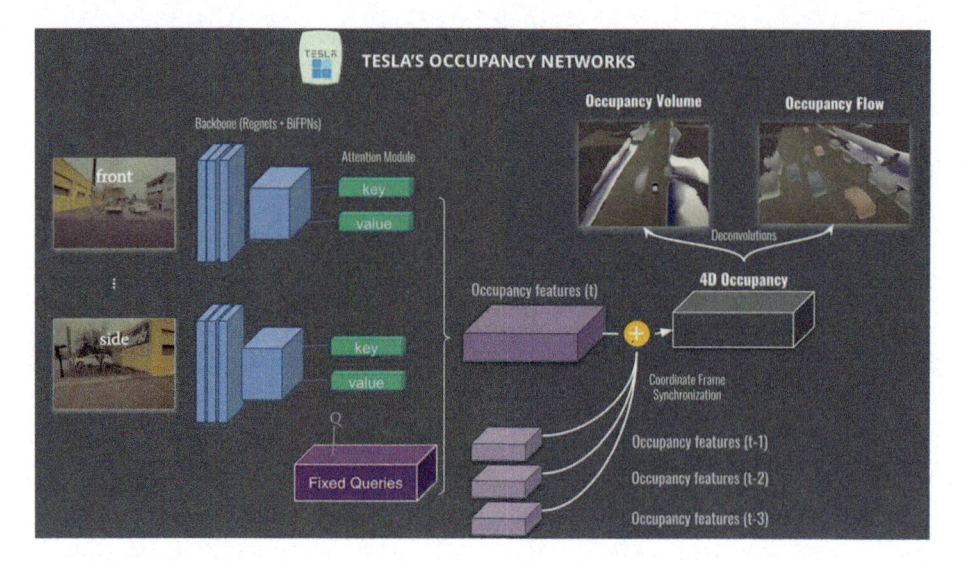

CHAPTER 6   MAPPING, LOCALIZATION, AND NAVIGATION

Figure 6-3. Tesla's Occupancy Network overview. Used with permission, source: `https://www.thinkautonomous.ai/blog/occupancy-networks/` *by Think Autonomous[39]*

While voxels have many use cases, one of their limitations is the high computational demand of reconstructing a scene in the high resolution that's needed for a voxel representation.

NeRF: Neural Radiance Fields

Neural Radiance Fields (NeRFs)[19] are used in the computer vision field because they allow the generation of detailed 3D scenes from 2D images. They offer valuable applications in robotics, particularly for localization, mapping, and understanding 3D environments. A NeRF is simply training a neural network to predict what an image of a scene would look like from a particular camera pose. The network is deliberately "overfit" to many pictures of a single scene and can then predict images for novel camera poses not in the training set.

CHAPTER 6 MAPPING, LOCALIZATION, AND NAVIGATION

NeRFs use a feed-forward network to generate detailed information about a scene. They use *rays*, which are lines in 3D space extending from a specific camera viewpoint into the scene in certain directions. These rays are lines in 3D space that project from a particular camera viewpoint into the scene along specific paths. The rendering process then merges this information into a 2D image. Since both components are differentiable, the entire system can be trained end-to-end. NeRFs define a 5D radiance field as an input of a location (x, y, z) and a 2D direction (specified by two angles in 3D space). As shown in Figure 6-3, the output from this function consists of volume density (opacity) and color (RGB value). The radiance field uses a feed-forward neural network with a 5D input to output the corresponding color and volume density information. An overview of the main steps in NeRFs is shown in Figure 6-4. We break down each of these steps in more detail next.

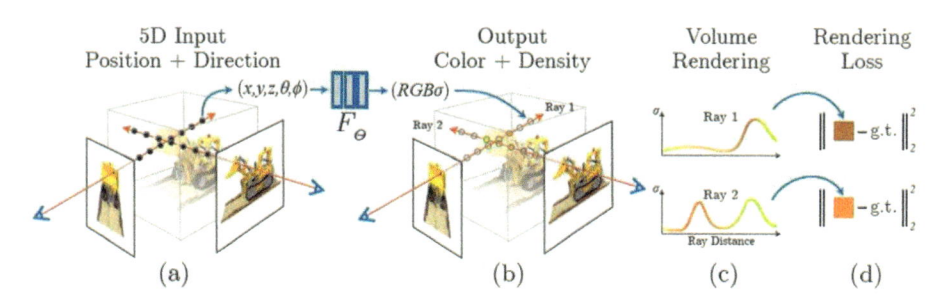

Figure 6-4. *Overview of NeRF and rendering. Used with permission, source: https://arxiv.org/pdf/2003.08934[19]*

To create a new view, camera rays are passed through the scene to get 3D points. These points, along with their 2D viewing directions, are input into the neural network to generate colors and densities. These are then combined using classical volume-rendering techniques into a 2D image.

CHAPTER 6   MAPPING, LOCALIZATION, AND NAVIGATION

The following code illustrates the MLP that can be used to convert the 5D input into RGB output:

```
# Model architecture

def init_nerf_model(D=8, W=256, input_ch=3, input_ch_views=3,
output_ch=4, skips=[4], use_viewdirs=False):
    relu = tf.keras.layers.ReLU()
    def dense(W, act=relu): return tf.keras.layers.Dense(W,
    activation=act)

    print('MODEL', input_ch, input_ch_views, type(
        input_ch), type(input_ch_views), use_viewdirs)
    input_ch = int(input_ch)
    input_ch_views = int(input_ch_views)

    inputs = tf.keras.Input(shape=(input_ch + input_ch_views))
    inputs_pts, inputs_views = tf.split(inputs, [input_ch,
    input_ch_views], -1)
    inputs_pts.set_shape([None, input_ch])
    inputs_views.set_shape([None, input_ch_views])

    print(inputs.shape, inputs_pts.shape, inputs_views.shape)
    outputs = inputs_pts
    for i in range(D):
        outputs = dense(W)(outputs)
        if i in skips:
            outputs = tf.concat([inputs_pts, outputs], -1)

    if use_viewdirs:
        alpha_out = dense(1, act=None)(outputs)
        bottleneck = dense(256, act=None)(outputs)
        inputs_viewdirs = tf.concat(
            [bottleneck, inputs_views], -1)   # concat viewdirs
```

```
        outputs = inputs_viewdirs
        # The supplement to the paper states there are 4 hidden
        layers here, but this is an error since
        # the experiments were actually run with 1 hidden
        layer, so we will leave it as 1.
        for i in range(1):
            outputs = dense(W//2)(outputs)
        outputs = dense(3, act=None)(outputs)
        outputs = tf.concat([outputs, alpha_out], -1)
    else:
        outputs = dense(output_ch, act=None)(outputs)

    model = tf.keras.Model(inputs=inputs, outputs=outputs)
    return model
```

This code has been taken without adaptation from [20].

NeRF computes loss by comparing the pixels of the rendered image to the ground truth images. Using this process, NeRFs can create detailed 3D reconstructions of physical spaces, which can be valuable for robots navigating complex environments. Additionally, the photorealistic images generated by NeRF capture the objects' geometry and appearance, improving the robots' ability to interpret scenes and navigate their surroundings accurately.

Semantic Mapping

Semantic mapping is creating a map so that objects and features in a scene are labelled with information that describes the scene. This can include identifying objects (car, dog), attributes about the objects (red, tall) and relationships (the vehicle is next to a tree). Traditionally, semantic mapping has been based on closed-vocabulary techniques, which means using a predefined set of labels. Due to large vision language models such

CHAPTER 6   MAPPING, LOCALIZATION, AND NAVIGATION

as CLIP[21], semantics are moving closer to open-vocabulary techniques. This allows for a more flexible and dynamic understanding of the scene and the ability to generate labels based on vocabulary learned from large image-text datasets.

Examples of methods that use this approach are Language Embedded Radiance Fields (LeRFs)[22], which use a neural approach to incorporate open-vocabulary semantics, and ConceptFusion, an explicit approach to semantic mapping.

Language Embedded Radiance Fields (LeRFs)

LeRF[22] integrates open-vocabulary semantic features into neural rendering, allowing detailed 3D scene representations to be created (see Figure 6-5). The main steps in LeRF are as follows:

- An input image is preprocessed into multiscale patches, which are smaller sections of an image extracted at different scales or resolutions. These patches are passed through a large vision language model such as CLIP to obtain CLIP features at different scales.

- The CLIP model is key in extracting visual and semantic information from the patches to generate feature maps at different scales.

- The neural network will take in spatial coordinates (x, y, z), view directions (θ, ϕ), and scale as inputs.

- Various parameters are outputted, such as RGB values, density, and DINO[23] features, which capture visual and semantic information from the input image, and CLIP features, which provide representations linking visual content with text descriptions.

CHAPTER 6 MAPPING, LOCALIZATION, AND NAVIGATION

- A loss function aligns the rendered features with semantic information from CLIP. This function is essential to ensure that the generated 3D scene has accurate semantic labels.

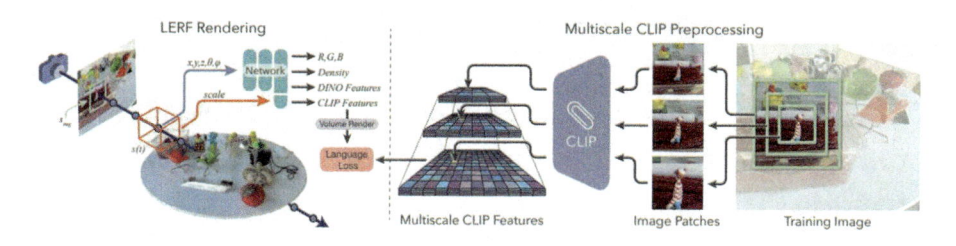

Figure 6-5. LeRF rendering and multiscale CLIP preprocessing. Used with permission, source: https://arxiv.org/pdf/2303.09553[22]

ConceptFusion

ConceptFusion[24] uses an explicit approach to semantic mapping that combines different levels of visual embeddings to create pixel-aligned embeddings that contain semantic information. ConceptFusion has a few main steps:

- Using an input image, masks and crops are generated. Masks identify different objects in the image, and crops are just smaller sections of the image that correspond to these masks.

- Two levels of embeddings are generated using a feature extractor that takes in the original image, masks, and crops. Global embeddings represent the overall context of the entire image, and region-level embeddings capture more detailed features of specific objects in the image.

CHAPTER 6   MAPPING, LOCALIZATION, AND NAVIGATION

- The global and region-level embeddings are combined using zero-shot pixel fusion, which aligns their features at the pixel level.

- The output of the zero-shot pixel fusion is a pixel-aligned embedding that captures the image's visual features and semantic information.

Given this learned embedding, you then have an image with labels that provide a semantic map of the entire scene. The pipeline of ConceptFusion is shown in Figure 6-6.

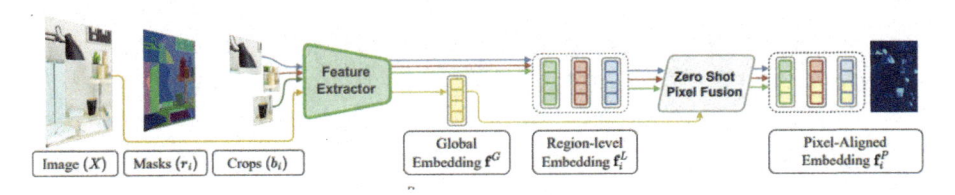

Figure 6-6. ConceptFusion generating pixel-aligned features. Used with permission, source: `https://concept-fusion.github.io/assets/pdf/2023-ConceptFusion.pdf` *[24]*

Localization

Understanding a robot's location is essential because it helps it track its path and determine its future actions. A typical approach to figuring out a robot's location involves using sensors such as LiDAR and odometry. For example, integrating LiDAR data with odometry can help track how the robot's position changes over time, creating a map of the robot's surroundings. An instance of such a localization system is Simultaneous Localization and Mapping (SLAM)[3]. However, SLAM can encounter issues when the robot is moved unexpectedly and encounters failure modes due to reliance on odometry. For this reason, using LiDAR or camera data with deep learning methods is another promising direction.

CHAPTER 6   MAPPING, LOCALIZATION, AND NAVIGATION

2D-to-2D Localization

2D-to-2D localization estimates the camera's position relative to a 2D map by matching images captured by the robot with features on the map. This method can use *explicit localization,* where the scene is represented by images tagged with geographical coordinates that can be directly compared and matched[25]. Alternatively, *implicit localization* uses a neural network that has been trained to understand and represent the scene. In this case, the camera's position is inferred based on the captured images without explicit geographical tags[25].

There are many 2D-to-2D localization methods, and while we don't cover them all here, Figure 6-7 provides a comprehensive overview for further reference.

	Model	Year	Agnostic	Performance (m/degree) 7Scenes	Cambridge	Contributions
Explicit Map	NN-Net [62]	2017	Yes	0.21/9.30	-	combine retrieval and relative pose estimation
	DeLS-3D [63]	2018	No	-	-	jointly learn with semantics
	AnchorNet [64]	2018	Yes	0.09/6.74	0.84/2.10	anchor point allocation
	RelocNet [65]	2018	Yes	0.21/6.73	-	camera frustum overlap loss
	CamNet [66]	2019	Yes	0.04/1.69	-	multi-stage image retrieval
	PixLoc [67]	2021	Yes	**0.03/0.98**	**0.15/0.25**	cast camera localization as metric learning
Relocalization in 2D Map / Implicit Map	PoseNet [68]	2015	No	0.44/10.44	2.09/6.84	first neural network in global pose regression
	Bayesian PoseNet [69]	2016	No	0.47/9.81	1.92/6.28	estimate Bayesian uncertainty for global pose
	BranchNet [70]	2017	No	0.29/8.30	-	multi-task learning for orientation and translation
	VidLoc [71]	2017	No	0.25/-	-	efficient localization from image sequences
	Geometric PoseNet [72]	2017	No	0.23/8.12	1.63/2.86	geometry-aware loss
	SVS-Pose [73]	2017	No	-	1.33/5.17	data augmentation in 3D space
	LSTM PoseNet [74]	2017	No	0.31/9.85	1.30/5.52	spatial correlation
	Hourglass PoseNet [75]	2017	No	0.23/9.53	-	hourglass-shaped architecture
	MapNet [76]	2018	No	0.21/7.77	1.63/3.64	impose spatial and temporal constraints
	SPP-Net [77]	2018	No	**0.18**/6.20	1.24/2.68	synthetic data augmentation
	GPoseNet [78]	2018	No	0.30/9.90	2.00/4.60	hybrid model with Gaussian Process Regressor
	LSG [79]	2019	No	0.19/7.47	-	odometry-aided localization
	PVL [80]	2019	No	-	1.60/4.21	prior-guided dropout mask to improve robustness
	AdPR [81]	2019	No	0.22/8.8	-	adversarial architecture
	AtLoc [82]	2019	No	0.20/7.56	-	attention-guided spatial correlation
	GR-Net [83]	2020	No	0.19/**6.33**	1.12/**2.40**	construct a view graph
	MS-Transformer [84]	2021	Yes	**0.18**/ 7.28	1.28/2.73	extend to multiple scenes with transformers

Figure 6-7. *Models used for 2D-to-2D localization. Used with permission, source:* `https://ieeexplore.ieee.org/stamp/stamp.jsp?tp=&arnumber=10260323[40]`

CHAPTER 6   MAPPING, LOCALIZATION, AND NAVIGATION

Figure 6-8 shows the typical 2D-to-2D-based localization architectures using an explicit map from RelocNet[26] and an implicit map from PoseNet[27]. In the case of explicit map-based localization, a new image is captured, and features are extracted. There is a preexisting database of features extracted from training images that are tagged with geographical coordinates. The features from the captured image are compared with those in the database to find the closest neighbor. The camera's position and orientation are determined by identifying the closest matching features in the database and then applying a coordinate transformation to map these features to the camera's pose. On the other hand, implicit map-based localization has another neural network module that takes the feature maps of the input image and directly predicts the camera's position and rotation. The entire pipeline is trained end-to-end to directly map from image features to camera pose.

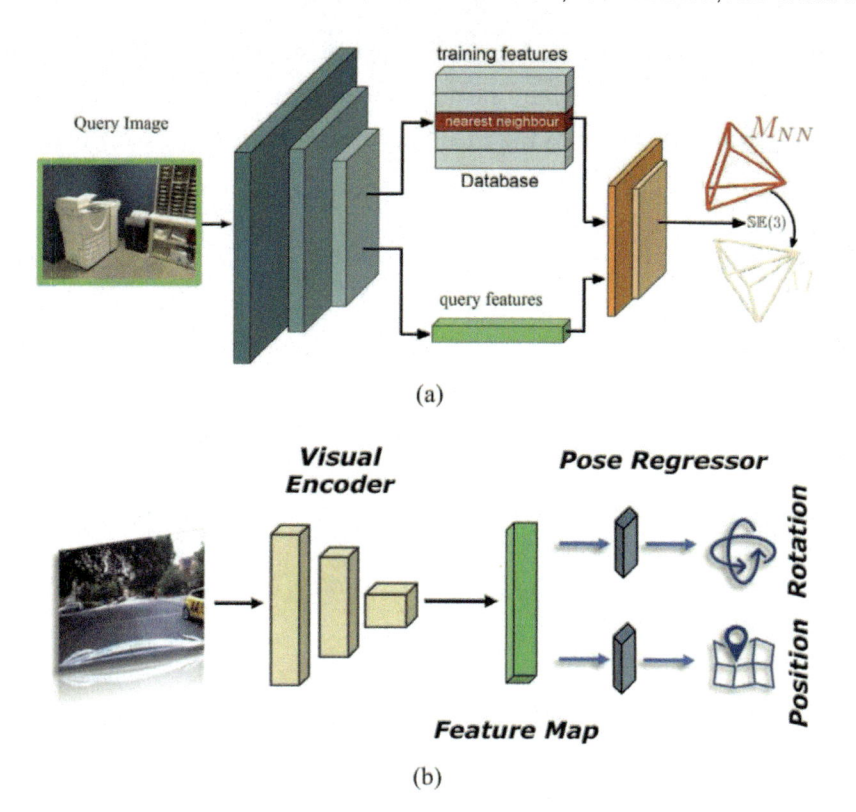

CHAPTER 6   MAPPING, LOCALIZATION, AND NAVIGATION

Figure 6-8. *Difference between explicit (a) RelocNet (https://www. robots.ox.ac.uk/~lav/Papers/balntas_etal_eccv2018/balntas_ etal_eccv2018.pdf)[26] and implicit (b) PoseNet (https://arxiv. org/abs/1505.07427)[27] map-based localization. Used both with permission, source: https://ieeexplore.ieee.org/stamp/stamp. jsp?tp=&arnumber=10260323[40]*

RelocNet (see Figure 6-9) determines a camera's position by comparing images. It learns to represent images so that similar images have similar representations using nearest-neighbor matching and continuous metric learning-based feature descriptors. The method incorporates information from overlapping parts of images

CHAPTER 6 MAPPING, LOCALIZATION, AND NAVIGATION

(camera frusta overlaps) to improve its accuracy. Additionally, RelocNet includes a pose regressor that fine-tunes the position estimate using geometric information, making the location prediction more accurate.

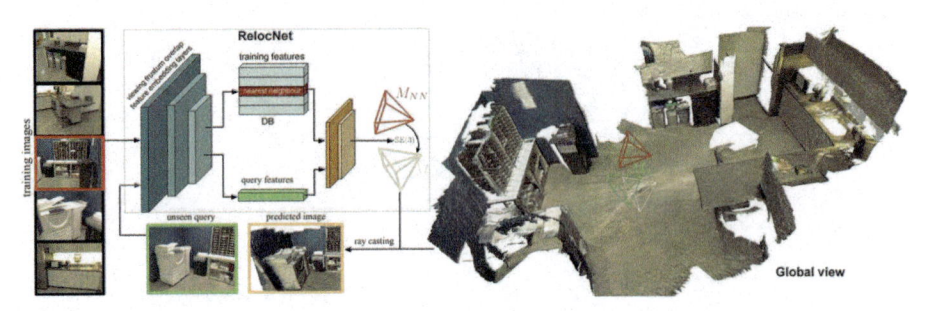

Figure 6-9. *Overview of RelocNet. Used with permission, source: https://openaccess.thecvf.com/content_ECCV_2018/papers/Vassileios_Balntas_RelocNet_Continous_Metric_ECCV_2018_paper.pdf[26]*

2D-to-3D Localization

Two common 2D-to-3D location methods are descriptor matching and scene coordinate regression, as shown in Figure 6-10. *Descriptor matching* determines a camera's position by comparing the image's high-level features and finer details with a database of pre-stored images to refine the match and calculate the exact position and orientation[25]. In contrast, *scene coordinate regression* uses a neural network to predict 3D coordinates for each pixel in the input image, assesses the confidence of these predictions, and then determines the camera pose based on these coordinates[25]. Both methods aim to determine the camera's location but use different techniques.

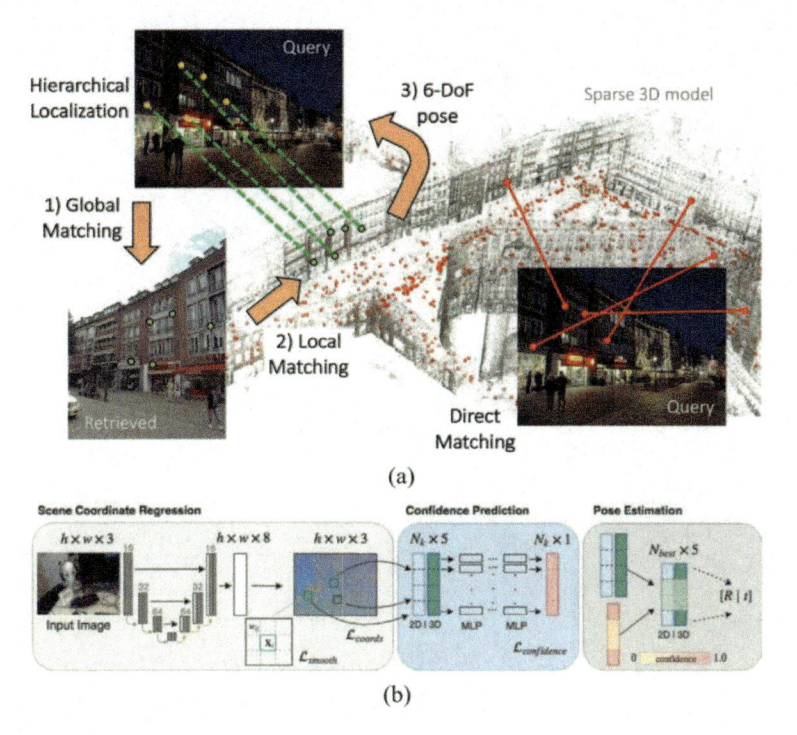

CHAPTER 6 MAPPING, LOCALIZATION, AND NAVIGATION

(a)

(b)

Figure 6-10. *Difference between descriptor matching (a), HF-Net (https://arxiv.org/pdf/1812.03506)[41], and scene coordinate regression (b), Confidence SCR (https://arxiv.org/abs/1805.08443)[42]. Used with permission, source: https://ieeexplore.ieee.org/stamp/stamp.jsp?tp=&arnumber=10260323[40]*

A summary of common 2D-to-3D localization methods is shown in Figure 6-11.

CHAPTER 6   MAPPING, LOCALIZATION, AND NAVIGATION

	Model	Year	Agnostic	Performance (m/degree) 7Scenes	Cambridge	Contributions
Relocalization in 3D Map / **Descriptor Based**	NetVLAD [91]	2016	Yes	-	-	differentiable VLAD layer
	DELF [113]	2017	Yes	-	-	attentive local feature descriptor
	InLoc [114]	2018	Yes	0.04/1.38	0.31/0.73	dense data association
	SVL [115]	2018	No	-	-	leverage a generative model for descriptor learning
	SuperPoint [116]	2018	Yes	-	-	jointly extract interest points and descriptors
	Sarlin et al. [117]	2018	Yes	-	-	hierarchical localization
	NC-Net [118]	2018	Yes	-	-	neighbourhood consensus constraints
	2D3D-MatchNet [119]	2019	Yes	-	-	jointly learn the descriptors for 2D and 3D keypoints
	HF-Net [120]	2019	Yes	0.042/1.3	0.356/0.31	coarse-to-fine localization
	D2-Net [121]	2019	Yes	-	-	jointly learn keypoints and descriptors
	Speciale et al [122]	2019	No	-	-	privacy preserving localization
	OOI-Net [123]	2019	No	-	-	objects-of-interest annotations
	Camposeco et al. [124]	2019	Yes	-	0.56/0.66	hybrid scene compression for localization
	Cheng et al. [125]	2019	Yes	-	-	cascaded parallel filtering
	Taira et al. [126]	2019	Yes	-	-	comprehensive analysis of pose verification
	R2D2 [127]	2019	Yes	-	-	learn a predictor of the descriptor discriminativeness
	ASLFeat [128]	2020	Yes	-	-	leverage deformable convolutional networks
	CD-VLM [129]	2021	Yes	-	-	cross-descriptor matching
	VS-Net [130]	2021	No	**0.024/0.8**	**0.136/0.24**	vote by segmentation
Scene Coordinate Regression	DSAC [131]	2017	No	0.20/6.3	0.32/0.78	differentiable RANSAC
	DSAC++ [132]	2018	No	0.08/2.40	0.19/0.50	without using a 3D model of the scene
	Angle DSAC++ [133]	2018	No	0.06/1.47	0.17/0.50	angle-based reprojection loss
	Dense SCR [134]	2018	No	0.04/1.4	-	full frame scene coordinate regression
	Confidence SCR [135]	2018	No	0.06/3.1	-	model uncertainty of correspondences
	ESAC [136]	2019	No	0.034/1.50	-	integrates DSAC in a Mixture of Experts
	NG-RANSAC [137]	2019	No	-	0.24/**0.30**	prior-guided model hypothesis search
	SANet [138]	2019	Yes	0.05/1.68	0.23/0.53	scene agnostic architecture for camera localization
	MV-SCR [139]	2019	No	0.05/1.63	0.17/0.40	multi-view constraints
	HSC-Net [140]	2020	No	0.03/0.90	**0.13**/**0.30**	hierarchical scene coordinate network
	KFNet [141]	2020	No	0.03/**0.88**	**0.13**/**0.30**	extends the problem to the time domain
	DSM [142]	2021	Yes	**0.027**/0.92	0.27/0.52	dense coordinates prediction

Figure 6-11.   2D-to-3D localization models. Used with permission, source: https://ieeexplore.ieee.org/stamp/stamp.jsp?tp=&arnumber=10260323[40]

3D-to-3D Localization

3D-to-3D localization, also known as LIDAR localization, involves determining the position and orientation of the camera by matching 3D point clouds generated by a LiDAR to a preexisting 3D map of the environment. This process usually means comparing the captured 3D image with the stored map to find the best fit[25]. This can be helpful for more complex environments where 2D mapping might fall short.

The process of 3D-to-3D localization is depicted in Figure 6-12. Using a LiDAR point cloud and a prebuilt map, PointNet[28] extracts relevant features. These features are then fed into a CNN for further feature extraction. Finally, the processed data is passed through Recurrent

Neural Networks (RNNs)[29] to refine the pose estimate, learn temporal relationships, and output the optimal camera pose.

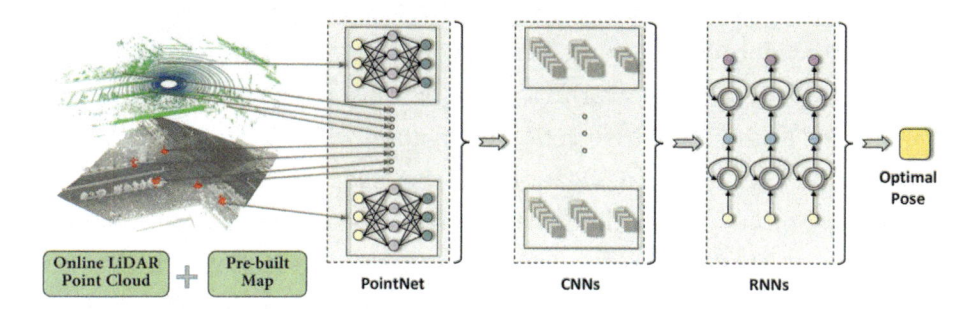

Figure 6-12. *3D-to-3D localization process. Used with permission, source: https://openaccess.thecvf.com/content_CVPR_2019/ papers/Lu_L3-Net_Towards_Learning_Based_LiDAR_Localization_for_Autonomous_Driving_CVPR_2019_paper.pdf[43]*

Figure 6-13 shows a summary of other 3D-to-3D localization methods.

Models	Agnostic	Contributions
LocNet [225]	No	convert 3D points into 2D matrix, search in global prior map
PointNetVLAD [226]	Yes	learn global descriptor from point clouds
Barsan et al. [227]	No	learn from LIDAR intensity maps and online point clouds
L3-Net [224]	No	extract feature by PointNet
PCAN [228]	Yes	predict the significance of each local point based on context
DeepICP [229]	Yes	generate matching correspondence from learned matching probabilities
DCP [230]	Yes	a learning based iterative closest point
D3Feat [231]	Yes	jointly learn detector and descriptors for 3D points

Figure 6-13. *3D-to-3D localization models. Used with permission, source: https://arxiv.org/pdf/2006.12567[25]*

We recommend checking out this paper[25], which we reference in this chapter, for further reading.

Navigation

Vision language models (VLMs)[30] are large models that can understand and process text and image data. Google's Gemini 1.5 Pro[31] is an

CHAPTER 6   MAPPING, LOCALIZATION, AND NAVIGATION

example of this, and it uses a Mixture-of-Experts architecture[32] to process long contexts. It can handle up to 1 million tokens in standard use and up to 10 million tokens in research settings. This translates to about one hour of video, eleven hours of audio, and over 30,000 lines of code.

Using VLMs can be powerful for robotics in solving navigation tasks. Specifically, Multimodal Instruction Navigation with Demonstration Tours (MINT) is a type of navigation task that uses recorded video demonstrations to guide robotic navigation. An example of this method is Mobility VLA[33], which uses a demonstration tour video of the environment and multimodal user instructions (images and language). The system has two main parts: a high-level policy, which uses the video and instructions to identify the goal, and a low-level policy, which uses this goal along with a pre-computed topological map to plan the robot's path and actions. An overview of Mobility VLA is shown in Figure 6-14.

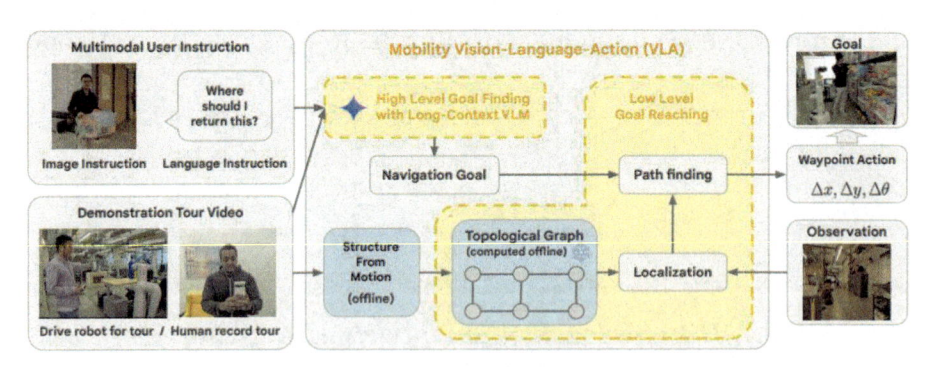

Figure 6-14. *Mobility VLA architecture. Used with permission, source:* `https://arxiv.org/pdf/2407.07775`*[33]*

The Mobility VLA architecture consists of the following main parts:

- **Multimodal user instruction:** The user provides instructions in the form of an image of the environment

CHAPTER 6   MAPPING, LOCALIZATION, AND NAVIGATION

or task that needs to be done. For example, an image of a shelf where an item needs to be placed. A language instruction consists of text instructions describing the task. For example, text saying "Place this box on the top shelf."

- **Demonstration tour video:** A recorded video where a human or a robot navigates through the environment and key locations and actions are captured. For example, a video showing someone walking through an office and pointing out where the desks, shelves, and doors are. This serves as a reference for the robot to understand the layout and important features of the environment.

A prompt (F, N, d, I) is prepared, which includes frames (images from the tour video), narratives (descriptions associated with each frame), distances (contextual information about the distance or location), and multimodal user instructions combining text and images. An example of this is shown in Figure 6-15.

```
You are a robot operating in a building and your task is to respond to the user
command about going to a specific location by finding the closest frame in the
tour video to navigate to.
These frames are from the tour of the building last year.
[Frame 1 Image f₁]
Frame 1. [Frame narrative n₁]
...
[Frame k Image f_k]
Frame k. [Frame narrative n_k]
This image is what you see now. You may or may not see the user in this image.
[Image Instruction I]
The user says: Where should I return this?
How would you respond? Can you find the closest frame?
```

Figure 6-15. *A demonstration of how a multimodal prompt is used by a Vision Language Model (VLM) to guide a robot in a navigation goal. Used with permission, source:* https://arxiv.org/pdf/2407.07775 *[33]*

CHAPTER 6   MAPPING, LOCALIZATION, AND NAVIGATION

- **Goal finding:** The VLM identifies specific frames in the video that correspond to the goal described by the input instructions. For example, the VLM could identify the frame showing the shelf where the box needs to be placed. This ensures that the robot understands what it needs to achieve based on context from the instructions and the video.

- **Navigation goal:** A high-level goal frame is identified by the VLM that matches the user's instruction. This goal could be reaching the shelf identified in the video.

- **Path finding:** The robot uses a pre-computed map, which is created from the demonstration tour video using a method classed structure-from-motion, which creates a 3D representation of the environment. Using this map, the robot finds the optimal path to reach the navigation goal.

This is where the low-level goal reaching policy can be used. As input, the specific frame identified by the high-level policy and the pre-constructed map are used. The robot captures a new image of its current view and then, using this the robot, determines its position within the topological graph. The robot checks if it has reached the goal vertex and if it has, the navigation goal is achieved. Otherwise, the robot computes the shortest path from its current position to the goal. The robot calculates the next action needed to move from its current position toward the first vertex in the path and then the robot performs the calculated action. This continues until the robot reaches the goal or the maximum number of steps is exceeded. See Figure 6-16.

CHAPTER 6   MAPPING, LOCALIZATION, AND NAVIGATION

Algorithm 1 Low-level Goal Reaching Policy

1: **Input:** goal frame index g, offline-constructed topological graph G.
2:
3: **while** timestep $\leq$ maximum steps **do**
4:     Get new camera observation image O
5:     Get start vertex v_s and robot pose T by localizing O in G
6:     **if** $v_s == v_g$ **then**
7:         Navigation goal reached, break
8:     **end if**
9:     Compute $S = [v_s, v_1, ..., v_g]$, the shortest path between v_s and v_g.
10:    Compute waypoint action a from the relative pose between T and v_1
11:    Execute a on robot
12: **end while**

Figure 6-16. This low-level policy computes waypoint actions at every timestep. Used with permission, source: https://arxiv.org/pdf/2407.07775 *[33]*

This method shows the power of using VLMs and leveraging their ability to capture long-contextual information for understanding complex environments and instructions. Using VLMs to process multimodal inputs can help robots get a more nuanced understanding of their tasks and surroundings.

Navigation and Exploration

Navigating new environments is a challenge for robots because they need to be able to reach a specific goal that is already defined and search for goals in new environments. These tasks are typically defined separately and solved using different models. NoMaD[34] is a technique that aims to build a unified approach that can solve both tasks and improve overall navigation performance. See Figure 6-17.

NoMaD uses a single diffusion policy that handles both goal-directed navigation and exploratory navigation:

- The robot takes in visual data from its surroundings over the most recent moments. Optionally, the robot can have a visual goal input indicating the goal location, which shows the robot where it needs to go.

CHAPTER 6   MAPPING, LOCALIZATION, AND NAVIGATION

- Two EfficientNet[35] encoders are used to process the visual data and goal input and turn them into tokens that can be used by a Transformer[9] decoder.

- The tokens are passed into a Transformer model, which uses the tokens to understand the context and make decisions about the robot's actions. The Transformer generates a context vector by polling information from the input tokens. This context vector represents the current understanding of the environment and the task.

- Goal masking is a method used to help the model understand whether it should be focusing on goal-directed navigation or exploratory navigation. It combines the information from the goal input with the context generated from the observations to make sure the model can adapt based on the task provided.

- The context vector is used in an action diffusion model, which predicts the future actions the robot should take. It uses a denoising process with ten steps to refine these action predictions. Throughout this process, the model computes the temporal distance between the current distance and the goal observation (if provided) to understand how far the robot is from the goal in terms of time.

- The action diffusion model outputs a sequence of future actions which guide the robot's movement to navigate the environment, whether that be searching for a goal or moving toward a known goal.

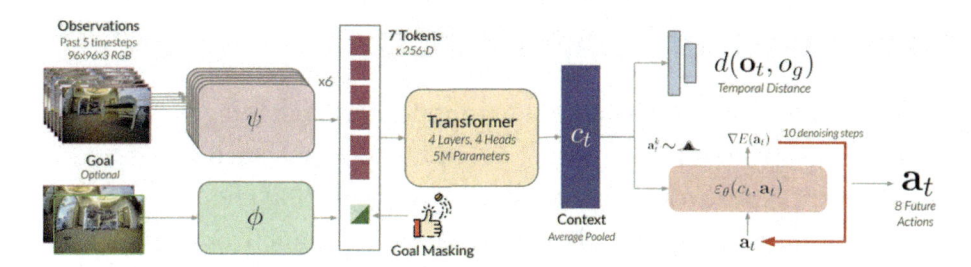

CHAPTER 6 MAPPING, LOCALIZATION, AND NAVIGATION

Figure 6-17. Architecture of NoMaD. Used with permission, source: https://arxiv.org/pdf/2310.07896 [34]

This model was tested in indoor and outdoor environments using the LoCoBot platform[36]. In exploratory mode, the robot identifies potential paths to find goals. In goal-directed mode, it navigates directly toward the specified targets. This demonstrates the ability of an approach like this to understand its environment, follow navigation routes, and make informed decisions at intersections. See Figure 6-18.

Figure 6-18. NoMaD's rollouts in indoor (top) and outdoor (bottom) environments. Yellow lines are future action samples from the undirected mode and blue lines indicate actions selected by the high-level planner. Used with permission, source: https://arxiv.org/pdf/2310.07896 [34]

CHAPTER 6   MAPPING, LOCALIZATION, AND NAVIGATION

Locomotion for Legged Robots

An important area of development in robotics research is legged locomotion—when a robot uses legs to move and navigate its environment. These robots are beneficial for inspection and exploration, where they must navigate challenging terrain and maneuver around obstacles. Recently, developments to bring down hardware costs, make simulation more realistic, and develop learning-based methods have pushed forward research in this field[37].

Hardware

Over the years, the hardware for quadrupedal robots has evolved to have improved performance and reduced cost. There are three types of actuations: hydraulic, electric motors, and series elastic actuators[37].

- **Hydraulic actuators:** These actuators are ideal for powerful and dynamic movements because they have high power-to-weight ratios. Regardless of this, they can be costly and require specific maintenance.

- **Electric motors:** These actuators are most commonly used in quadrupedal robots. In the past, these actuators were used alongside other components, like high-ratio gears to improve torque. Recently, high torque-density actuators have been developed that allow for direct joint control with fewer or no additional parts. These developments have reduced complexity and improved performance.

- **Series elastic actuators:** These actuators have an elastic component that absorbs impact and allows for more precise control.

CHAPTER 6 MAPPING, LOCALIZATION, AND NAVIGATION

More recently, the field has been focusing on proprioceptive actuators, which offer higher torque density and control. These actuators use sensors to measure the force and position of the joints, which allows the control to be more adaptive. These actuators will ultimately enable quadrupedal robots to perform more complex tasks. Figure 6-19 shows how quadrupedal robots have evolved over time.

Figure 6-19. *How quadruped robots have evolved overtime. Used with permission, source:* `https://arxiv.org/pdf/2406.01152` *[37]*

Simulation

Simulation is essential to test control algorithms for quadrupedal robots before deploying them in the real world. Recent simulations use rigid body models, which can more accurately handle collisions by considering friction, elastic interactions (two objects collide and bounce off each other without loss of energy in the system), and inelastic interactions (when two objects collide, they may deform, stick together, or lose some kinetic energy)[37]. GPU acceleration has also sped up simulations and the training process for more complex tasks using a mixture of real-world data from sensor input and simulated data.

CHAPTER 6   MAPPING, LOCALIZATION, AND NAVIGATION

MDP Formulation

Markov Decision Processes (MDPs) are used in RL as a framework for decision-making. For locomotion in legged robots, MDPs help design control policies.

The dynamics represent how the robot's state (position, speed, etc.) changes over time. The state evolved based on the robot's current state and the action it takes. This can be represented as follows:

$$D \equiv (\mathcal{S}, \mathcal{A}, \mathcal{P}, \mathcal{R})$$

Equation 6-1

This equation represents the state of the robot at the next time step, which is determined by the current state, the action taken, and the change over a small time step (dt). The change in the state due to actions is modeled using a probability function:

$$p(s_{t+1} | s_t, a_t)$$

Equation 6-2

This gives the likelihood of moving to the next state from the current state after taking an action.

Learning-based Methods

A few of the common learning-based methods used to solve MDP for locomotion policies[37] are shown in Figure 6-20.

CHAPTER 6   MAPPING, LOCALIZATION, AND NAVIGATION

- **End-to-end learning:** The robot learns how to move by treating the entire task as one big problem, often formulated as a control problem by MDPs. Oftentimes, RL algorithms like TRPO (Trust Region Policy Optimization) and PPO (Proximal Policy Optimization), which are described in Chapter 7, are used to solve this. The intuition is that these algorithms help the robot learn by making small, safe updates in movements.

- **Curriculum learning:** The goal of curriculum learning is to mimic the way humans learn by slowly increasing the difficulty of tasks over time. This method helps the robot build up skills over time, which can make it a helpful technique in complex environments where learning slowly with increasing difficulty in tasks can be useful.

- **Hierarchical learning:** The complex tasks are broken down into smaller, simpler tasks. A high-level policy decides on the larger actions, such as choosing where to step, and the low-level policies handle the specific details like moving the robot's legs. This can make it easier to handle complicated behaviors by focusing on specific actions one part at a time.

- **Privileged learning**: A "teacher" policy is trained in a simulation where it has full access to all needed information. This policy guides a "student" policy, which operates on more limited information, similar

297

to real-world conditions. The "student" policy learns to leverage the information it has by interrupting a history of observations to make decisions. This approach can be helpful to bridge the gap between a simulation and the real world by helping the robot perform tasks in the real world using a fully informed, simulated model.

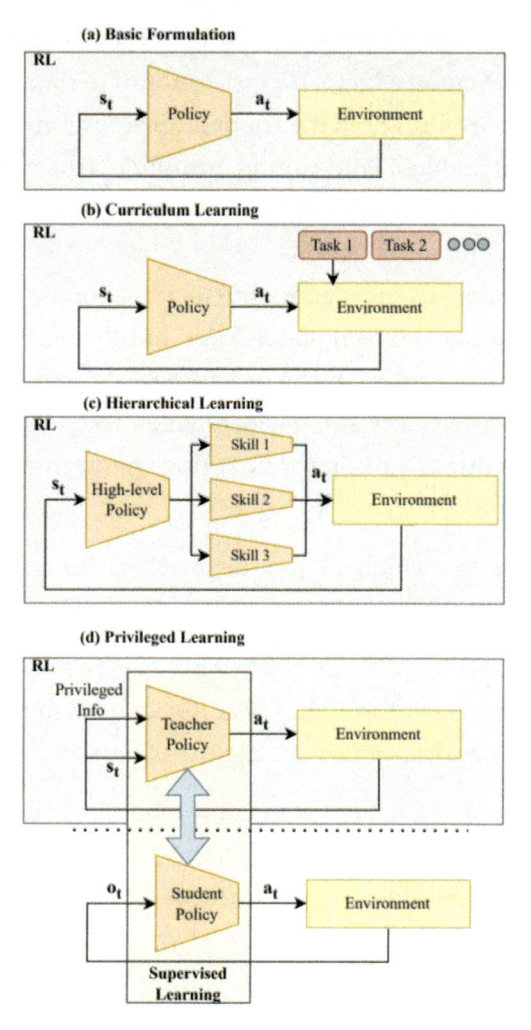

CHAPTER 6 MAPPING, LOCALIZATION, AND NAVIGATION

Figure 6-20. *Popular learning frameworks. Used with permission, source: https://arxiv.org/pdf/2406.01152[37]*

CHAPTER 6   MAPPING, LOCALIZATION, AND NAVIGATION

A future direction is using foundation models, which are large models that have been trained on a large, diverse amount of data. For legged robots, foundation models help the robot interpret its environment and plan a sequence of actions. Fine-tuning foundation models with robot-specific data, such as for manipulation tasks, can also be a useful direction for directly controlling robot actions.

One of the challenges in legged robot locomotion is the need for benchmarks to evaluate control policies and hardware. The Barkour benchmark[38] aims to solve this by providing a standardized obstacle course to measure the agility and performance of legged robots (see Figure 6-21). The course is designed after dog agility competitions and includes four main obstacles:

- Pause tables, which are platforms where the robot must stop and hold its position for a specified time.

- Weave poles, which are a series of poles that the robot must maneuver through. This directly tests the robot's ability to make precise and agile movements.

- An A-frame, which is a steep ramp that the robot must climb up and down, allowing a way to evaluate how well it can handle inclined surfaces.

- A board jump, which is a wide gap that the robot must leap over, evaluating the ability of the robot to jump.

This course evaluates different movements, like running, moving sideways, climbing, and jumping. A specific area and allocated time is used for each obstacle for fair and consistent evaluation.

CHAPTER 6   MAPPING, LOCALIZATION, AND NAVIGATION

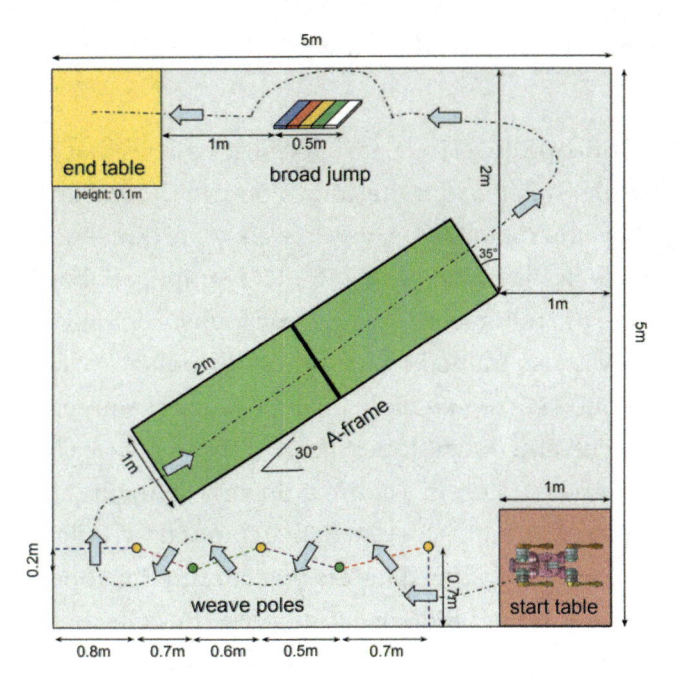

Figure 6-21.   The design for the Barkour course. Used with permission, source: https://arxiv.org/pdf/2305.14654[38]

A score measures how quickly and accurately the robot completes the course. The scoring starts at 1.0, and points are deducted for each failed or skipped obstacle or for taking longer than the allotted time.

$$R_{\text{agility}} = 1.0 - \max\left(\frac{t_{\text{run}} - t_{\text{allotted}}}{t_{\text{allotted}}}, 0\right) - \text{penalties}$$

Equation 6-3

Here, the time allocated is the standard time based on dog agility competitions. Penalties include a deduction of 0.1 for each failed or skipped obstacle and 0.01 for each second over the allotted time. See Figure 6-22.

CHAPTER 6   MAPPING, LOCALIZATION, AND NAVIGATION

To tackle the benchmark, there are two main phases to establishing the baseline:

- **Training phase**: There are specialist policies that include walking, jumping, and slope. In the walking policy, the robot learns how to walk in all directions. This policy is trained using RL. The jumping policy is where the robot learns to jump over obstacles and this policy is also trained on RL. The slope policy is where the robot learns to handle inclined slopes by practicing going up and down slopes. Each of these policies is trained separately in a simulation environment. After training, the robot's performance under each policy is recorded, creating datasets that capture the robot's behavior under different conditions. The collected dataset is used to train a Locomotion Transformer, which is a generalist policy that integrates the learned behaviors into a single, diverse policy that can adapt to the various tasks on the course.

- **Deployment phase**: The evaluation has two parts: the evaluation with specialist policies and the evaluation with the generalist policy. For the specialist policies, a navigation controller, which is a high-level controller, determines the overall path the robot should take throughout the course. This is decided based on the robot's position and an elevation map of the course. Based on the commands from the navigation controller, the robot switches between specialist policies (walking, jumping, and slope) to navigate through the course. Similar to the specialist approach, the navigation controller guides the robot through the course in the evaluation of the generalist policy. Instead of

CHAPTER 6   MAPPING, LOCALIZATION, AND NAVIGATION

switching between different policies, the robot uses the Locomotion Transformer policy to dynamically adjust the movements based on the environment and its own state.

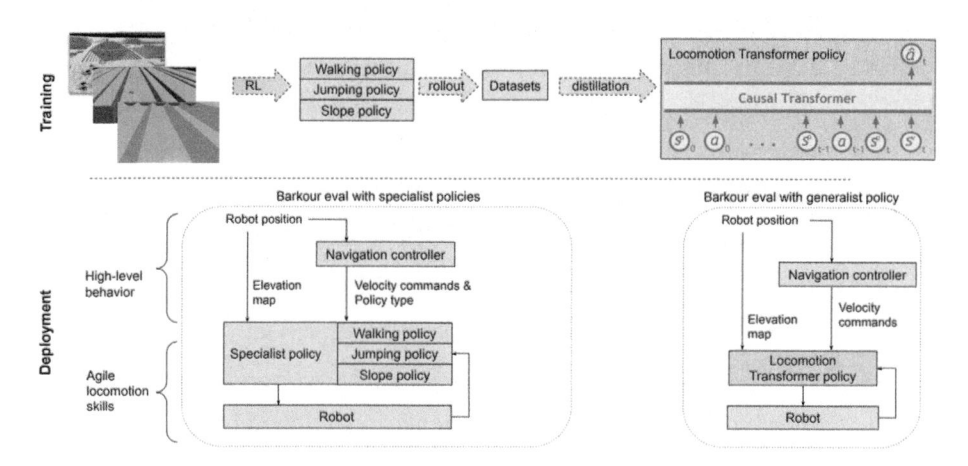

Figure 6-22.   Methods used to establish a baseline for the Barkour benchmark. Used with permission, source: `https://arxiv.org/pdf/2305.14654`*[38]*

Summary

During this chapter, you learned that:

- Deep learning can learn complex patterns from high-dimensional sensor data and adapt to changing environments, but it requires large, labeled datasets and computational resources. Traditional methods are more interpretable and effective with less data but need manual feature engineering.

CHAPTER 6   MAPPING, LOCALIZATION, AND NAVIGATION

- Most robots use an RGB camera and IMU sensors and some incorporate LiDAR for 3D mapping. The choice of sensors and estimation techniques varies based on the environment and task.

- Various representations for defining mapping tasks exist in robotics. Geometric mapping creates 3D maps with depth data, while voxel representation divides 3D spaces into cubes to indicate occupancy. NeRF (Neural Radiance Fields) generates 3D scenes from 2D images. Semantic mapping uses models like CLIP to label objects in scenes with examples of open-vocabulary approaches like LeRF and ConceptFusion.

- Localization methods include 2D-to-2D localization, which compare captured images to pre-tagged maps, with methods like RelocNet and PoseNet using neural networks to predict a robot's position. 2D-to-3D localization matches image features or uses scene-coordinate regression to estimate a robot's position. 3D-to-3D localization uses LiDAR to compare point clouds to a 3D map.

- Some navigation approaches include Mobility VLA, which uses a combination of demonstration tours and multimodal instructions to guide robots. NoMaD unifies goal-directed and exploratory navigation using a diffusion policy that adapts to varying tasks by combining visual data and goal inputs.

- Advances in hardware, simulation, and learning methods (e.g., curriculum and hierarchical learning) have enabled robots to navigate rough terrain with growing research in legged locomotion.

CHAPTER 6   MAPPING, LOCALIZATION, AND NAVIGATION

The next chapter covers reinforcement learning's role in robotics, focusing on how agents learn from interactions to improve performance, tackles challenges like reward design and sample efficiency, and applies techniques like sim-to-real transfer and model-free methods like PPO. It also highlights emerging trends, like integrating LLMs and learning without manually engineered or explicit reward functions.

References

[1] Aqel, Mohammad OA, et al. "Review of visual odometry: types, approaches, challenges, and applications." *SpringerPlus* 5 (2016): 1-26.

[2] Zhang, Ji and Sanjiv Singh. "LOAM: LiDAR odometry and mapping in real-time." *Robotics: Science and systems*. Vol. 2. No. 9. 2014.

[3] https://www.mathworks.com/discovery/slam.html

[4] https://www.mathworks.com/help/vision/ug/what-is-structure-from-motion.html;jsessionid=bad2 9d217299a9c65ea6e531733c

[5] https://thekalmanfilter.com/kalman-filter-explained-simply/

[6] https://github.com/commaai/rednose

[7] Wang, Shuzhe, et al. "Dust3r: Geometric 3D vision made easy." Proceedings of the IEEE/CVF Conference on Computer Vision and Pattern Recognition. 2024.

[8] Dosovitskiy, Alexey, et al. "An image is worth 16x16 words: Transformers for image recognition at scale." *arXiv preprint arXiv*:2010.11929 (2020).

[9] Vaswani, Ashish. "Attention is all you need." *arXiv preprint arXiv*:1706.03762 (2017).

[10] Padkan, Nazanin, et al. "Evaluating Monocular Depth Estimation Methods." International Archives of the Photogrammetry, Remote Sensing and Spatial Information Sciences 48.1 (2023): 137-144.

[11] Silberman, Nathan, et al. "Indoor segmentation and support inference from RGBD images." Computer Vision–ECCV 2012: 12th European Conference on Computer Vision, Florence, Italy, October 7-13, 2012, Proceedings, Part V 12. Springer Berlin Heidelberg, 2012.

[12] https://www.cvlibs.net/datasets/kitti/

[13] He, Kaiming, et al. "Deep residual learning for image recognition." Proceedings of the IEEE Conference on Computer Vision and Pattern Recognition. 2016.

[14] Huang, Gao, et al. "Densely connected convolutional networks." Proceedings of the IEEE Conference on Computer Vision and Pattern Recognition. 2017.

[15] Wagner, Joshua. "Chapter 8 Attention and Self-Attention for NLP." Modern Approaches in Natural Language Processing, 8 Sept. 2020, slds-lmu.github.io/seminar_nlp_ss20/attention-and-self-attention-for-nlp.html.

[16] Khan, Faisal, Saqib Salahuddin, and Hossein Javidnia. "Deep learning-based monocular depth estimation methods—a state-of-the-art review." *Sensors* 20.8 (2020): 2272.

CHAPTER 6   MAPPING, LOCALIZATION, AND NAVIGATION

[17] Mescheder, Lars, et al. "Occupancy networks: Learning 3D reconstruction in function space." Proceedings of the IEEE/CVF Conference on Computer Vision and Pattern Recognition. 2019.

[18] https://youtu.be/6x-Xb_uT7ts?feature=shared

[19] Mildenhall, Ben, et al. " Nerf: Representing scenes as neural radiance fields for view synthesis." *Communications of the ACM* 65.1 (2021): 99-106.

[20] https://github.com/bmild/nerf/blob/master/run_nerf_helpers.py

[21] Radford, Alec, et al. "Learning transferable visual models from natural language supervision." International Conference on Machine Learning. PMLR, 2021.

[22] Kerr, Justin, et al. "LeRF: Language embedded radiance fields." Proceedings of the IEEE/CVF International Conference on Computer Vision. 2023.

[23] Caron, Mathilde, et al. "Emerging properties in self-supervised vision transformers." Proceedings of the IEEE/CVF International Conference on Computer Vision. 2021.

[24] Jatavallabhula, Krishna Murthy, et al. "ConceptFusion: Open-set multimodal 3D mapping." *arXiv preprint arXiv*:2302.07241 (2023).

[25] Chen, Changhao, et al. "A survey on deep learning for localization and mapping: Towards the age of spatial machine intelligence." *arXiv preprint arXiv*:2006.12567 (2020).

[26] Balntas, Vassileios, Shuda Li, and Victor Prisacariu. "RelocNet: Continuous metric learning relocalisation using neural nets." Proceedings of the European Conference on Computer Vision (ECCV). 2018.

[27] Kendall, Alex, Matthew Grimes, and Roberto Cipolla. "PoseNet: A convolutional network for real-time 6-DoF camera relocalization." Proceedings of the IEEE International Conference on Computer Vision. 2015.

[28] Qi, Charles R., et al. "PointNet: Deep learning on point sets for 3D classification and segmentation." Proceedings of the IEEE Conference on Computer Vision and Pattern Recognition. 2017.

[29] https://www.geeksforgeeks.org/introduction-to-recurrent-neural-network/

[30] Noyan, Merve and Edward Beeching. "Vision Language Models Explained." *Hugging Face,* 11 Apr. 2024, huggingface.co/blog/vlms.

[31] Pichai, Sundar and Demis Hassabis. "Introducing Gemini: Our Largest and Most Capable AI Model." Google, 6 Dec. 2023, blog.google/technology/ai/google-gemini-ai/#sundar-note.

[32] https://huggingface.co/blog/moe

[33] Chiang, Hao-Tien Lewis, et al. "Mobility VLA: Multimodal Instruction Navigation with Long-Context VLMs and Topological Graphs." *arXiv preprint arXiv:*2407.07775 (2024).

[34] Sridhar, Ajay, et al. "NoMaD: Goal masked diffusion policies for navigation and exploration." 2024 IEEE International Conference on Robotics and Automation (ICRA). IEEE, 2024.

CHAPTER 6   MAPPING, LOCALIZATION, AND NAVIGATION

[35] Tan, Mingxing and Quoc Le. "EfficientNet: Rethinking model scaling for convolutional neural networks." International Conference on Machine Learning. PMLR, 2019.

[36] http://www.locobot.org/

[37] Ha, Sehoon, et al. "Learning-based legged locomotion; state of the art and future perspectives." *arXiv preprint arXiv*:2406.01152 (2024).

[38] Caluwaerts, Ken, et al. "Barkour: Benchmarking animal-level agility with quadruped robots." *arXiv preprint arXiv*:2305.14654 (2023).

[39] "A Look at Tesla's Occupancy Networks." *Think Autonomous,* 12 Sept. 2022, www.thinkautonomous.ai/blog/occupancy-networks/.

[40] Chen, Changhao, et al. "Deep learning for visual localization and mapping: A survey." IEEE Transactions on Neural Networks and Learning Systems (2023).

[41] Sarlin, Paul-Edouard, et al. "From coarse to fine: Robust hierarchical localization at large scale." Proceedings of the IEEE/CVF Conference on Computer Vision and Pattern Recognition. 2019.

[42] Bui, Mai, et al. "Scene coordinate and correspondence learning for image-based localization." *arXiv preprint arXiv*:1805.08443 (2018).

[43] Lu, Weixin, et al. "L3-net: Towards learning based LiDAR localization for autonomous driving." Proceedings of the IEEE/CVF Conference on Computer Vision and Pattern Recognition. 2019.

CHAPTER 7

Reinforcement Learning and Control

Chris Paxton is a roboticist who has worked for FAIR labs at Meta and NVIDIA research. He earned his PhD in Computer Science in 2019 from the Johns Hopkins University in Baltimore, Maryland, focusing on using learning to create powerful task and motion planning capabilities for robots operating in human environments. His work won the ICRA 2021 Best Paper Award on Human-Robot Interaction and was nominated for best systems paper at CoRL 2021, among other things. His research focuses on using language, perception, planning, and policy learning to make robots into general-purpose assistants.

The dominant paradigm for learning in general these days is *supervised learning:* taking a known set of data and fitting some large model to it, which can be used for downstream applications. But this leaves many questions unanswered: where is the data coming from? If the model is not good enough—and no model is ever good enough for everything—how will it improve, and which data is necessary for it to improve? It would be great if the robots could collect their own data and improve on their own. Studying how to do this is the main idea of reinforcement learning.

The dream of reinforcement learning is to build intelligent systems that can learn as humans do. They collect data through their interactions with the world, intelligently choosing which goals to attempt and which skills

CHAPTER 7 REINFORCEMENT LEARNING AND CONTROL

to employ in different situations so as to improve their underlying skill set and, potentially, their model of the world.

In many ways, reinforcement learning seems like a true prerequisite for real, robust embodied intelligence. Agents deployed in the real world must be able to recover and learn from their mistakes; in general, people do not want a robot that makes the same mistake over and over again. In addition, reinforcement learning is perhaps the clearest route to true *superhuman* intelligence. It is optimizing some underlying objective and collecting its own data, so it can surpass the performance of even domain experts[12] and constantly surprise its creators. See Figure 7-1.

Figure 7-1. A Boston Dynamics spot robot trained to walk over rough terrain via reinforcement learning. Modern robots are often taught to walk via reinforcement learning, and Boston Dynamics is far from the only company to do so. Used with permission, source: https://bostondynamics.com/blog/starting-on-the-right-foot-with-reinforcement-learning/ by Boston Dynamics[10]

CHAPTER 7 REINFORCEMENT LEARNING AND CONTROL

In spite of its high-profile successes in the research community, reinforcement is a still-emerging and constantly-changing area of study. Unlike in the case of supervised learning, which has dramatic high-profile successes in applications like computer vision and language, reinforcement learning still has a lot to prove at scale, both in robotics and outside of it. However, it has many clear applications, from learning how to move around in the world[10][11][22][23], to how to navigate in various environments[24][25], to how to grasp and manipulate objects[25][26][27][33], and it has even found applications in large language models[33][34].

Reinforcement learning has an unusual place in the milieu of machine learning methods. It sits equal to supervised and unsupervised learning as one of the three basic machine learning paradigms. *Supervised learning* involves taking labeled data and fitting a model to it, and *unsupervised learning* involves taking whatever data exists in the world and using it to learn a generally-useful model. *Reinforcement learning* tends to be much more goal-oriented, and as such involves making many more assumptions.

Perhaps the clearest example is what reinforcement learning is used for in robotics. It is generally *not* used to build general-purpose systems or build the foundations for generalist algorithms. It's used to excel at a specific task, and is often accelerated by pretrained vision or vision-language backbones.

This has been famously summed up by Prof Yann Lecunn in his cake metaphor. As applied to robotics, this means:

1. Self-supervised learning (a variant of unsupervised learning) is used to train the models used as a visual backbone.

2. Demonstration data is used to get the policy started, to avoid local minima and dramatically accelerate policy learning.

3. Finally, reinforcement learning can be used to achieve truly superhuman performance, outperforming the experts within whatever domain the robot is operating in.

CHAPTER 7   REINFORCEMENT LEARNING AND CONTROL

The particular ratios of these components vary. In many cases, a clearer *problem setup* and strong engineered priors can substitute for the learned models preceding reinforcement learning, which creates many of the difficulties.

This chapter includes a brief overview of the ideas of reinforcement learning, looks at several widely used methods, and goes over common modern applications of reinforcement learning, both to machine learning in general and to robotics in particular.

Reinforcement Learning Basics

Reinforcement learning is generally formalized as a *Markov Decision Process*, or MDP (see Figure 7-2). A Markov Decision Process is a model of the form (S, A, T, R), where:

1. S is a *state space*, a representation of the world such as the poses of objects a robot might need to manipulate.

2. A is the *action space*, representing which actions the agent can take from each state.

3. T is the *transition* model, in the form of the conditional probability distribution $P(S_{t+1} | S_t, A_t)$, meaning it represents how likely it will transition from one state to the next given a particular action.

4. R is the *reward function,* representing the immediate value of transitioning from state S_t to state S_{t+1}.

CHAPTER 7   REINFORCEMENT LEARNING AND CONTROL

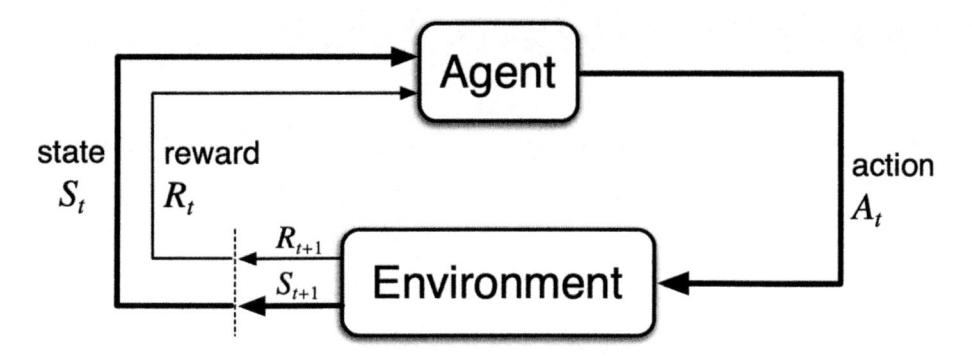

Figure 7-2. *An illustration of the basic reinforcement learning loop. An agent—such as a robot—takes actions from its action set A, which results in new states in S and observed rewards, R. Used with permission, source:* `http://incompleteideas.net/book/RLbook2020.pdf` *[44]*

From any given state $s \in S$, within the Markov Decision Process, the agent's goal is to choose the best possible action a, such that the agent maximizes accumulated reward over its lifetime. The choice of a is determined by the policy $a = \pi(s)$; the goal of reinforcement learning is then to learn the policy function π, which itself is often represented by a neural network.

Immediately, you can see how this might cover a wide range of problems: the actions might be displaying different types of ads to a user browsing a website, or they might be controllable degrees of freedom of a robot arm. Positive rewards can be given for click-through, or when a grasp was successfully executed, or when a mobile robot is getting closer to its destination.

315

CHAPTER 7   REINFORCEMENT LEARNING AND CONTROL

Solving a Markov Decision Process

Assume for now that you have access to the transition function T, and that you might easily visit all possible states $s \in S$. Starting from state s_0, you might simply choose the best next state given the transition probabilities:

$$\pi(s) = argmax_a \sum_{s'} P(s,s'|a) R(s,s') + \gamma V(s')$$

Where $P(s, s' | a)$ is the probability of transitioning from state s to state s' given action a, $R(s, s')$ is the reward associated with that transition, γ is some discount factor, and $V(s')$ is the *value* of state s'. The value is the expected reward-to-go from this point onward. To put it in other words, the *policy* computes the action that you expect will lead to the highest overall reward.

For a finite, fixed-size Markov Decision Process, you can solve this via dynamic programming by performing Bellman value iteration[1][2][19]:

$$V(s)_{i+1} = max_a \sum_{s'} P(s,s'|a) R(s,s') + \gamma V_i(s')$$

for the i-th iteration of the algorithm, until it converges. Then the best action can be chosen simply. This is the foundation of reinforcement learning, and you will see how it gets built upon for more complex applications later in this chapter.

Considerations

When applying this to robotics, there are many obvious issues. First, these algorithms mostly work on finite, relatively small, discrete spaces where it's reasonable to compute probabilities and iterate over all possible states in the state space. Robotics state spaces tend to be large and *continuous*, meaning that direct application of methods like Bellman value iteration

CHAPTER 7   REINFORCEMENT LEARNING AND CONTROL

are not trivial. In addition, there's a great deal of difficulty *scaling* these methods, and a great deal of work goes into designing the underlying MDP, with its state, actions, and rewards[1].

Rewards can take a lot of forms; the principal concern in robotics is that they may be *sparse*, meaning that it is often hard to observe a meaningful, informative reward at a transition from S_t to S_{t+1}.

Consider the example of the maze in Figure 7-2. Imagine a *dense reward function* for an agent solving a maze, which represents the reward as a penalty, given a:

$$-\|p-g\|_2$$

Where $p = (x, y)$ is the robot's position and $g = (x_g, y_g)$ is the position of the other end of the maze. This has a few advantages: it is easy to compute, it's easy to specify, it's relatively generic, and it will be useful in all mazes and in many related tasks like navigation. However, it's very greedy: if the solution involves a lot of backtracking, this reward function may lead the agent astray, because it will *penalize* the agent for moving away from the goal, even if there is no way to get closer!

Consider instead a *sparse* reward function. This is even more generic: you just give one point to the agent if it reaches the end, and no points otherwise. This sort of approach is ideal for most applications, but is much harder to solve, as the reward provides essentially no information along the way.

In practice, most applications of reinforcement learning thrive in the first case, where there is a dense reward function, and fail in the latter case. Design of the reward function becomes a major part of many reinforcement learning applications.

CHAPTER 7   REINFORCEMENT LEARNING AND CONTROL

Model-Free vs Model-Based RL

One fundamental concern with all reinforcement learning, particularly in robotics, is *sample complexity*. How much data does your method need, in order to work well at scale? Can it solve arbitrarily complex problems?

Take a "traditional" RL algorithm, represented as the Markov Decision Process. It explores an environment—like the maze—based on either sampling from a stochastic policy, or by adding some noise to policy outputs. Then, based on these experiences, the algorithm will directly optimize the underlying policy. This is called *model-free reinforcement learning* for reasons we get to in a moment.

At first glance, this seems great. It relies on making very few assumptions about the world; in principle, such a method will work in any domain, for any application. However, it quickly becomes apparent that under some realistic assumptions—sparse rewards like in the maze example, or high-dimensional problems like controlling a high-DoF robot arm—suddenly you might need a very large amount of information to properly characterize the underlying space.

There are many potential solutions to this, from the ever-popular reward shaping—manually designing a system of rewards to encourage the agent to succeed—to the more principled. But fundamentally one major way you might improve reinforcement learning performance is via *planning*, hallucinating different possible futures that you believe will be optimal, instead of actually executing them. This requires having an accurate *world model*; hence, model-based reinforcement learning. See Figure 7-3.

CHAPTER 7 REINFORCEMENT LEARNING AND CONTROL

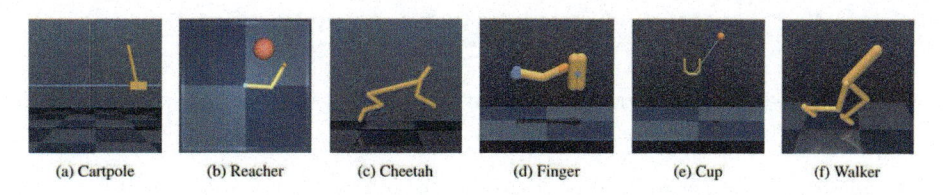

Figure 7-3. *Examples of visual domains used in PlaNet[16]. These are common reinforcement learning domains used for model development. Model-based RL algorithms like PlaNet are capable of more efficient learning, but sometimes they don't scale as well. Used with permission, source:* `https://proceedings.mlr.press/v97/hafner19a/hafner19a.pdf`*[16]*

Of course, using a world model to generate different possible futures is easier said than done. For this to be useful, that model has to be able to be evaluated offline efficiently, which is often very difficult in complex domains! In addition, certain environments can be nearly impossible to simulate (a bin full of deformable objects, for example), and even if they *can* be simulated, this might not be possible at faster than real time. A recent push that's shown impressive results has been model-based reinforcement learning with learned models. Works such as AlphaGo[12] and PlaNet[16] showed that in some domains, a useful model can be learned, which then provides all the advantages of model-based RL without needing this model.

For the most part, the dream of model-based RL has not arrived. The most common means of performing reinforcement learning as of 2024 is Proximal Policy Optimization (PPO), a highly scalable, model-free algorithm[6]. PPO has shown great results in a wide range of different robotics problems, from learning dexterous robot skills[35][36] to navigation[24] to mobile manipulation[31]. However, there remain many challenges and the debate is not truly solved.

With that context, the next section goes into more detail on these two main families of reinforcement learning methods.

CHAPTER 7 REINFORCEMENT LEARNING AND CONTROL

Model-Free Reinforcement Learning

As a reminder, a reinforcement learning problem is formalized as a Markov Decision Process (S, A, T, R), with states S, actions A, rewards R, and state-transitions T. In model-free reinforcement learning, you neither have access to nor attempt to estimate the transition probability T.

This is useful because learning and properly estimating T can be expensive and difficult, and T is often very hard to model—sometimes harder to model than simply estimating the *value* of a particular state in the abstract! The value $V(s)$ for states $s \in S$ represents the reward-to-go from a particular state s, meaning the reward that will be accumulated for all states the agent expects to visit from this point forward. To put it simply, this can be a simpler value to learn, since it's often a direct quality of the state itself—if you do not want to fall, teetering on the edge of a cliff is obviously bad—and it requires estimating only the value itself, a single quantity, instead of predicting a full state, which could be extremely high dimensional. See Figure 7-4.

Figure 7-4. OpenAI used Proximal Policy Optimization (PPO), a model-free reinforcement learning method, to train a policy that could defeat Dota 2 world champions. Used with permission, source: https://www.vox.com/2019/4/13/18309418/open-ai-dota-triumph-og [45]. Note: This book is in no way affiliated with or endorsed by OpenAI.

CHAPTER 7   REINFORCEMENT LEARNING AND CONTROL

As a result, model-free methods have historically been more tractable and more scalable than their model-based competitors. Some of the most storied reinforcement learning approaches have been model-free, including Deep Q Learning[20], which brought deep learning into the mainstream with exciting results on playing Atari games; AlphaGo, which could play Go at a level exceeding the best humans in the world[12]; and Proximal Policy Optimization[6], a general method used to play the video game Dota 2 at a competitive level[7].

Q Learning

The Q function is simply defined as $Q(s, a)$ for a state $s \in S$ and an action $a \in A$, where $Q(s, a)$ is the expected value (reward-to-go) from the given state, if the specified action is taken. If you successively take the action that maximizes this function's return value from each state, you would presumably solve the problem in an optimal way.

The difficulty, then, is learning the Q function in the first place. The core of Q learning is a modified version of the Bellman value iteration algorithm from the "Solving a Markov Decision Process" section, modified to use the Q function as follows:

$$Q_{i+1}(s,a) = (1-\alpha)Q_i(s,a) + \alpha \ R(s,s') + \gamma \ Q_i(s,a)$$

where $0 <= \alpha <= 1$ is a learning rate. Note the absence of the transition probability in the original version of the algorithm. However, this formulation is more practical in many ways, being able to handle problems where the transition probabilities are unknown or difficult to compute, allowing the agent to learn optimal behavior through direct interaction with the environment.

Deep Q Learning

In 2014, Google DeepMind released Deep Q Learning, which is a variant of Q learning modified to use a convolutional neural network as a function

321

CHAPTER 7   REINFORCEMENT LEARNING AND CONTROL

approximator[20]. This opens up some interesting problems, particularly that it is very susceptible to oscillations or changes in the parameters underlying the neural net.

Instead, they propose deep reinforcement learning over a *replay buffer*. As the agent explores, data points are added to a memory, and when computing an update to the policy, a random batch of observations (s, a, r, s') are sampled from memory. Then the current target Q value can be computed and a gradient descent update can be performed on the neural network weights.

Successful for the first time at performing a wide variety of tasks from pixels alone, this innovation fueled an explosion in reinforcement learning research. However, Deep Q Learning has some limitations when applied to robotics: it still operates over *discrete* action spaces, where robot actions are generally continuous. It is susceptible to a wide range of hyperparameters that need to be tuned, and it's often unreliable. This leads to a second family of methods that are less restricted: policy gradient methods.

Policy Gradient Methods

Deep Q Learning has a good number of weaknesses, however. The fact that you're still approximating these Q values iteratively introduces a lot of potential for instability when learning, as small changes in the policy can lead to dramatic changes in evaluation, especially in response to changing states. While you're optimizing Q(s, a), it's important to remember what you're actually optimizing is still

$$Q_{i+1}(s,a|\theta) = (1-\alpha) Q_i(s,a|\theta) + \alpha\ R(s,s') + \gamma\ Q_i(s,a)$$

where θ is some set of neural network weights. Instead, policy gradient methods aim to optimize the policy parameters \\θ so as to optimize the expected return. The expected return is given as

$$J(\theta) = E\left\{ \sum_{k=0}^{H} \gamma^k r_k \right\}$$

CHAPTER 7 REINFORCEMENT LEARNING AND CONTROL

Where r_k is the reward at the k-th step, gamma is a discount factor, and so on. In other words, the *return* is the discounted sum of all future expected rewards.

Now, instead of updating Q, assume you have some policy with parameters θ, where θ could for example be the parameters of a neural network. You might then take the derivative with respect to these policy parameters and optimize them directly[1]:

$$\theta_{k+1} = \theta_k + \alpha_k \nabla_{\theta_0} J(\pi_0)\big|_{\theta_0=\theta_k}$$

Again for policy parameters θ. This gives you a formulation where, now, you just need to compute this derivative of the return $J(\theta)$ and can directly optimize the policy—hence, policy gradients.

You can then express the gradient $\delta J(\theta)$[2] as:

$$\nabla_\theta J(\pi_\theta) = \mathbb{E}_{\tau \sim p_{\pi_\theta}(\tau)} \left[\sum_{t=0}^{H} \gamma^t \nabla_\theta \log \pi_\theta(a_t|s_t) \underbrace{\left(\sum_{t'=t}^{H} \gamma^{t'-t} r(s_{t'}, a_{t'}) - b(s_t) \right)}_{\text{return estimate } \hat{A}(s_t, a_t)} \right]$$

Where

$\tau \sim p_{\pi_\theta}(\tau)$ indicates that you are sampling a trajectory from the policy

$b(s_t)$ is the *baseline* reward-to-go

$\hat{A}(s_t, a_t)$ is the return estimate—an estimate of what the expected return given (s, a) would be out to the horizon

The term $\hat{A}$ can be computed in various ways; often it is estimated by a second neural network, called the *critic*.

Overall, policy gradient methods let you theoretically directly optimize what you really care about, which is to say the policy parameters. While there are many ways of computing the return and therefore the gradients, this section focuses on just a couple options.

CHAPTER 7 REINFORCEMENT LEARNING AND CONTROL

Trust Region Policy Optimization

Of course, as elegant as it sounds, there are some serious issues with this idea when applied naively. Gradient-based optimization is nice, but susceptible both to being caught in local minima, and to instability due to taking steps that are too large. In addition, merely *computing* the gradients with respect to the thousands or even millions of parameters in a deep neural network can be problematic.

Trust Region Policy Optimization (TRPO)[3] aims to solve this problem by constraining the Kullback-Leibler (KL) divergence between the distributions of the old and new policies during policy updates. It does this while optimizing a surrogate advantage function between the old and new policies[4].

$$\theta_{k+1} = \arg\max_{\theta} \mathcal{L}(\theta_k, \theta) \quad \text{s.t.} \quad \bar{D}_{\text{KL}}(\theta\|\theta_k) \leq \delta$$

In other words, TRPO maximizes the *advantage*—the estimated improvement—between an old and a new policy at each step of the learning process, while constraining the KL divergence between the old and the new policies.

In practice, this is *still* too difficult to solve, so TRPO makes some approximations, particularly a Taylor expansion of the objective and constraint:

$$\mathcal{L}(\theta_k, \theta) \approx g^T(\theta - \theta_k)$$

$$\bar{D}_{KL}(\theta\|\theta_k) \approx \frac{1}{2}(\theta - \theta_k)^T H(\theta - \theta_k)$$

CHAPTER 7   REINFORCEMENT LEARNING AND CONTROL

This results in an approximate constrained objective function[2][5]:

$$\theta_{k+1} = \arg\max_{\theta} g^T(\theta - \theta_k) \quad \text{s.t.} \quad \frac{1}{2}(\theta - \theta_k)^T H(\theta - \theta_k) \leq \delta$$

Which can be solved analytically. However, you're *still* not done here, because the Taylor series expansion above may have introduced an error, which means that the original constraint is not fully satisfied. Therefore, you add a backtracking line search to the original analytical solution

$$\theta_{k+1} = \theta_k + \alpha^j \sqrt{\frac{2\delta}{g^T H^{-1} g}} H^{-1} g,$$

Where α in (0, 1) is the backtracking coefficient and j is the smallest nonnegative integer such that there is a positive advantage and the KL divergence constraint is satisfied.

TRPO was able to substantially outperform Deep Q Learning on Atari domains from vision inputs[3], meaning it was a notable step forward for reinforcement learning. However, TRPO has a few notable issues:

1. As obvious, the math is fairly complex—it requires solving a constrained optimization problem, for example. This makes it harder to implement.

2. It's computationally inefficient, requiring computing things like the KL divergence.

Fortunately, TRPO was soon followed by a method that resolved these issues: Proximal Policy Optimization.

CHAPTER 7   REINFORCEMENT LEARNING AND CONTROL

Proximal Policy Optimization

Finally, the most widely used reinforcement learning of the current day—Proximal Policy Optimization (PPO)—introduced by OpenAI in 2017[6].

The idea behind PPO is the same as TRPO: you need a way to substantially update policies with huge numbers of parameters without causing performance to collapse. TRPO solved this by applying a constraint with complex second-order methods. PPO adds some extra constraints but results in an overall simpler optimization problem.

The core equation being solved by PPO is this:

$$\theta_{k+1} = \arg\max_{\theta} \mathbb{E}_{s,a \sim \pi_{\theta_k}} \left[L(s, a, \theta_k, \theta) \right],$$

Where at each step, you are finding the θ that *maximizes* L(s, a, θ, θ_k) such that

$$L(s, a, \theta_k, \theta) = \min \left(\frac{\pi_\theta(a \mid s)}{\pi_{\theta_k}(a \mid s)} A^{\pi_{\theta_k}}(s, a),\ g(\epsilon, A^{\pi_{\theta_k}}(s, a)) \right)$$

Where g is given by

$$g(\epsilon, A) = \begin{cases} (1+\epsilon)A & \text{if } A \geq 0 \\ (1-\epsilon)A & \text{if } A < 0 \end{cases}$$

A is the advantage.

So, what does all this mean? Let's break it down:

1. If the advantage is positive, the g term will be $(1 + \varepsilon)$ A, and the min() in L means that there is no advantage to moving very far away from the old policy, π_θ.

2. If the advantage is negative, the objective increases if the action becomes *less* likely, again constrained to stay near the old policy.

CHAPTER 7 REINFORCEMENT LEARNING AND CONTROL

Therefore, you have a substantially simpler problem, which can be repeatedly optimized by several (stochastic, minibatch) gradient descent steps, before a new batch is sampled. See Figure 7-5.

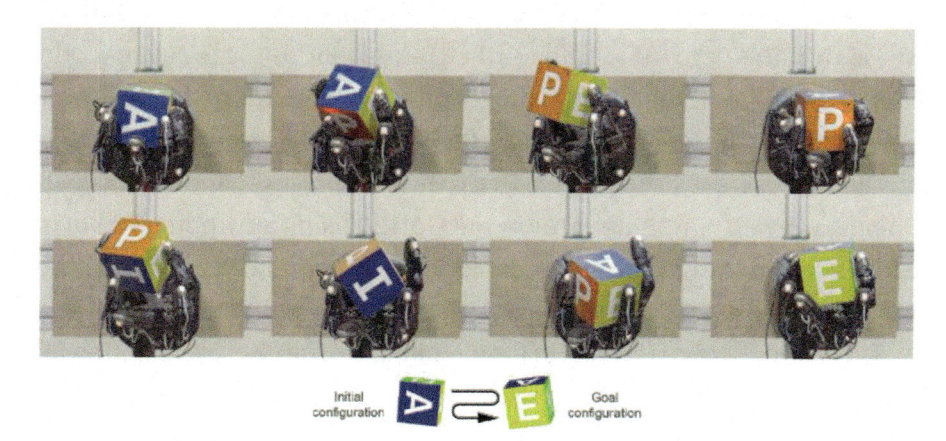

Figure 7-5. A five-fingered robotic hand trained using PPO to manipulate cubes into any configuration. PPO is one of the more widely employed reinforcement learning algorithms at present. Used with permission, source: https://journals.sagepub.com/doi/pdf/10.1177/0278364919887447 *[9]. Note: This book is in no way affiliated with or endorsed by OpenAI.*

In practice, PPO is an extremely effective algorithm, being used for everything from playing games like Dota[7] to a wide range of robotics tasks, such as quadruped locomotion[8] and dexterous manipulation[9].

Model-Based Reinforcement Learning

Model-based reinforcement learning differs in that it presumes a *model* of the world exists, i.e. that you will be able to approximate the state probability distribution P(s, s' | a) during learning. This is less commonly used in practice than model-free methods, but it's useful to understand

CHAPTER 7   REINFORCEMENT LEARNING AND CONTROL

why this probability distribution—this *world model*—is such a compelling addition to the reinforcement learning framework, and indeed the robot learning framework in general.

Model-based reinforcement learning does not have as many widely-used algorithms across the field as PPO. Intuitively, this is because model-based RL requires that model of the world; and often, these assumptions can be built into the algorithm itself. Perhaps the most significant line of work in this space is from Google DeepMind, which used model-based RL to defeat world champions at the game Go[12], using a combination of tree search—which is to say, traditional planning over which states to explore—and neural networks to learn which areas to perform that tree search in.

This kind of approach was expanded in follow-up works, resulting in MuZero, a model-based system which uses planning to determine which states are worth exploring[14]. See Figure 7-6. It learns what actions to take, as well as what states will *result* from those actions. By making these types of predictions, MuZero can solve much more challenging, long-horizon tasks, like the aforementioned Go. Model-based reinforcement learning methods from this family—AlphaProof and AlphaGeometry2—achieved silver medal status at the International Math Olympiad in 2024[15], showing how capable model-based methods can be at solving specific extremely challenging problem sets.

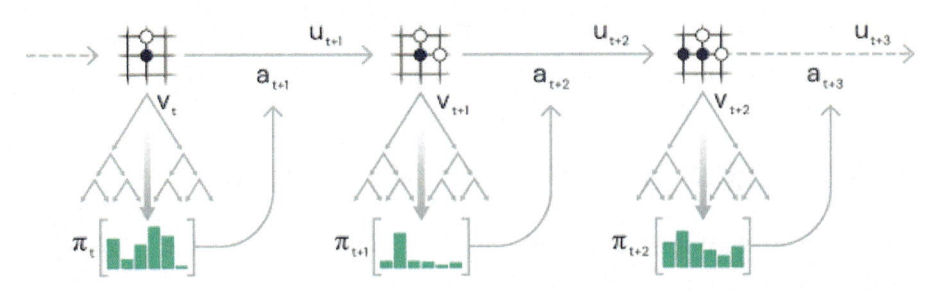

Figure 7-6.   MuZero uses planning to determine where to explore next. It learns a dynamics function that's used to model how the future state will change and what rewards will result. Used with permission, source: https://deepmind.google/discover/blog/muzero-mastering-go-chess-shogi-and-atari-without-rules/ [14].

CHAPTER 7 REINFORCEMENT LEARNING AND CONTROL

Robotics: Model-Based RL for Continuous Control

There have been a few attempts to develop model-based RL methods that can be applied to robotics. These share some similar ideas to the MuZero method:

1. In addition to learning a policy, one must learn a *simulator*.

2. This simulator is essentially giving you the transition function T(s, a) -> s' from the original reinforcement learning simulator.

3. It allows you to *plan* and compute which sequences of actions will give you an optimal return before actually executing them.

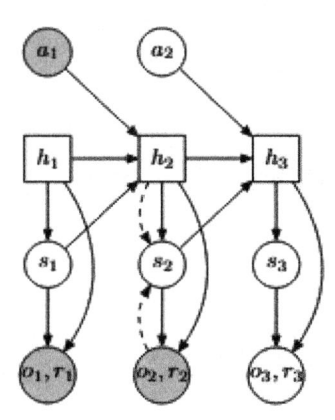

Figure 7-7. *The Recurrent State Space Model from PlaNet. Model-based RL methods can use search and a learned world model to focus their exploration in more useful regions of the state space. Used with permission, source:* https://proceedings.mlr.press/v97/hafner19a/hafner19a.pdf *[16].*

CHAPTER 7　REINFORCEMENT LEARNING AND CONTROL

One such method is the Deep Planning Network (PlaNet) proposed in [16]. PlaNet uses a Recurrent State Space Model (RSSM) to predict future latent states given action sequences. See Figure 7-7. Predicting *latent* spaces instead of full observations means that the method does not need to decode and generate a full image to perform a "rollout" of a future plan; this makes the overall method more efficient and faster.

The core idea of PlaNet[16] and, relatedly, Dreamer[17], is that they make predictions in some *latent* space. See Figure 7-8. In this latent space, you use data to learn not just the policy π but some transition function as well:

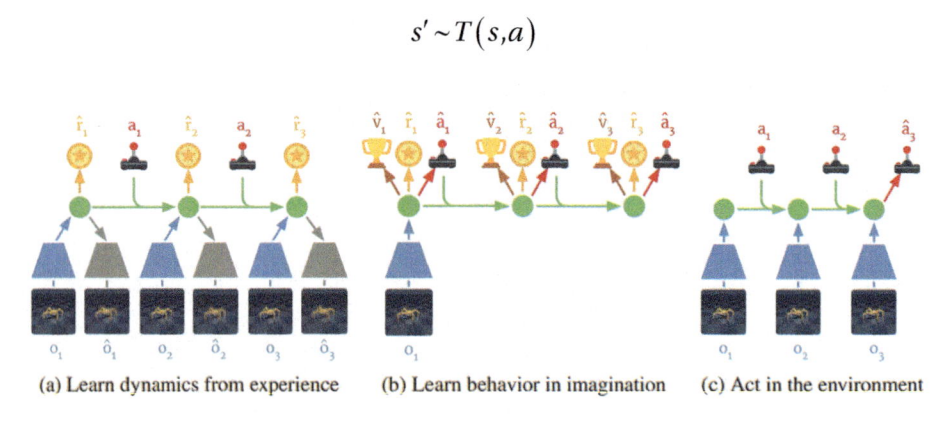

$$s' \sim T(s,a)$$

Figure 7-8. *Model learning approach from Dreamer. Used with permission, source:* https://arxiv.org/pdf/1912.01603 [17]

This tells you that, given state s and action *a*, you will end up in a state s' that results from this state. Models like PlaNet assume deterministic state dynamics (i.e., $s' = T(s, a)$). This transition function is often referred to as a *world model*.

However, this full world modeling challenge is extremely difficult. You can compare this to generative AI techniques for video generation, like SORA[21]—on the outside, learning such a world model is an even more difficult problem than generating a video! This is because the robot's world model must be an extremely accurate simulator of the world. Models like

CHAPTER 7 REINFORCEMENT LEARNING AND CONTROL

the Universal Policy[30] use video generation as a "world model" to predict futures for robot actions. This can be used for directly planning out which actions to take, but also provides an example for how such a world model can be learned and what it entails.

Model-based methods, to summarize, have an incredible promise: to use planning and foresight to choose better actions and explore more efficiently, solving a core issue facing model-free reinforcement learning. This has allowed them to solve incredibly difficult problems like the International Math Olympiad. However, there's no clear "best" approach yet, and there are few model-based methods that can be widely applied in robotics without substantial work—even more than the usual reward and environment engineering necessary for model-free methods. However, this is an exciting and promising area of research.

Offline Reinforcement Learning

Reinforcement learning is often solved as an inherently *online* process, meaning that it is performed as an agent interacts with a (real or simulated) world. This is inherently limiting in the age of big data; we want algorithms that can be applied on very large datasets collected in myriad ways. Sometimes human data will be useful; sometimes autonomous exploration; other times, it will even be necessary to use heuristic policies to get a good start in certain difficult domains. In this area, previous reinforcement learning approaches often fail.

However, the underlying reinforcement learning problem formulation is still a valuable one. One line of work looks more intensely at offline, off-policy reinforcement learning methods—those that can be applied on large, precomputed datasets—in many ways, these datasets are most reflective of the realities of robot learning problems[2].

CHAPTER 7   REINFORCEMENT LEARNING AND CONTROL

Take Decision Transformer[13] as an example, shown in Figure 7-9. It follows the classic reinforcement learning structure, with actions $a \in A$, states $s \in S$, and even a reward function $R(s, s')$. However, you do not learn any of these via interacting with the environment, as in the original reinforcement learning formulation. Instead, it uses a large offline dataset, and has reward labels as a part of the data itself.

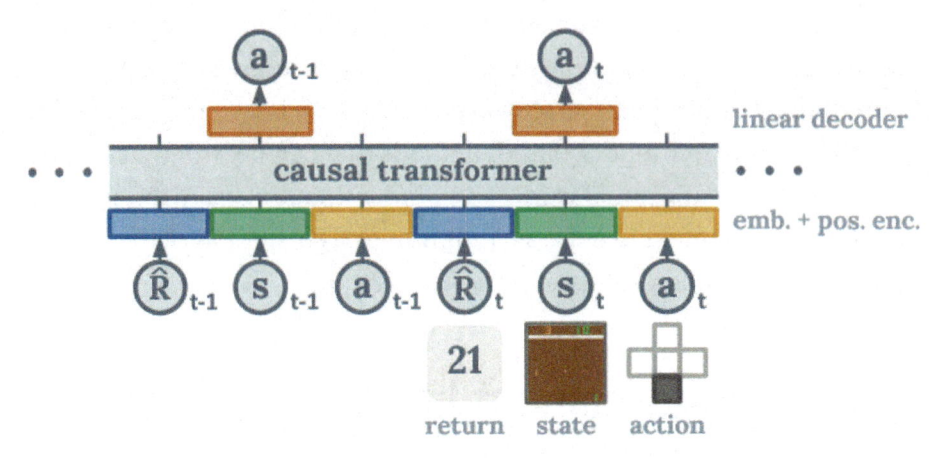

Figure 7-9. Decision Transformer is an example of an offline reinforcement learning method, which works similarly to supervised techniques used for large language models. Used with permission, source: https://proceedings.neurips.cc/paper_files/paper/2021/file/7f489f642a0ddb10272b5c31057f0663-Paper.pdf [13].

CHAPTER 7   REINFORCEMENT LEARNING AND CONTROL

Applications and Challenges

Figure 7-10. Deploying RL skills for mobile robot navigation in a variety of real-world home environments. Reinforcement learning skills have been tested on robot learning problems and in many cases are competitive with or even exceed traditional robotics methods. Used with permission, source: https://www.science.org/doi/10.1126/scirobotics.adf6991[24]

Robotics has long been one of the main motivating examples for reinforcement learning. On the face of it, it's the perfect application: we do not know how to solve many important robotics problems, but we know what these problems look like when they succeed.

However, applying reinforcement learning to robotics problems has a couple huge issues. The biggest is, as always, the data. Robotics data is fairly rare and expensive to collect, but reinforcement learning methods can require hundreds of thousands or even millions of steps[12][14]. There are two ways forward: train in simulation[10][24][25][28][29] or use a very large number of robots to collect data at scale in the real world[26][27][32]. See Figure 7-10.

Using a large number of robots is obviously very expensive and comes with a lot of headaches—a fleet of robots must be maintained, data is difficult to collect, and policy rollouts are quite expensive. However, sim-to-real, despite its great appeal, has real shortcomings. Works that do sim-to-real testing have been able to show some impressive results, and for certain things like quadruped gait learning and humanoid gait learning, it is quickly becoming standard[25]. For more complex semantic tasks, however, sim-to-real often cannot transfer at all—instead, works learning semantic skills like object search use detectors trained on supervised learning and a mixture of traditional robot control for the best and most generalizable results[24]. Sim-to-real transfer, however, is a fast moving and thrilling area, and through better real-to-sim, incredibly impressive policies can be trained for dynamic, reactive, and robust robot skills such as soccer playing[28][29].

Scaling Up RL in the Real World

The gold standard, however, is real-world data collection. A large amount of work has been done on scaling up real-world data collection, particularly spearheaded by Google's robotics lab in California[26][27], which used an "arm farm" of seven robots in their MT-Opt work. See Figure 7-11.

CHAPTER 7   REINFORCEMENT LEARNING AND CONTROL

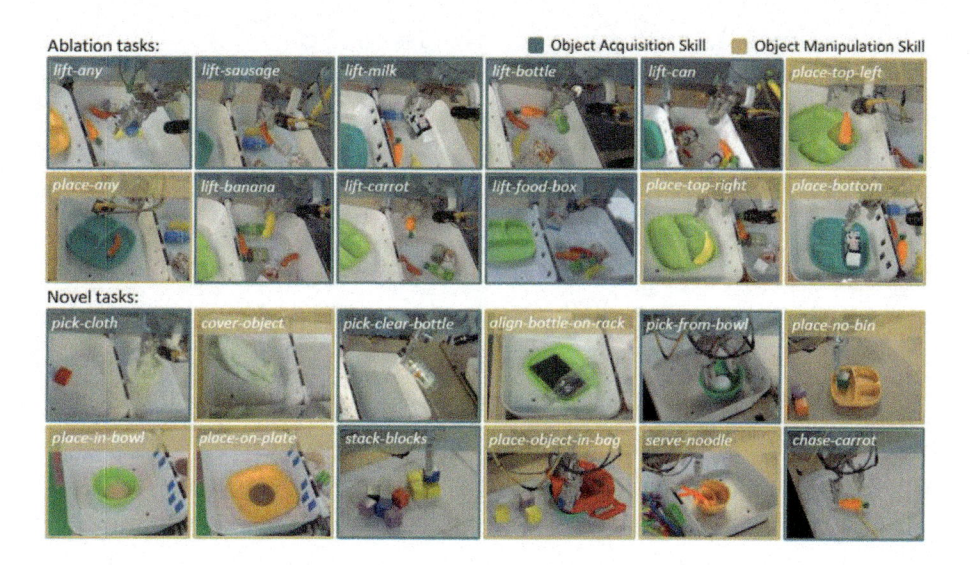

Figure 7-11. *Examples of manipulation tasks from MT-Opt. Training large numbers of manipulation skills on large numbers of real robots could potentially be a route to generally useful robots. Used with permission, source: https://arxiv.org/pdf/2104.08212 [26]*

In particular, MT-Opt looks at how to train general-purpose robot policies. This means that it does something called *multitask* reinforcement learning, where the policy is additionally parameterized by some goal (which can be thought of as a part of the state in the traditional RL formulation from the beginning of the chapter). In this case, an extra token is passed to indicate which task is being executed.

Then the model can be set up to train a large number of skills all at once. MT-Opt codifies how you can perform reinforcement learning at scale in such a setting:

1. Set up a system so tasks can be easily reset or can reset themselves.

2. Train success classifiers to determine when tasks succeed.

CHAPTER 7   REINFORCEMENT LEARNING AND CONTROL

3. Crucially, make sure you have tasks that can be accomplished easily enough that you get a positive reward signal quickly, before building to harder tasks.

The team was able to train robots to perform a wide variety of tasks, doing things like aligning, rearranging, and moving objects from one location to another. All in all, they trained their system to perform 12 different tasks using a fleet of seven robot arms.

Importantly, training all of these tasks at once as part of a multitask policy actually helps *accelerate* single-task learning in a few ways. Many aspects of visual feature learning are going to be shared across different settings, which means that these do not need to be learned over and over again. Things like approaching the surface of the table or avoiding collisions are shared across all policies. In addition, entire sub-skills might be shared: nudging an object out of the way, for example, to access a different one, might appear in multiple skills. It might even be possible to *learn* this nudging skill as a part of some task where it is actively being rewarded, which will then enable the robot to learn other skills that would be otherwise impossible with a relatively sparse reward function.

This was then extended as AW-Opt[27], a method that allows robots to learn using both online and offline, sub-optimal data by using some ideas from advantage-weighted regression. The advantage here is that it helps address a key shortcoming of on-robot reinforcement learning: that data is still hard to collect, even with a fleet of robots, and that *bad* data is much more common than *good* data.

CHAPTER 7 REINFORCEMENT LEARNING AND CONTROL

Figure 7-12. Large-scale mobile robot trash sorting from Google and Everyday Robots. Twenty-three robots were deployed performing trash sorting over the course of two years. Used with permission, source: https://arxiv.org/pdf/2305.03270[32]

Finally, it's worth noting a dramatic experiment building upon these works. Researchers at Everyday Robots deployed a fleet of 23 mobile manipulators, shown in Figure 7-12, to perform trash-sorting tasks. Over two years, they used reinforcement learning at scale to collect vast amounts of data—9,527 hours of robot experiences—and use them to train policies for sorting across nine different trash-distribution scenarios. The robots had to place objects correctly into compost, recycling, or garbage bins. This is perhaps the largest real-world RL undertaking performed, where robots had to sequentially execute pick-and-place tasks for a very wide variety of objects, understand what those objects were, and move them into different locations. To do this, they had both a simulation and a "teaching" environment in which data could be collected more efficiently.

These methods show promise, but also still show common RL pitfalls: data is hard to get, RL policies often do not *generalize* well outside of what they saw during training and so training environments are carefully constructed, and collecting real data with useful rewards is difficult. The next section looks at cases where this is less relevant: training mostly in simulation.

CHAPTER 7   REINFORCEMENT LEARNING AND CONTROL

Learning to Walk

Figure 7-13. Boston Dynamics trains its robots to walk in simulation. Used with permission, source: https://bostondynamics.com/blog/starting-on-the-right-foot-with-reinforcement-learning/ [10]

Reinforcement learning has one spectacular success story: walking skills for mobile robots. Walking has long been the quintessential robot skill. This has become the de facto way quadruped locomotion is done, with companies like Unitree, Boston Dynamics[10], and Anybotics[22] beginning to ship products trained with reinforcement learning. See Figure 7-13.

These robots commonly use depth sensors to determine the geometry of the world around them. This problem is perfect for the reinforcement learning case: unlike complex problems like trash sorting or pick-and-place, feedback is frequent, reward functions are valuable, and it's easy to reset and recover. RL methods like PPO and TRPO are extremely effective in these kinds of scenarios, where feedback to the policy—and thus, the ability to estimate the necessary gradient—is readily available.

CHAPTER 7 REINFORCEMENT LEARNING AND CONTROL

Recently, researchers have moved toward fast, dynamic motions trained using reinforcement learning; training in simulation and testing in the real world allows less efficient model-free algorithms like PPO[6] and TRPO[4][5][6] to be deployed on real hardware[11][23].

Robots Playing Soccer

Training robots to play soccer using deep RL[28]. Robot skills can be learned in simulation and transferred to real robots. This has even been extended to the multi-agent reinforcement learning case[29].

Bringing these together, as a case study in robot reinforcement learning, consider recent work on having robots play soccer. This is a interesting task for robots because:

1. It involves *multiple* intelligent agents.

2. Robots must learn using a high-dimensional, partially-observable observation space—images.

3. The robots themselves are relatively high degree-of-freedom, being small humanoids, and they need to be able to run and recover when they inevitably fall.

They train using maximum a posteriori policy optimization (MPO), an off-policy actor-critic algorithm, and they generate large amounts of data in simulation to accomplish this. They separately train skills like scoring goals and getting up off the ground if the robot falls[28], before combining them—a common strategy in longer horizon tasks also seen elsewhere[31]. They then train an agent policy that has to compete against increasingly difficult opponents.

Their agent is capable of *zero-shot* sim-to-real transfer, which is to say that it works without any real-world data. This is accomplished through adding perturbations and domain randomizations to the simulation environment. They randomized the floor friction and applied random forces to the agents as they moved[28], without which zero-shot sim to real transfer was not possible. This was taken further with random changes to lighting and color saturation[29] in order to improve visual transfer.

In the end, this resulted in robot agents that could play soccer against each other, kick the ball, and score goals.

Reinforcement Learning and Large Language Models

Large language models (LLMs) have made a dramatic impact on the robotics space, as well as on the wider tech world, and part of the magic that has made LLMs so appealing is a technique called Reinforcement Learning from Human Feedback (RLHF). As LLMs are highly relevant to robotics, it's important to discuss the reinforcement learning use case associated with them[33].

Reinforcement Learning from Human Feedback (RLHF)

RLHF is performed *after* training an LLM on a large corpus of data. After training an LLM on all useful (supervised) data, you perform two steps:

1. Train a reward model.
2. Use this reward model to update the original LLM.

As noted, getting a reward function is the most difficult part of applying reinforcement learning to many real-world problems. LLMs are no exception. To learn the reward function, human annotators must rank the outputs of the original LLM (this is the human feedback part of RLHF). Then, these rankings are used to fine-tune a second language model, which could be the same architecture as the first one, or, as is often the case, a smaller model.

CHAPTER 7 REINFORCEMENT LEARNING AND CONTROL

Then, given a model that can predict reward from text outputs, you now have the reinforcement learning problem. The state, s, is the original text, the action *a* the context, and the reward is the newly-trained reward function estimator. From here, you can compute policy gradient steps as per the PPO algorithm[6], modifying the weights of a copy of the original language model.

What is the result? A model that produces text more like whatever it is that humans prefer! In general, models trained with RLHF have a strongly preferable style to those trained without it, which is particularly important for a chatbot or an AI research assistant!

There are some weaknesses, however. Gathering this human data is extremely expensive[33], and you cannot use your reward function too much either. You generally cannot take that many gradient steps or do too many iterations of RLHF without the model beginning to find loopholes that it can take advantage of. If the LLM can find adversarial examples that fool the reward function, performance will begin to drop again[34]. Also, because it's based on human preferences, it can focus overly on style transfer and other visual aspects over substance[34]. Finally, there are many open questions about how to set up the RL algorithm, and there are many potential improvements to PPO[6] suitable for RLHF.

Direct Preference Optimization (DPO)

Direct Preference Optimization (DPO) simplifies the RLHF process by removing the need for a separate reward model[37]. Instead, DPO treats the problem as a classification task. The model is directly fine-tuned using human preference data, which more directly aligns the models' outputs with what humans prefer. The process has these main steps[38]:

- It starts with supervised fine-tuning (SFT) where the model is fine-tuned on a labeled dataset. The labels represent the preferred responses and the output of the model is aligned with specific guidelines.

CHAPTER 7   REINFORCEMENT LEARNING AND CONTROL

- After SFT, the model is further refined using preference data, which consists of pairs of outputs ranked by human annotators. In DPO, a preference loss function is defined that reflects how well the model's outputs align with these human rankings. Since this loss is a function of the policy itself, you don't need a separate reward model.

- The model is optimized by reducing the preference loss such that, for each pair of outputs, the model learns to produce the one that humans prefer more often.

Overall, DPO can be more stable and efficient than RLHF, as it eliminates the need for a lot of sampling and hyperparameter tuning. For this reason, it can be a useful method for fine-tuning LLMs with human preferences without the overhead of traditional RL methods.

Reinforcement Learning from AI Feedback (RLAIF)

RLAIF automates the collection of human preference data using an off-the-shelf LLM to generate preference labels[39]. Instead of relying on binary labels, RLAIF uses the log probabilities of different preference outputs, essentially creating a preference distribution using a softmax function.

This approach reduces the time and cost usually associated with collecting human feedback and has shown to perform similarly to RLHF, especially when using large models like Google's PaLM[40].

As LLMs become more integrated into robotics, it can be useful to understand how human preference data can be leveraged. This is especially useful in applications where robots need to interact with humans and abide by their needs/preferences. RLHF, DPO, and RLAIF all provide paths to fine-tune LLMs to align better with human behavior. However, each of these methods has its own tradeoffs depending on the application, LLMs being used, and so on, and thus further research is needed to refine these methods for real-world robotics applications.

CHAPTER 7   REINFORCEMENT LEARNING AND CONTROL

Challenges in RL for Robotics

Although there have been some successful applications of RL for real-world robotics, it is important to note some of the key challenges that the field still faces. Many of these challenges have been discussed throughout the chapter. The following is a consolidated summary of these challenges, along with potential solutions:

- **Sample efficiency**: RL algorithms often require a large number of samples or interactions with the environment to learn useful policies. In robotics, collecting these samples in the real world can be costly and time-consuming. Improving sample efficiency is important for reducing the amount of data required for learning while still achieving good performance. One way to do this is relying more heavily on simulation data while improving Sim2Real methods, such that the gap between simulated and real-world environments is minimized. These ideas were covered in Chapter 5.

- **Transfer learning**: Generalizing RL policies from one robot or environment to another is a difficult challenge. What works well in a simulated environment or on one robot may not be directly transferable to a different robot or real-world setting. Developing transferable RL algorithms that can adapt to new conditions is critical for practical use. Using techniques like meta-learning, where you train RL agents to learn how to learn, and domain randomization could be helpful in improving performance in new environments.

- **Scalability**: As robotic tasks become more complex and involve high-dimensional state and action spaces, scalability becomes a concern. Some RL algorithms

343

that work well for simple tasks may not scale to handle the complexities of real-world robotics applications. A few directions for developing scalable RL algorithms use hierarchical RL, where complex tasks are broken down into simpler subtasks, and developing more efficient exploration strategies that focus on promising areas of the state space.

- **Reward design**: Designing suitable reward functions that guide the learning process is often one of the hardest tasks in RL. Misaligned or poorly defined rewards can lead to suboptimal policies or to your agent not learning anything useful. Thus, developing reward functions that accurately represent the task's objectives and constraints is an important aspect of RL algorithm design. A few strategies here include using Inverse RL (IRL), where the reward function can be learned based on observed expert behavior, incrementally shaping the reward function, or having multiple objectives in the reward design.

Emerging Trends in RL for Robotics

Reinforcement learning is making a comeback in robotics, after the field has been dominated by imitation learning approaches for years.

- **Sim-to-real**: Generating the data for RL training is difficult; even the best RL methods, like TRPO and PPO, require very large amounts of training data, which is often impractical to collect in the real world. Research like [28] and [29] has showed how we can apply reinforcement learning to exciting problems, thanks to clever use of simulation.

CHAPTER 7   REINFORCEMENT LEARNING AND CONTROL

- **Integrating large language models**: Including large language models, large vision-language models, and turning them into *action* models that can act in real environments may require reinforcement learning, wherein AI agents repeatedly interact with virtual environments or systems[41].

- **Learning without manual reward functions**: The reward function is one of the most difficult parts of a reinforcement learning problem. Works like Eureka[42] have looked into using LLMs and AI agents to design and tune reward functions; works like VIP[43] have looked at extracting these reward functions from video. Finding ways to use Internet-scale data to reduce or remove the need for reward tuning will be crucial to scaling RL for real-world robotics.

Conclusions

Reinforcement learning is a powerful and general-purpose framework for describing robotics problems, and it has the great promise of allowing us to automatically learn skills for a wide variety of problems for which it is hard to collect data, and to improve on expert performance.

Reinforcement learning has previously been applied to multi-step manipulation tasks, to locomotion, and to language models. It's a powerful toolkit that appears in real robots products like quadrupedal robots and AI chatbots.

However, there are great open questions to resolve. We need robotics RL algorithms that are more capable of generalizing to unseen objects and environments. Leveraging more offline data, whether through mixing in imitation learning as per AW-Opt[27], offline RL methods[2][17], or pretraining[30][33], will be a large part of what makes this successful, given the limitations of collecting all data on-site[26][27][32].

Future work in reinforcement learning will surely take advantage of learning in simulation and sim-to-real transfer, which continues to produce incredible results[24][28][29] and circumvents some of the biggest data efficiency problems facing reinforcement learning. All in all, it seems certain that the robots of the future will use reinforcement learning to become faster, stronger, and smarter—the question is how.

Summary

In this chapter, you learned the following ideas and concepts:

- Supervised learning relies on labeled data, but it has limitations. Reinforcement learning addresses these challenges by allowing robots to autonomously collect data and learn as they interact with their environment.

- The ultimate goal of RL is to create intelligent systems that learn like humans, by interacting with their environment, setting goals, and refining their skills. RL is essential for achieving robust embodied intelligence in robots by letting them learn from mistakes and improve.

- The basics of RL, including its formulation as a Markov Decision Process (MDP), and the differences between model-free and model-based methods.

- Projects like Google's MT-Opt and AW-Opt improve learning by training robots on multiple tasks, allowing for better generalization and faster single-task learning because of shared visual features and sub-skills.

CHAPTER 7 REINFORCEMENT LEARNING AND CONTROL

- Reinforcement Learning from Human Feedback (RLHF) is used in large language models to fine-tune outputs based on human preferences, but it faces drawbacks like high data collection costs and potential adversarial behavior. Alternative approaches include Direct Preference Optimization (DPO) and Reinforcement Learning from AI Feedback (RLAIF), which can be more efficient methods for aligning LLMs with human preferences without explicit reward models.

- Real-world RL struggles with data scarcity, poor generalization, and reward design difficulties that can in some ways be addressed by methods like sim-to-real transfer.

- In the future, RL in robotics will likely focus on improving sample efficiency, transfer learning, scalability, and reward design, using more offline data and sim-to-real methods.

The next chapter explores self-driving technology as a key application of robotics, focusing on its economic potential and the software and hardware frameworks that enable perception, prediction, planning, and safety.

References

[1] Peters, Jan. "Policy gradient methods." *Scholarpedia* 5.11 (2010): 3698.

[2] Levine, Sergey, et al. "Offline reinforcement learning: Tutorial, review, and perspectives on open problems." *arXiv preprint arXiv:*2005.01643 (2020).

[3] Schulman, John, et al. "Trust region policy optimization." *International Conference on Machine Learning.* PMLR, 2015.

[4] Jayakody, Dilith. "Trust Region Policy Optimization (TRPO) - A Quick Introduction." 3 Apr. 2023, `dilithjay.com/blog/trpo#objective-function`.

[5] TRPO Docs, "Spinning Up OpenAI." `https://spinningup.openai.com/en/latest/algorithms/trpo.html`

[6] Schulman, John, et al. "Proximal policy optimization algorithms." *arXiv preprint arXiv:*1707.06347 (2017).

[7] Berner, Christopher, et al. "Dota 2 with large scale deep reinforcement learning." *arXiv preprint arXiv:*1912.06680 (2019).

[8] Tsounis, Vassilios, et al. "DeepGait: Planning and control of quadrupedal gaits using deep reinforcement learning." *IEEE Robotics and Automation Letters* 5.2 (2020): 3699-3706.

[9] Andrychowicz, Marcin, OpenAI, et al. "Learning dexterous in-hand manipulation." *The International Journal of Robotics Research* 39.1 (2020): 3-20.

[10] "Starting on the Right Foot with Reinforcement Learning." Boston Dynamics. `https://bostondynamics.com/blog/starting-on-the-right-foot-with-reinforcement-learning/`

[11] Bellegarda, Guillaume, et al. "Robust high-speed running for quadruped robots via deep reinforcement learning." *2022 IEEE/RSJ International Conference on Intelligent Robots and Systems (IROS).* IEEE, 2022.

CHAPTER 7 REINFORCEMENT LEARNING AND CONTROL

[12] AlphaGo https://deepmind.google/technologies/alphago/

[13] Chen, Lili, et al. "Decision Transformer: Reinforcement learning via sequence modeling." *Advances in Neural Information Processing Systems* 34 (2021): 15084-15097.

[14] Schrittwieser, Julian, et al. "Mastering Atari, Go, chess, and shogi by planning with a learned model." *Nature* 588.7839 (2020): 604-609.

[15] "AI achieves silver-medal standard solving International Mathematical Olympiad problems." Google DeepMind. https://deepmind.google/discover/blog/ai-solves-imo-problems-at-silver-medal-level/

[16] Hafner, Danijar, et al. "Learning latent dynamics for planning from pixels." International Conference on Machine Learning. PMLR, 2019.

[17] Hafner, Danijar, et al. "Dream to control: Learning behaviors by latent imagination." *arXiv preprint arXiv*:1912.01603 (2019).

[18] "Reinforcement Learning. Geeks for Geeks." https://www.geeksforgeeks.org/what-is-reinforcement-learning/#

[19] Bellman, Richard. "A Markovian decision process." *Journal of Mathematics and Mechanics* (1957): 679-684.

[20] Mnih, Volodymyr, et al. "Human-level control through deep reinforcement learning." *Nature* 518.7540 (2015): 529-533.

CHAPTER 7   REINFORCEMENT LEARNING AND CONTROL

[21] SORA. OpenAI. https://openai.com/index/sora/

[22] "Superior Robot Mobility – Where AI Meets the Real World." Anybotics. https://www.anybotics.com/news/superior-robot-mobility-where-ai-meets-the-real-world/

[23] Hwangbo, Jemin, et al. "Learning agile and dynamic motor skills for legged robots." *Science Robotics* 4.26 (2019): eaau5872.

[24] Gervet, Theophile, et al. "Navigating to objects in the real world." *Science Robotics* 8.79 (2023): eadf6991.

[25] Tang, Chen, et al. "Deep Reinforcement Learning for Robotics: A Survey of Real-World Successes." *arXiv preprint arXiv*:2408.03539 (2024).

[26] Kalashnikov, Dmitry, et al. "Mt-opt: Continuous multi-task robotic reinforcement learning at scale." *arXiv preprint arXiv*:2104.08212 (2021).

[27] Lu, Yao, et al. "Aw-opt: Learning robotic skills with imitation and reinforcement at scale." *arXiv preprint arXiv*:2111.05424 (2021).

[28] Haarnoja, T., Moran, B., Lever, G., Huang, S. H., Tirumala, D., Humplik, J., ... and Heess, N. (2024). "Learning agile soccer skills for a bipedal robot with deep reinforcement learning." *Science Robotics*, 9(89), eadi8022. https://arxiv.org/pdf/2405.02425

[29] Tirumala, D., Wulfmeier, M., Moran, B., Huang, S., Humplik, J., Lever, G., ... and Heess, N. (2024). "Learning Robot Soccer from Egocentric Vision with Deep Reinforcement Learning." *arXiv preprint arXiv*:2405.02425. https://arxiv.org/pdf/2405.02425

[30] Du, Y., Yang, S., Dai, B., Dai, H., Nachum, O., Tenenbaum, J., ... and Abbeel, P. (2024). "Learning universal policies via text-guided video generation." *Advances in Neural Information Processing Systems, 36.* https://proceedings.neurips.cc/paper_files/paper/2023/file/1d5b9233ad716a43be5c0d3023cb82d0-Paper-Conference.pdf

[31] Yenamandra, S., Ramachandran, A., Yadav, K., Wang, A., Khanna, M., Gervet, T., ... and Paxton, C. (2023). "HomeRobot: Open-vocabulary mobile manipulation." *arXiv preprint arXiv:2306.11565.*

[32] Herzog, A., Rao, K., Hausman, K., Lu, Y., Wohlhart, P., Yan, M., ... and Levine, S. (2023). Deep RL at scale: Sorting waste in office buildings with a fleet of mobile manipulators. *arXiv preprint arXiv:2305.03270.* https://rl-at-scale.github.io/assets/rl_at_scale.pdf

[33] Nathan Lambert, Louis Castricato, Leandro von Werra, and Dahoas Alex Havrilla. https://huggingface.co/blog/rlhf

[34] Andrej Karpathy. X. https://twitter.com/karpathy/status/1821277264996352246

[35] Andrychowicz, O. M., Baker, B., Chociej, M., Jozefowicz, R., McGrew, B., Pachocki, J., ... and Zaremba, W. (2020). "Learning dexterous in-hand manipulation." *The International Journal of Robotics Research, 39*(1), 3-20. https://journals.sagepub.com/doi/pdf/10.1177/0278364919887447

[36] Yu, C. and Wang, P. (2022). "Dexterous manipulation for multi-fingered robotic hands with reinforcement learning: A review." *Frontiers in Neurorobotics, 16,* 861825. https://www.ncbi.nlm.nih.gov/pmc/articles/PMC9083362/#B6

CHAPTER 7   REINFORCEMENT LEARNING AND CONTROL

[37] Rafailov, Rafael, et al. "Direct preference optimization: Your language model is secretly a reward model." *Advances in Neural Information Processing Systems* 36 (2024).

[38] https://toloka.ai/blog/direct-preference-optimization

[39] Lee, Harrison, et al. "RLAIF: Scaling Reinforcement Learning from Human Feedback with AI feedback." *arXiv preprint arXiv*:2309.00267 (2023).

[40] Chowdhery, Aakanksha, et al. "PaLM: Scaling language modeling with pathways." *Journal of Machine Learning Research* 24.240 (2023): 1-113.

[41] Zhai, Y., Bai, H., Lin, Z., Pan, J., Tong, S., Zhou, Y., ... and Levine, S. (2024). "Fine-Tuning Large Vision-Language Models as Decision-Making Agents via Reinforcement Learning." *arXiv preprint arXiv:2405.10292*.

[42] Ma, Y. J., Liang, W., Wang, G., Huang, D. A., Bastani, O., Jayaraman, D., ... and Anandkumar, A. (2023). "Eureka: Human-level reward design via coding large language models." *arXiv preprint arXiv:2310.12931*.

[43] Ma, Y. J., Sodhani, S., Jayaraman, D., Bastani, O., Kumar, V., and Zhang, A. (2022). "ViP: Towards universal visual reward and representation via value-implicit pre-training." *arXiv preprint arXiv:2210.00030*.

[44] Sutton, Richard S. and Andrew G. Barto. *Reinforcement Learning: An Introduction.* MIT press, 2018.

[45] Piper, Kelsey. "AI Triumphs against the World's Top Pro Team in Strategy Game Dota 2." *Vox*, 13 Apr. 2019, www.vox.com/2019/4/13/18309418/open-ai-dota-triumph-og

CHAPTER 8

Self-Driving Vehicles

This chapter covers autonomous driving technology and explains how autonomous cars are built. The goal of this chapter is to help you understand how such a system is designed and built and what metrics are used to evaluate the performance of various components.

Economic Opportunity

Self-driving cars present a massive economic opportunity, as the transportation industry is roughly worth 7 trillion dollars[1]. The transportation industry is also divided into highway driving for freight and other operations, which forms the backbone of the supply chain, and urban driving for transporting humans.

Historically, self-driving innovation picked up pace during the DARPA urban challenge[2] funded by the United States Department of Defense in 2007. Several teams competed in this challenge, which eventually spun out multiple startups, including Zoox[3] arising out of Stanford, Waymo[4] led by folks from Berkeley and Carnegie Mellon University (CMU), Argo AI[5] taking birth in CMU, and Cruise Automation[6], which was later acquired by General Motors (GM). After the COVID-19 pandemic and the associated recession-inspired funding crunch, there was a widespread consolidation in the market, with Amazon acquiring Zoox, Uber selling off ATG to Aurora[7], Cruise acquiring Voyager[8], Apple acquiring drive.ai[9], and several other startups going bust.

CHAPTER 8   SELF-DRIVING VEHICLES

There are two prominent business models for self-driving. One is the personal car model, pursued by Tesla[10] and Comma AI[11], where self-driving cars are sold to individual customers. Given that use of personal cars spans a wide geographical area which can be costly to map in high definition, companies betting on this business model invest in as little mapping as possible. Since the cars are sold to individuals, there is a requirement to keep the cost of self-driving, both the software and the sensor suite, affordable. The second model is the robotaxi business model, where companies run self-driving cars as robotaxis, akin to Uber and Lyft, in a geofenced area that has been thoroughly mapped and tested. This is pursued by Waymo and Cruise. Usually, the cost of a taxi is amortized across its entire lifespan, allowing them to be slightly more expensive than personal vehicles. These two business models also lead to slightly different technical philosophies:

1. **Sensors and redundancy:** Due to cost concerns, personal vehicle self-driving business models tend to rely less on expensive sensors like LiDARs, leading to vision-only self-driving cars supported by multiview cameras. In recent years, however, LiDAR prices have fallen drastically, as shown in Figure 8-1.

2. **Style of intervention and liability:** In the personal vehicle business model, since the vehicle is owned and operated by the customer, intervention in tricky scenarios, or what is called *disengagement* (when an autonomous system is deactivated in favor of human driving manually), is provided by the driver at the wheel. When accidents result from poorly behaved self-driving, the liability falls on the customer. Contrary to this, in a robotaxi business model, intervention is provided by teleoperation and remote humans, allowing a car to fully remove the driver inside.

CHAPTER 8   SELF-DRIVING VEHICLES

LiDAR Price & Growth Rate

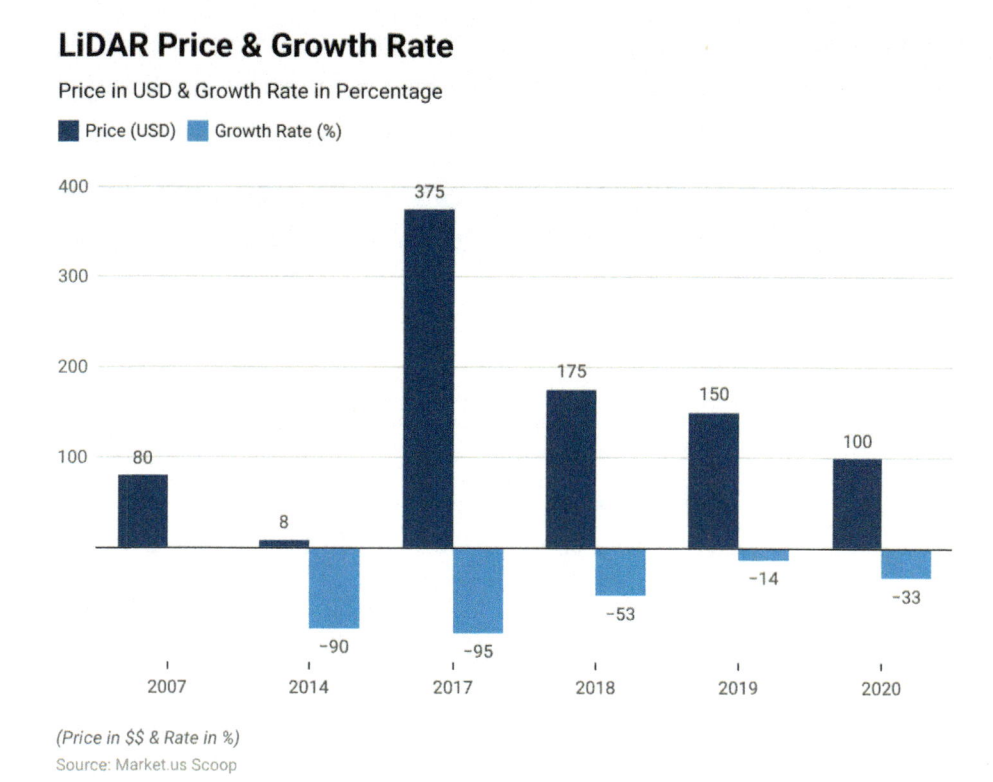

Price in USD & Growth Rate in Percentage

- Price (USD)
- Growth Rate (%)

Year	Price (USD)	Growth Rate (%)
2007	80	—
2014	8	-90
2017	375	-95
2018	175	-53
2019	150	-14
2020	100	-33

(Price in $$ & Rate in %)
Source: Market.us Scoop

Figure 8-1. *Decreasing cost of LiDAR. Used with permission, source: https://scoop.market.us/lidar-statistics/[19]*

Starting in 2023, Cruise and Waymo obtained permits to charge customers for self-driving taxi rides in San Francisco city as well as in parts of the Bay Area[12].

System Design

A self-driving system, in its simplest interpretation, is designed with inputs being the sensor inputs and outputs being acceleration, steering, and the angle of the car. The goal of the system is to successfully go from point A

CHAPTER 8   SELF-DRIVING VEHICLES

to point B in the map while reducing costs like time and without violating traffic rules or other constraints. The high-level architecture and building blocks of these systems are shown in Figure 8-2.

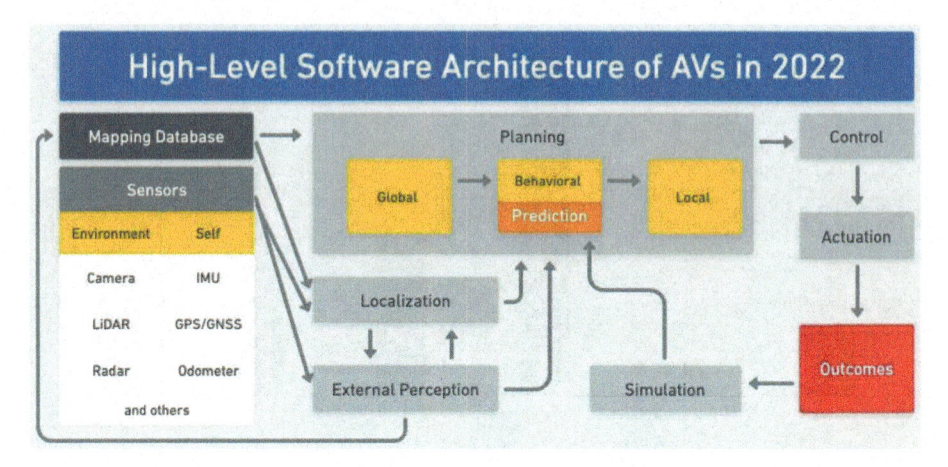

Figure 8-2. *Self-driving system architecture. Used with permission, source:* `https://medium.com/@justinmilner/a-visual-guide-to-the-software-architecture-of-autonomous-vehicles-390b1744cbd6`[20]

For urban and highway driving, a car is expected to have a response frequency of upwards of 10 Hz, which means that the entire stack end to end and running on device/remote needs to deliver an output at least once every 100 milliseconds.

The main components of these systems are:

1. Perception: Sensor suite, perception algorithms, and localization

2. Prediction

3. Planning and Control

Many of these components were covered in depth in previous chapters, so this chapter only covers the self-driving-related details driving the design of these systems.

CHAPTER 8 SELF-DRIVING VEHICLES

End-to-End Self-Driving (E2E)

Several companies approach or have done work in pushing for self-driving systems that are an end-to-end neural net. Comma AI, Wayve[13], and Waabi[14] are a few examples. An end-to-end neural planner usually takes sensors and maps as inputs and outputs either a lowest cost trajectory or current actions at each time step. Compared to the classical design, the E2E planner is a single-neuron neural network, with highly interpretable intermediate outputs for safety verifications. Compared to traditional planners, E2E planners could be more unsafe, since a neural network has full control over the actions of the car. Due to safety concerns, it's harder to productionize this technology. While most self-driving cars on the road today are classical planners, the continued rise of deep learning points to E2E planners becoming a tangible possibility in the future. Figure 8-3 shows an early E2E neural planner proposed by Uber ATG Research group.

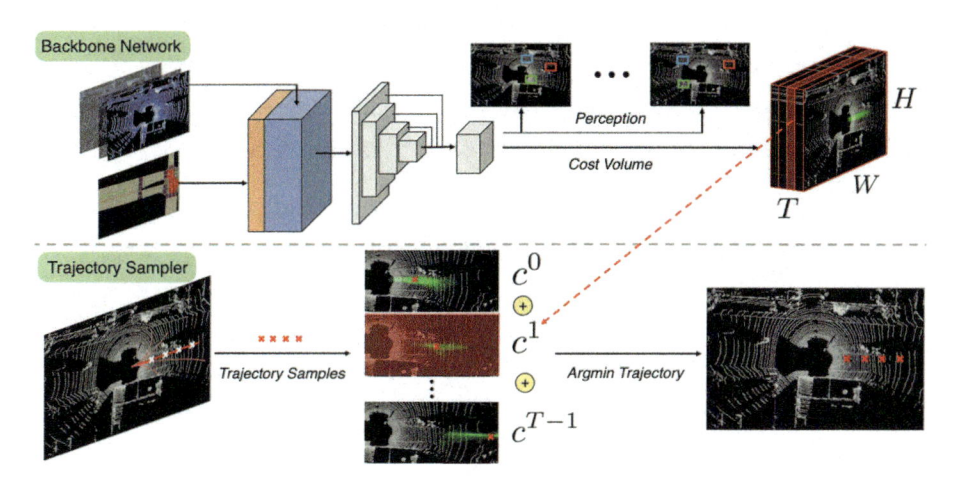

Figure 8-3. End-to-end interpretable neural planner. Used with permission, source: https://arxiv.org/pdf/2101.06679[21]

CHAPTER 8   SELF-DRIVING VEHICLES

Perception

The goal of self-driving perception is to build an environment model around the vehicle that is aware of all static, dynamic objects and any contextual scene information required to make a driving decision.

- Static objects include nonmoving entities like traffic lights, lane lines, buildings, trees, and so on. Static entities from perception can be used to localize on a preloaded map by matching static artifacts observed by the vehicle to artifacts saved in a preloaded high-definition map. They can also be used to determine if the map is wrong, such as if there's construction or if the map is outdated, to either dynamically replan or request remote assistance.

- Dynamic objects include moving agents like people, other vehicles like cars, bikes, and trucks, live animals, and so on. Intentions of objects are also detected using tracking and specific intent prediction models.

- Contextual scene information includes relationships between objects. Here are a few examples:

    - A biker and their bike move together
    - Which lanes contain which objects
    - Which animal is leashed to which person
    - Whether a trailer truck is an articulated object or not

CHAPTER 8 SELF-DRIVING VEHICLES

In order to assist good perception of the environment, three stages are involved:

1. **Sensing:** For self-driving cars, the main consideration for the sensing suite is to ensure that there are very few blind spots, and that resolution of camera and LiDAR images is dense. Modern self-driving cars are also equipped with audio sensors (to detect oncoming emergency vehicles, for example), IMUs for odometry, radar for velocity estimation, and so on. Figure 8-4 shows an example sensor suite on a car. Usually, surround view cameras are used for near-range information, such as onboarding passengers and curbs, and long-range dense LiDARs are used for far-view information.

CHAPTER 8   SELF-DRIVING VEHICLES

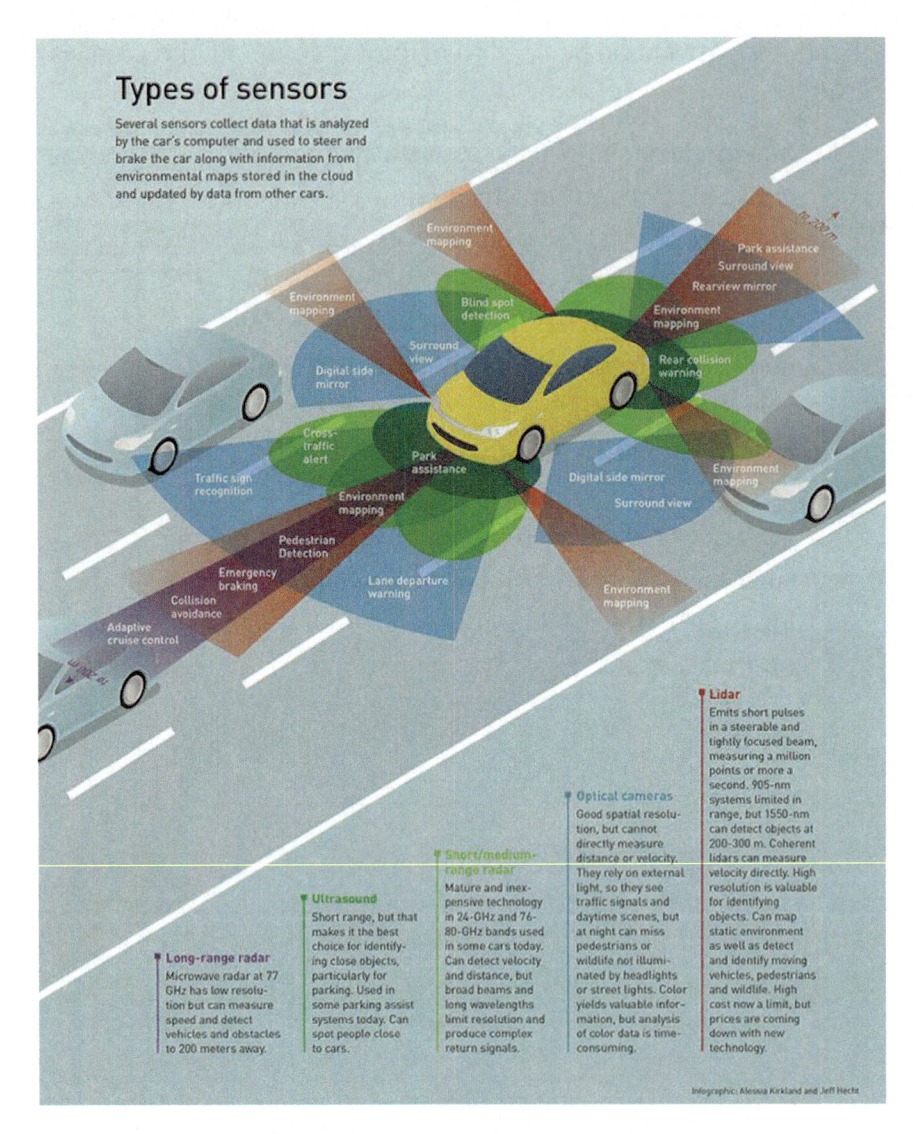

Figure 8-4. *"Types of sensors" from the article by J. Hecht, "LiDAR for Self-Driving Cars," Optics & Photonics News 29(1), 26-33 (2018), created by Alessia Kirkland, Senior Manager and Creative Director of OPN. Used with permission, source:* https://www.optica-opn.org/home/articles/volume_29/january_2018/features/lidar_for_self-driving_cars/ *[22] Reprinted/adapted with permission from © Optica Publishing Group.*

2. **Computer vision:** 2D and 3D computer vision algorithms are applied on fused or unfused sensory images to obtain relevant information from the scene and objects, as discussed earlier in this section. Chapter 2 discusses the type of algorithms used to detect this information. Figure 8-5 shows typical computer vision output where objects have been detected by bounding boxes.

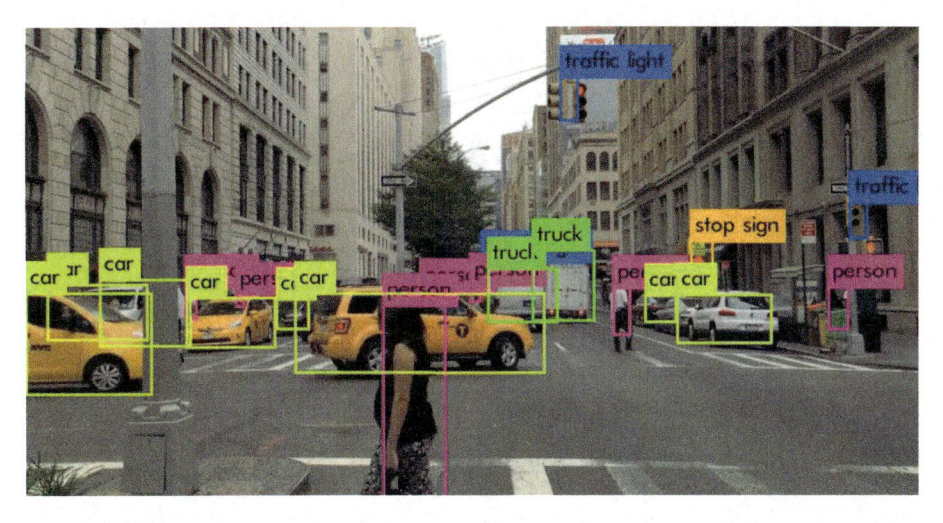

Figure 8-5. *Typical computer vision output. Used with permission, source: screenshot from* https://youtu.be/YmbhRxQkLMg?feature=shared*[23]*

3. **Sensor fusion:** The outputs from various sensors are combined to remove ghost objects, improve confidence of true objects, and combine object features. For example, LiDAR depth is a lot more precise than stereo depth, so once an object is classified using a camera image, its depth attribute is updated from its corresponding instance in the LiDAR image.

CHAPTER 8   SELF-DRIVING VEHICLES

Prediction

It is very important for a self-driving car to anticipate the actions of other agents. This is done via intent prediction. In state-of-the-art self-driving software stacks, intent prediction is done via large neural nets. Waymo's ChauffeurNet[15] paper proposed one of the first intent-prediction systems, with inputs involving maps, traffic lights, speed limit information, and other agents, as well as past trajectories to predict future trajectory, where all inputs were transformed to images for processing by convolutional neural nets. These inputs are shown in Figure 8-6.

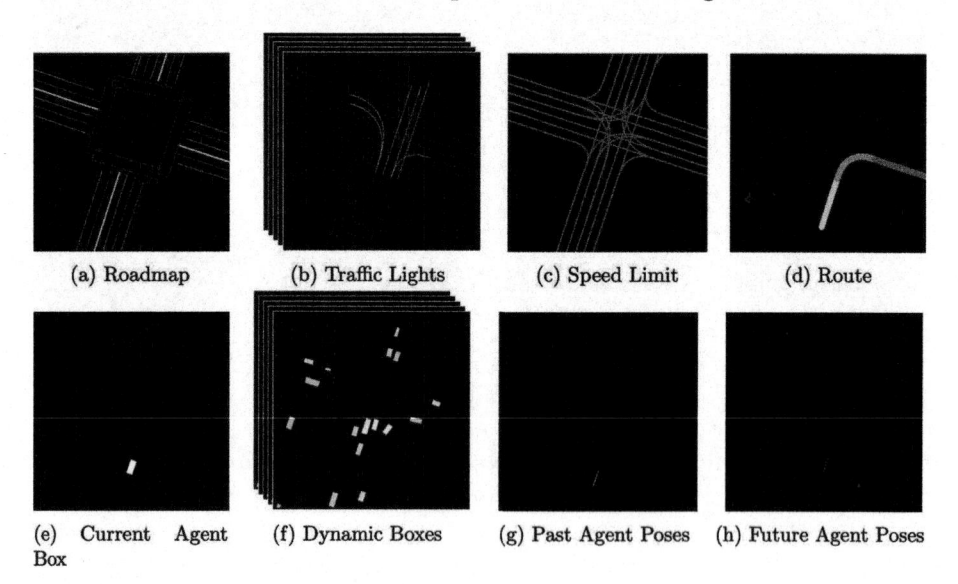

(a) Roadmap   (b) Traffic Lights   (c) Speed Limit   (d) Route

(e) Current Agent Box   (f) Dynamic Boxes   (g) Past Agent Poses   (h) Future Agent Poses

Figure 8-6. *Predicting future agent pose. Used with permission, source: https://arxiv.org/pdf/1812.03079.pdf[15]*

Waymo's ChauffeurNet architecture predicts the future actions of other road agents using a combination of convolutional and recurrent neural networks. The core of the model is a FeatureNet that processes inputs (maps, traffic lights, etc.) and transforms them into features. These features are then fed into an AgentRNN, which predicts driving parameters like heading, speed, and waypoints. Additionally, a Road Mask Net and

Perception RNN are used to predict the drivable area and detect other agents, respectively. The model itself is trained of various loss functions—heading loss, speed loss, and collision loss—which focus on minimizing differences between the model's prediction and actual outcomes.

The architecture also has a memory component, which keeps track of past agent locations. This improves overall model prediction of future positions and trajectories. By using convolutional neural networks for image processing and recurrent neural networks for handling sequence data, ChauffeurNet can leverage different types of data to understand the driving environment and predict future actions of agents in complex traffic scenarios.

While ChauffeurNet predicts the future of the driver, similar logic can be applied to predict the intents of other agents as well.

Intent prediction is crucial for driving situations like:

1. Traffic light intersections, such as four-way and two-way stops, where multiple agents indicate their intentions through hand signals, lights, and other ways.

2. During lane change and merging situations, where it is important to know if other parties are giving you the right of way or not.

Planning

Given an agent's surroundings, such as its location on the map, the static and dynamic obstacles around it, its destination, and relevant traffic rules, planning is the problem of finding an optimal trajectory toward the safe achievement of reaching the destination. Costs involved during planning are multifold:

1. **Collision cost:** An accident with any agent or static obstacle has among the highest costs in planning and is to be avoided with priority.

2. **Traffic rule violation:** Violating speed limits, traffic lights, HOV lane rules, and other traffic rules constitutes another cost. Compared to collision costs, these are lower, which means that a violation of traffic rules or speed limits may be incurred to avoid a collision.

3. **Distance and time cost:** Reaching a destination within optimal path and time.

The most common type of planners in self-driving are search-based planners. There are also neural planners and reinforcement-learned planners in research settings. The main downsides of neural planners are non-interpretability of decision making as well as lack of control over final actions chosen by the car.

The following code segment shows how a search-based planner (A* search used in the example here) can be utilized. Note that this is a highly simplified version of a planner. Lane segments refer to elements in the map, and segments of lanes that may be occupied by an agent. For this example, the map is interpreted as a graph of lane segments, with a start and destination. Note that this code segment is modified based on A* search code provided here[16].

```
// A* search algorithm for self-driving to go from lane segment
start // to lane segment destination
1. Initialize an to_explore with the first lane segment where
   the ego is, and set its f = 0
2. Initialize the explored list
3. while the to_explore list is not empty
   a) find the lane segment with the least f on the to_explore
      list, call it "q"
   b) pop q off the to_explore list
```

CHAPTER 8 SELF-DRIVING VEHICLES

   c) generate q's successors, ie, lane segments that can be traversed to from q and set their parents to q
   d) for each successor
      i) if successor is the goal, stop search
      ii) else, compute both **g** and **h** for successor
         successor.**g** = q.**g** + cost to go from q to successor
         successor.**h** = cost to go from goal to successor distance (eg Manhattan distance cost + other costs discussed above)
         successor.**f** = successor.**g** + successor.**h**
      iii) if a node with the same position as successor is in the to_explore list which has a lower **f** than successor, skip this successor
      iV) if a node with the same position as successor is in the explored list which has a lower **f** than successor, skip this successor otherwise, add the node to the to_explore list end (for loop)
   e) push q on the explored list end (while loop)

Similarly, an intersection may be modelled as follows:

```
// Modeling a traffic intersection in C++

class Intersection
{
    public:
    Type type_of_intersection; // type of intersection,
                                       say traffic
                              // light, 4 way stop, 2 way stop
    LaneSegment[] lanesegments; // list of lane segments
                                       that are
                              // part of the intersection
                                       and their
```

```
                                    // relationships to each other
    Crosswalk[] crosswalks;         // crosswalks that are
                                       part of the
                                    // intersection and agent IDs
                                    // belonging to them
    TrafficLights[] trafficlights;  // traffic light
                                       elements and
                                    // their states that are
                                    // part of the intersection
    Agent[] order_of_arrival;       // order of arrival
                                       of agents
    Agent[] agents;                 // all agents and their
                                       attributes
                                    // that are relevant to the
                                    // intersection
};
```

Safety

Engineering safety is a core concept when building robust robotics products. This is especially crucial with self-driving systems, as vehicles operate among people, sometimes at high speeds, and mistakes can be deadly.

AI Safety and Systems

Previous safety standards were designed with traditional systems in mind and don't cater as well to AI-based self-driving systems. For example, ML systems can be unpredictable, use incomplete data, and are often difficult to interpret, which makes it hard to guarantee their safety using existing standards.

A key failure mode of ML systems is breaking down when interacting with people or their environments in unexpected ways. Since ML methods rely heavily on the completeness of the data, ensuring that important scenarios and data is accounted for is crucial. If data is missing for important scenarios that the vehicle may encounter, its behavior could be unexpected and more likely to break down.

A group in the ISO (the International Organization for Standardization) [17] is focused on developing standards to ensure that AI systems being developed are safe and reliable for applications that involve human interactions, like autonomous vehicles. To account for the unique challenges that ML systems face in autonomous vehicles, we can also draw on principles from the AI safety community. The goal of AI safety is to ensure that AI systems act fairly and don't discriminate in any way. A few key ideas of AI safety, as described in [18], include ensuring that AI is fair (doesn't discriminate), transparent (easy to understand), and secure (protected from hackers).

Safety Considerations

This section discusses a few aspects of general safety to consider when it comes to the hardware and algorithm design of self-driving cars. Many of these points are focused on companies building self-driving cars that will be used publicly, but these ideas are similarly applicable in research settings.

Sensor Reliability

Self-driving cars can use many sensors, including cameras, radar, and LiDAR, to understand their environment. To ensure that sensors work reliably, it can be helpful to use multiple types of sensors to cover various weaknesses, perform regular calibration to ensure that sensors perform

CHAPTER 8 SELF-DRIVING VEHICLES

well, and use sensor-fusion techniques to validate information from different sensors. That way, if a single sensor is spooked, others can balance it out.

Algorithmic Robustness

The algorithms used in self-driving cars must be able to work in a wide range of driving scenarios, whether it be empty streets or complex urban environments. To catch edge cases and measure adaptation, these algorithms should be tested in varying real-world and simulated environments. It's also important to implement fail-safe methods and monitoring methods to detect and address any anomalies or edge cases that the vehicle encounters that are not accounted for in algorithm design.

Cybersecurity

Using sensors and other technologies in self-driving cars makes them more vulnerable to cyberattacks, which can compromise safety and privacy. To protect against these attacks, it's important to have encryption methods in place to protect data and establish detection systems to monitor for and respond to potential threats. Companies should conduct regular security audits and collaborate with cybersecurity experts to identify and mitigate data-related risks.

Ethical Decision-Making

It's important that companies have clear ethical guidelines that dictate how ethical decisions should be made by the self-driving car and ensure that these guidelines are reflected in how the vehicle handles edge cases. To build trust in the company and with the public, having clear documentation and communicating openly about these topics is important.

Human-Machine Interaction

The interaction between self-driving cars and the people riding in them should be intuitive and user-friendly. When designing these vehicles, it can be useful to have clear visual and auditory signals to inform passengers about the vehicle's actions and goals. This can include designing controls and interfaces that are easy to understand and operate, and allowing passengers to take over control in any emergency cases. Making sure these human-machine interactions are straightforward can be extremely important for public trust and to maintain safety.

Fail-Safe Mechanisms

Self-driving cars must have mechanisms in place to handle unexpected failures safely. This includes developing comprehensive emergency protocols that allow the vehicle to safely stop or return to a secure state if any issues come up. Having manual or backup control systems that can take over if the primary system fails can be useful during deployment to ensure safe operation continues even if a malfunction happens.

Data Privacy

Given the vast amount of data that's collected by self-driving cars, protecting user privacy is important. When training models or sharing data, anonymizing that data can be useful to prevent the identification of individuals. Developing standards about data usage and privacy policies, and sharing these standards publicly can build trust with the people using self-driving cars.

CHAPTER 8   SELF-DRIVING VEHICLES

Summary

This chapter covered the following points:

- There is a huge economic opportunity to build self-driving cars, as the transportation industry is worth $7 trillion.

- The system design for these vehicles can be broken into perception, prediction, and planning algorithms, along with hardware considerations. There is a growing interest in end-to-end neural planners, but they face challenges in safety and commercialization.

- The goal of perception in self-driving is to build an environment model that is aware of static, dynamic objects, and contextual scene information around the vehicle.

- Prediction focuses on anticipating the actions of other agents in the environment using large neural networks.

- Planning involves finding an optimal trajectory for the vehicle to achieve its goal safely while minimizing various costs, including collision costs, traffic rule violation costs, and time costs.

- Safety is a critical aspect of self-driving cars, including AI safety, ML system robustness, and general safety considerations for companies and research.

The next chapter explores common tasks in industrial robotics, including peg-in-hole assembly, welding, and warehouse automation. It discusses various robot types, key system design considerations for hardware, the use of software (such as CNNs for grasping), scaling reinforcement learning for grasping, as well as performance and safety metrics.

References

[1] https://www.datamintelligence.com/research-report/transportation-industry-market

[2] https://www.darpa.mil/about-us/timeline/-grand-challenge-for-autonomous-vehicles

[3] https://zoox.com/

[4] https://waymo.com/

[5] https://www.bloomberg.com/profile/company/1489336D:US

[6] https://www.getcruise.com/

[7] https://investor.uber.com/news-events/news/press-release-details/2020/Aurora-is-acquiring-Ubers-self-driving-unit-Advanced-Technologies-Group-accelerating-development-of-the-Aurora-Driver/default.aspx

[8] Korosec, Kirsten. "Cruise Acquires Self-Driving Startup Voyage." *TechCrunch*, 15 Mar. 2021, techcrunch.com/2021/03/15/cruise-acquires-self-driving-startup-voyage/.

[9] Leswing, Kif. "Apple Bought Autonomous Vehicle Start-up Drive.Ai." CNBC, 25 June 2019, www.cnbc.com/2019/06/26/apple-buys-autonomous-vehicle-start-up-driveai.html.

[10] https://www.tesla.com/

[11] https://www.comma.ai/

[12] Templeton, Brad. "Waymo and Cruise Get Permits for Full-Day Robotaxi for Money in SF, LA, Silicon Valley." *Forbes Magazine,* 10 Aug. 2023, www.forbes.com/sites/bradtempleton/2023/08/10/waymo-and-cruise-get-permits-for-full-day-robotaxi-for-money-in-sf-la-silicon-valley/.

[13] https://wayve.ai/

[14] https://waabi.ai/

[15] Bansal, Mayank, Alex Krizhevsky, and Abhijit Ogale. "ChauffeurNet: Learning to drive by imitating the best and synthesizing the worst." *arXiv preprint arXiv*: 1812.03079 (2018).

[16] Belwariar, Rachit. "A* Search Algorithm." *GeeksforGeeks,* 30 July 2024, www.geeksforgeeks.org/a-search-algorithm/.

[17] https://www.iso.org/home.html

[18] Burton S., Gauerhof L., Sethy B. B., Habli I., Hawkins R. (2019). "Confidence arguments for evidence of performance in machine learning for highly automated driving functions," in *Computer Safety, Reliability, and Security,* Vol. 11699 of Lecture Notes in Computer Science, eds A. Romanovsky, E. Troubitsyna, I. Gashi, E. Schoitsch, and F. Bitsch (Cham: Springer International Publishing;), 365–377. 10.1007/978-3-030-26250-1_30.

[19] Pangarkar, Tajammul. "LiDAR Statistics 2024 by New Light Detection Technology." Market.us Scoop, 10 July 2024, scoop.market.us/lidar-statistics/.

[20] Milner, Justin. "A Visual Guide to the Software Architecture of Autonomous Vehicles." *Medium*, 25 Sept. 2022, medium.com/@justinmilner/a-visual-guide-to-the-software-architecture-of-autonomous-vehicles-390b1744cbd6.

[21] Zeng, Wenyuan, et al. "End-to-end interpretable neural motion planner." Proceedings of the IEEE/CVF Conference on Computer Vision and Pattern Recognition. 2019.

[22] Hecht, Jeff. "LiDAR for Self-Driving Cars." Optica-Opn.org, *Optics & Photonics News,* 2018, www.optica-opn.org/home/articles/volume_29/january_2018/features/lidar_for_self-driving_cars/.

[23] Redmon, Joseph. "YOLO in New York." YouTube, 12 Nov. 2017, youtu.be/YmbhRxQkLMg?feature=shared.

CHAPTER 9

Industrial Robotics

Industrial robots perform tasks in industrial and manufacturing settings such as assembly, welding, and packing. These tasks are repetitive and labor-intensive and require a level of precision and consistency. By using deep learning, we can improve a robot's ability to perceive and understand their environment and perform these complex tasks. Using sensor data, robots can recognize objects, plan paths, and execute on tasks. Deep learning can especially be important, as it allows robots to adapt to their changing, unstructured environment and perform tasks in predictive maintenance and operations alongside humans.

Deep learning techniques are being used by several businesses in manufacturing and industrial settings. One is Amazon, which uses computer vision methods to deploy a fleet of robots in their fulfillment centers. According to Amazon, "In 2022, 1 billion packages, or one-eighth of all the orders we delivered to customers worldwide, were sorted by Robin, one of Amazon's robotic-handling systems."[1] By creating a collaborative environment between robots and humans, they have seen an increase in efficiency, a reduction in errors, and an increase in how many orders they can process in their fulfillment centers. Before diving into which deep learning methods are used, the following section talks about which tasks are commonly performed in manufacturing settings.

CHAPTER 9   INDUSTRIAL ROBOTICS

Common Tasks

Manufacturing and industrial robotics are employed in a wide range of tasks to automate processes, improve efficiency, and increase precision. A few common tasks in manufacturing and industrial robotics include pick-and-place, peg-in-hole, and welding.

Pick-and-Place

In industrial automation, picking up an object from one position and placing it in another is known as "pick-and-place." In fields such as automotive, electronics, and logistics, this task is extremely important for a robot to complete, as it helps place components on assembly lines and generally manipulate heavy items. A pick-and-place system typically comprises a robotic arm or manipulator with end effectors like grippers, vacuum cups, or magnetic attachments. For placement and control, it also has sensors, a vision system, and a controller. Commonly, FANUC's robotic arms are used in automotive assembly[2], ABB's robots are used for electronics manufacturing, and Amazon's robotics are used in fulfillment centers[3]. Figure 9-1 is an illustration of a pick-and-place robot system.

CHAPTER 9　INDUSTRIAL ROBOTICS

Figure 9-1. Pick-and-place system by Covariant AI. Used with permission, source: https://covariant.ai/insights/automation-upgraded-robotic-sorter-induction/[31]

Peg-in-Hole

In the peg-in-hole task, an item must be carefully inserted into the correct hole or container. Electronics production, automobile assembly, and aircraft manufacturing all depend on this work. For example, in electronics manufacturing, you may need to place chips on circuit boards, which requires extreme precision that robots like ABB's IRB series[4] can help with. In the automotive industry, aircraft manufacturers use robots like FANUC's M-20iA[5] or KUKA's KR Quantec robots[6] to assemble components. The components of a peg-in-hole system include a robotic arm, end effectors (usually a gripper or other specialized tool), sensors, and a vision system for alignment and positioning. Figure 9-2 depicts a system with a peg-in-hole robot as an example.

377

CHAPTER 9   INDUSTRIAL ROBOTICS

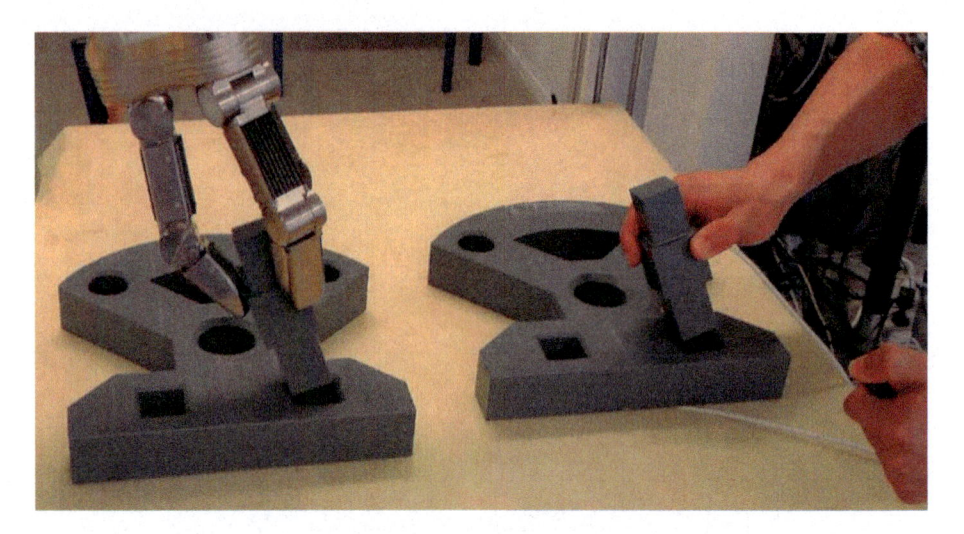

Figure 9-2.   *Robot performing peg-in-hole task. Used with permission, source:* https://www.semanticscholar.org/paper/ Robot-Learning-from-Demonstration-in-Robotic-A-Zhu-Hu/ d77d2925eee76dbb41d7c2fbda138b2f7beeec62 *[32]*

Welding

Robotic welding machines are used to weld materials together by melting and then combining them using heat. This is an important task in sectors including aircraft, construction, and the automobile industry. In the aircraft industry, you need precise and high-quality welds and in the construction sector, robots are used for tasks such as steel beam welding. In the automotive industry, manufacturers like Tesla[7], General Motors[8], and Toyota[9] use robots to weld car bodies. This can speed up manufacturing time and, in some cases, improve the quality of the welds. Common robots used for these tasks are the KUKA's KR Quantec[6], FANUC ARC Mate series[10], and ABB's IRB 6700 robots[11]. A welding robot system incorporates a robotic arm with a torch or tool, a welding power source, sensors, and a vision system for tracking. Figure 9-3 shows an illustration of a welding robot.

CHAPTER 9   INDUSTRIAL ROBOTICS

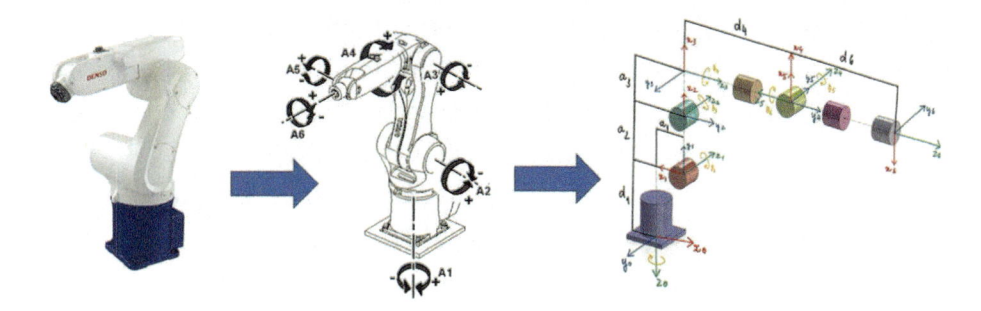

Figure 9-3. Welding robot example showing mechanical structure and joint rotations. Used with permission, source: `https://journals.sagepub.com/doi/full/10.1177/16878132241260525`[33]

Warehouse Tasks

Outside of improving specific industrial processes, robots can also be used to improve overall warehousing duties. Robots can do inventory movement and fulfillment, inventory tracking, and scanning, which can help modern warehouses and distribution centers operate as efficiently as possible.

- **Inventory movement and fulfillment:** Effective management and handling of items within a warehouse or distribution center are necessary for inventory movement and fulfillment tasks. To fulfill client orders and maintain stock levels, robots are used for choosing, packaging, sorting, and delivering products. Typically, automated guided vehicles (AGVs)—which are mobile robots that have conveyors, shelves, grippers, and sensors—are used for inventory transportation and fulfillment. They can be linked to a centralized command system that controls orders and inventory information.

- **Inventory tracking and scanning**: Robots are used for inventory tracking and scanning to monitor and control inventory levels, ensure accuracy, and track the movement of items throughout the warehouse. These robots have various sensors, including cameras, barcode scanners, RFID readers, and computer vision systems for monitoring, recognizing objects, navigation, and obstacle avoidance.

Common Robots

Robots used in industrial settings come in many shapes and forms, depending on the environment and tasks being completed[12].

Standalone Industrial Robots

- **Articulated robots:** These robots have rotary joints and between three to six degrees of freedom. For this reason, they are flexible and can rotate in multiple directions. They have a base, arm, wrist, end effector, control system, and sensors/actuators, which allow for controlling movement. Commonly these types of robots are used in assembly and welding. An example of a robot like this is the Kuka KR Quantec series,[6] which are commonly used in manufacturing for welding car bodies.
- **Parallel robots:** These robots have three arms, which are connected to a base platform using universal joints. The arms themselves are placed in a parallelogram configuration, which allows them to move quickly. For

these reasons, parallel robots are ideal for packing and higher precision assembly. ABBs FlexPicker robots are examples of these; they have been used in food processing plants[13] for pick-and-place and to prepare frozen pizzas[14].

- **Gantry robots:** With three linear axes of control placed at 90-degree angles to each other, these robots are known for being simple to control and useful when space is limited. Gantry robots consist of linear actuators, guide rails, and a control system. Güdel's TMF Gantry robots[15] are an example of this; they are commonly used in automotive assembly lines to manipulate heavy parts.

Collaborative Robots (Cobots)

As it says in the name, these robots work with humans and are designed with safety in mind. They are useful in tasks where flexibility and working on ground with humans is important.

- **Assembly:** These cobots can perform tasks such as screwing, fitting, and joining parts, often working alongside human workers.
- **Material handling:** They assist in moving, sorting, and packaging materials, which reduces the physical strain on human workers and increases throughput.
- **Pick-and-place:** These cobots are used to pick items from one location and place them in another, which is useful in logistics and warehousing.

- **Quality inspection:** Using vision systems, these cobots can inspect products for defects. They are very effective at performing routine tasks, which frees up human operators for other needed duties.

  - **Safety:** Using sensors, these cobots can detect contact with humans or objects and stop or slow down in these scenarios to avoid injury. They are often designed to apply little force, which can be helpful in cases where there is unintended contact with a human.

  - **Hardware:** Force sensors are used to detect force and torque applied by the cobot and adjust movement as needed. Cameras and other sensors mentioned in Chapter 2 are used to recognize humans and objects, navigate in their environment, and perform tasks such as quality inspections. Oftentimes, there is some GUI or easy-to-use interface for humans to easily program or interact with the robot in case of changes to the environment or tasks.

A popular cobot is Universal Robots' UR series[16], which is often used in smaller spaces. It is lightweight compared to other robots, is fast, and can tend to different applications easily. These robots are commonly used for welding and stacking objects.

Mobile Robots

There are two common types of mobile robots: Autonomous Guided Vehicles (AGVs) and Autonomous Mobile Robots (AMRs). The main difference is that, while AGVs follow a fixed, predetermined path using tape, wires, or rails in the physical environment, AMRs can use sensors like cameras and LiDAR to navigate their environment more dynamically, without the need for physical guides.

- **AGVs**: These are mobile robots that follow a fixed path and predetermined routes to transport materials from one location to another. They often require modifying the physical environment using guide tracks or markers for them to operate. They are commonly used in warehouses/distribution centers, where you are following the same path to transport an object from one location to another. They are made up of navigation systems that can navigate their environment and then follow those predefined paths, drive motors for movements, and various sensors to detect any obstacles. An example of these are Amazon's Robotics, which usually transport items within their fulfillment centers[17]. These AGVs follow a predefined path to receive and deliver inventory items and reduce repetitive tasks that humans need to do.

- **AMRs**: These robots use cameras, LiDARs, and algorithms to navigate their environment in a more flexible manner. AMRs can be used in logistics and manufacturing to transport materials and pick-and-place items and deliver them within facilities. AMRs have navigation systems with mapping, localization abilities, sensors for detecting and avoiding obstacles, and path planning. Fetch Robotics' AMRs[18] are commonly used in warehouses for picking and transporting items autonomously.

CHAPTER 9   INDUSTRIAL ROBOTICS

Humanoids

Humanoid robots are designed to complete tasks like a human would, with a head, torso, arms, and legs. They are made up of actuators that help move the robots limbs and joints, sensors that are used to provide feedback on the environment, control systems that process sensor data, manage actuators that move, and end effectors that interact with items to perform specific tasks like gripping or manipulating items. Recently the company Figure built humanoid robots and signed an agreement with BMW[19] to use their robots in automotive manufacturing. Similarly, companies like Tesla are building humanoid robots to work alongside humans in their factories[20].

For more information on the hardware and software design of these humanoid robots, we recommend reading Chapter 10.

Market Opportunity

According to a McKinsey report, "The overall automation market is growing rapidly: at least some expert sources expect robot shipments to increase by up to 50 percent each year through 2030, with warehouse automation growing by more than 10 percent per year"[21]. The industrial robotics market worldwide was valued at 54 billion in 2023 and is expected to grow to 142.8 billion[22]. The growth of this market and the applications breakdown are shown in Figure 9-4.

CHAPTER 9 INDUSTRIAL ROBOTICS

Global Industrial Robotics Market
Size, by application, 2022-2032 (USD Billion)

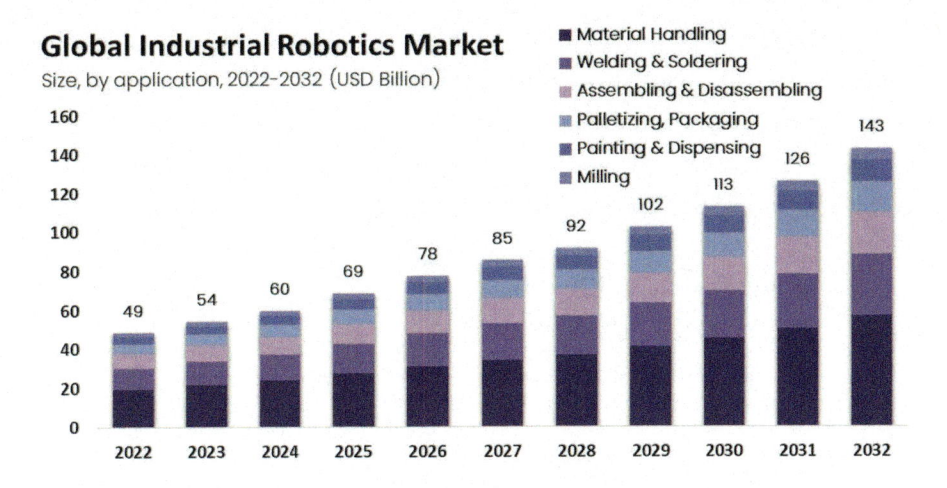

- Material Handling
- Welding & Soldering
- Assembling & Disassembling
- Palletizing, Packaging
- Painting & Dispensing
- Milling

Year	2022	2023	2024	2025	2026	2027	2028	2029	2030	2031	2032
USD Billion	49	54	60	69	78	85	92	102	113	126	143

Figure 9-4. Growth of industrial robotics market and main applications in welding, soldering, assembling, and so on. Used with permission, source: `https://market.us/report/industrial-robotics-market/`*[34]*

There are a few reasons for growing market interest, including:

- **Decreasing prices:** Due to lower production costs in regions with cheaper labor, robots have become more accessible.

- **Robots:** There is an increase in the variety of robots and improved software with greater precision and mobility, which have allowed robots to be deployed across a wider range of manufacturing industries.

- **Labor:** Rising labor costs globally have also made the investment in robotics more attractive because companies are seeking to automate to maintain competitiveness and address skill shortages.

CHAPTER 9   INDUSTRIAL ROBOTICS

According to recent data from Statista's Market Insights, ABB leads the market with a 21 percent global market share in 2022, KUKA and Japan's Kawasaki make up 9 percent share, and other important players are Fanuc with an 8 percent share, Mitsubishi, and Yaskawa, who have 5 percent of the market each[23]. A breakdown of this is shown in Figure 9-5.

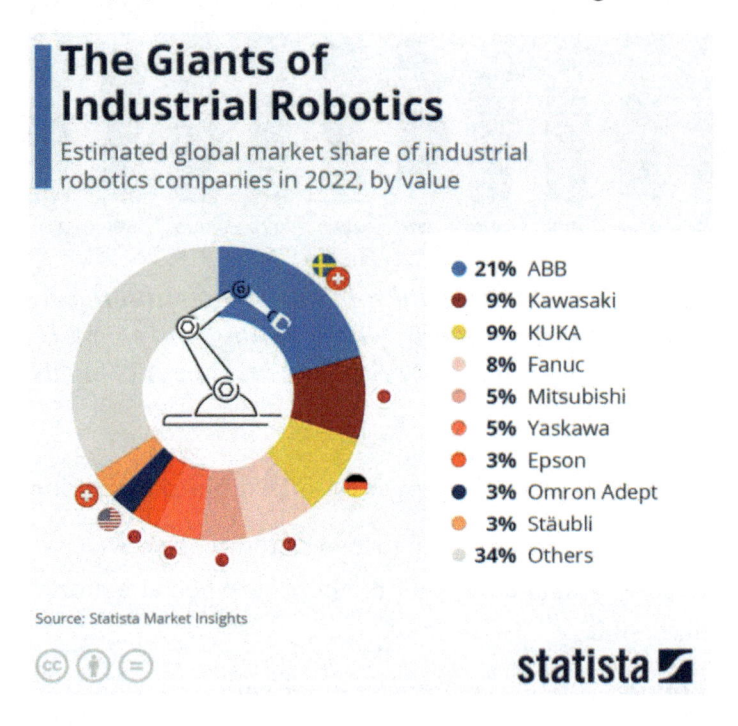

Figure 9-5. *Companies that produce robots commonly used in industrial tasks. Used with permission, source:* https://www.statista.com/chart/32239/global-market-share-of-industrial-robotics-companies/ *by Statista[23]*

These robotic manufacturing companies design the arms or platforms that are deployed by other companies. There are also companies like Boston Dynamics, which leverages deep learning methods for their robots, including their robot, Spot[24], which can be used for industrial inspections.

CHAPTER 9 INDUSTRIAL ROBOTICS

System Design for Pick-and-Place Robots

Now that you are familiar with the common tasks in industrial robotics, this section explains how to design a system to accomplish these tasks, specifically the task of pick-and-place. The pick-and-place task is essential because it is a specific task that is usually part of accomplishing larger manufacturing goals, such as assembling components, packaging products, and sorting materials.

The most common pick-and-place robots[25][26] include:

- **Robotic arm**: This five-axis robotic arm is used for most common pick-and-place tasks that involve picking up and dropping an object in a single plane. To rotate objects, you need a six-axis arm or seven-axis arm to turn objects before dropping/placing them somewhere.

- **Cartesian**: These robots can move in x, y, and z axes and use linear actuators to control belts, balls, and so on, which can help with positioning. These robots are often used for material handling, CNC operations, electronics, and food because they are reliable and provide more precise manipulation.

- **Delta**: They have three or four arms that are connected on a base and provide speed with pick-and-place tasks.

Hardware Components

The main components of pick-and-place robots include the arm, end effectors, control system, and sensors[25].

CHAPTER 9 INDUSTRIAL ROBOTICS

- **Arm:** The arm itself can range from five-axis designs for basic tasks, to six-axis and seven-axis arms, which can help perform more complex rotations and orientation changes. The arm has a combination of motors and actuators, which provide control and sensors used for managing pressure/force.

- **End effectors:** An end effector is attached to the end of the robot arm and is the part of the arm that is used to grip, hold, and manipulate objects using picking and placing. A few common end effectors include:

  - *Vacuum grippers:* These grippers create a vacuum that allows you to lift objects. This gripper is ideal for objects that are relatively flat, such as glass and metal sheets. These grippers are commonly used in electronics and packing for these reasons.

  - *Mechanical grippers:* These can be two or three finger grippers that are usually designed to handle different shapes and sizes of objects. Often these grippers are used for picking and placing in automotive parts and consumer electronics. As they are able to manipulate different objects, they are commonly used for gripping and manipulation.

- **Control system:** This is the "brain" of the pick-and-place robot because it is used to manage the movements and actions that the arm takes. Inputs are taken from sensors and commands are sent to actuators to execute actions. Some common controllers include:

CHAPTER 9   INDUSTRIAL ROBOTICS

- *PID:* These controllers continuously adjust the robot's movement based on feedback from the environment. They use three components: the proportional component, which corrects for differences when the robot goes off path by applying a correction proportional to the error, the integral component, which removes common recurring errors by accumulating past errors and adjusting the path to correct this offset, and the derivative component, which predicts change and adjusts position so that you don't miss the target position.

- *ML:* Neural network based controllers are trained on a larger dataset of robot movement and outcomes. The robot can learn various different pick-and-place actions and ideally adapt to any environment that it operates in. In reality, training these models requires extensive data and can be difficult to generalize. A lot of research is working to improve this through simulation data and agent-based learning through RL.

- **Sensors:** Vision sensors and force sensors are typically used in pick-and-place tasks. Vision sensors help identify and locate objects before picking them up and force sensors ensure that the robot grips objects with an appropriate amount of force.

All these components interact together to accomplish a pick-and-place task. Sensors are used initially to understand the environment, control systems provide a command to execute on the task based on where the initial and target location are sensed, and sensors are used during grasping to ensure enough grip is applied to execute the task.

CHAPTER 9 INDUSTRIAL ROBOTICS

Software Components

Over the years, there have been many deep learning methods developed to perform grasping. Why is deep learning even good for a task like pick-and-place?

Given the repetitive nature of pick-and-place tasks, many demonstrations can be collected. This provides a diverse dataset to train deep learning models. These models need to generalize to various object morphologies and environments, which is challenging for traditional methods. For example, handling soft objects requires being delicate to avoid damage, and deep learning models can learn the specific ways to grasp such objects and ensure they land correctly in their designated places. This adaptability and precision make deep learning an ideal approach for improving the efficiency and accuracy of pick-and-place tasks in changing and different visual environments. A common approach is using Convolutional Neural Networks (CNNs), which was explained in Chapter 2 to execute grasping tasks.

Convolutional Neural Networks for Grasping

A popular method that uses CNNs for grasping was presented in this paper[27]. As input, they take in an image that is captured before the robot attempts to grasp any object (without gripper) and an input image that is captured at some specific time t during the grasping process. These images are combined and processed using multiple layers of convolutional filters. The idea is for each layer to extract different features from the image, including edges, shapes, textures of the environment, and specific objects to grasp. The filter sizes are fairly small (3x3, 5x5, etc.) and many are applied per layer. After convolutional layers, there are pooling layers to reduce the size of feature maps and fully connected layers toward

CHAPTER 9 INDUSTRIAL ROBOTICS

the end that takes in the flattened feature maps as a vector and makes predictions. Here, the network outputs a probability indicating the success of grasping an object. An important aspect of this approach is that it takes in a proposed motor command (describing planned movement of the robotic arm) and considers this alongside the input images of the scene to determine the success of the grasp. This architecture is shown in Figure 9-6.

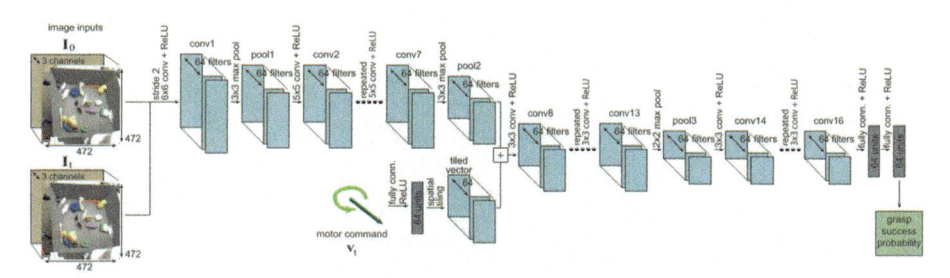

Figure 9-6. *Architecture used for CNN for robot grasping in [27]. Used with permission, source: https://journals.sagepub.com/doi/full/10.1177/0278364917710318[27]*

This method uses training data from real physical robots that attempt grasps, records images, poses, and movements at each time step, and then evaluates the success of the grasp to produce labeled training samples. Each sample includes the image, the current pose to the final pose, and the overall grasp success. The setup for collecting grasping training data is shown in Figure 9-7.

CHAPTER 9   INDUSTRIAL ROBOTICS

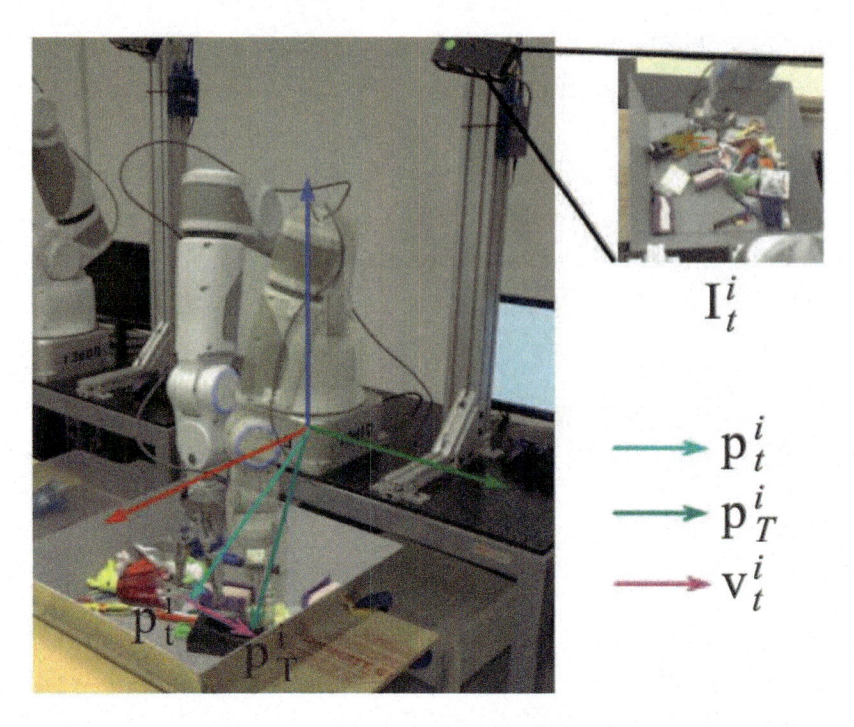

Figure 9-7. Grasp setup includes multiple timesteps, which correspond to an image and a pose. Used with permission, source: https://journals.sagepub.com/doi/full/10.1177/0278364917710318[27]

Another important aspect of this method is the serving mechanism. It takes the trained network from the previous step to determine motor commands, which will improve the likelihood of grasping success. Conceptually, this part of the method is optimizing the commands through a sampling-based approach, making sure while staying within constraints, it is adapting the gripper's position based on predicted success probabilities. In the end, this approach is able to effectively grasp objects, even in cluttered environments.

A lot of work has built on this or taken other directions using CNNs for grasping, but fundamentally all of these do the same thing—they use CNNs to process images of the scene, extract relevant features, predict the optimal motor commands for grasping, and then evaluate the probability of these commands to successfully guide the robot's actions.

Scaling RL for Robotic Grasping

Reinforcement Learning (RL) has shown success in individual robotic applications like grasping and manipulation. One of the key challenges in scaling up these systems is ensuring that they can reliably be used in everyday life. A paper by Google[28] tackled this problem by getting 23 RL-enabled robots over two years in their offices to sort waste/recycling and then collect this data. Their system combined deep RL from real-world data with simulation data to accomplish grasping. Although this was done in the context of trash sorting, it acts as a case study of how ideas you've learned throughout this book on simulation, RL, and perception methods can be combined to solve real-world robotic manipulation challenges.

Their pipeline uses three main components:

- A scripted policy where they detect objects, plan the grasp pose, and generate the trajectories to reach the object as a way to start collecting data in simulation and the real world.

- The RL training starts in simulation and simulation-to-real methods are used to bridge the gap between simulation and the real world.

- They start training on tasks with varying levels of difficulty, with sorting trash being the hardest task. Initially the tasks start out easier and there's higher success, which gradually turns into harder tasks. By quickly introducing easier tasks at the start of the training, the aim is for the model to learn and be able to adapt to more challenging tasks.

CHAPTER 9   INDUSTRIAL ROBOTICS

PI-QT-Opt[29] is used to train the final policy on the full dataset that is collected from simulations and real world. The Q-function architecture takes in an RGB image of the scene that has unsorted objects and the masks of the shapes of the objects in the scene are generated. Features are extracted from the images and masks using convolutional layers. The features from the masks and the original image are combined. One encoder, called the forward encoder, is used to process the current image and the background encoder processes the next image as a way to predict future states. An LSTM layer is used to keep context overtime on observations and actions. The Q-function uses the current state and action to predict future rewards and uses the most recent six time steps. The final output is given as a predicted value, which tells you the expected success of a particular action. This architecture is shown in more detail in Figure 9-8.

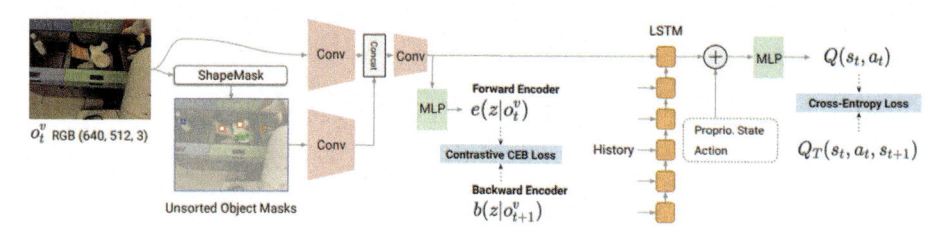

Figure 9-8. Overview of PI-QT-Opt system used to sort trash. Used with permission, source: https://rl-at-scale.github.io/assets/rl_at_scale.pdf[28]

An important aspect highlighted in this work is the need for large-scale data collection and the combination of online and offline data to improve model performance for robotic grasping.

Multimodal Grasping System

An interesting direction to build a foundation model for robotics was presented by Covariant AI, in their method called RFM-1[30]. This method is particularly suited for manufacturing, logistics, and warehouse

CHAPTER 9 INDUSTRIAL ROBOTICS

applications, as it can adapt to changes and collaborate and communicate with human workers. An important aspect is that Covariant has been able to collect diverse data of their robots performing tasks in challenging dynamic environments with many different object shapes and types, as shown in Figure 9-9. A few unique aspects of RFM-1 include:

- **Multimodal:** The model itself is a transformer that tokenizes different modalities such as text, sensor data, and camera data into a shared space, which allows for next-token prediction. This means the model can take in any input modality and output in any modality as well.

- **Physics understanding:** Using a technique that allows for learning world models, the method can gain a deeper understanding of physics, which can help with performing realistic actions and reacting to changes in the environment. This is done by generating videos where input tokens of an initial image and robot actions are used to predict future video tokens.

- **Language re-programming:** A big challenge in robotics is re-programming robots when changes occur. Oftentimes, this is a difficult task that only the person who initially programmed the robot can successfully do. Through natural language commands, a person can instruct a robot to do a task and the robot can ask for help when it needs it. This back-and-forth dialogue is useful in the real world, where environmental changes require changing the task or re-programming the robot in any way.

CHAPTER 9   INDUSTRIAL ROBOTICS

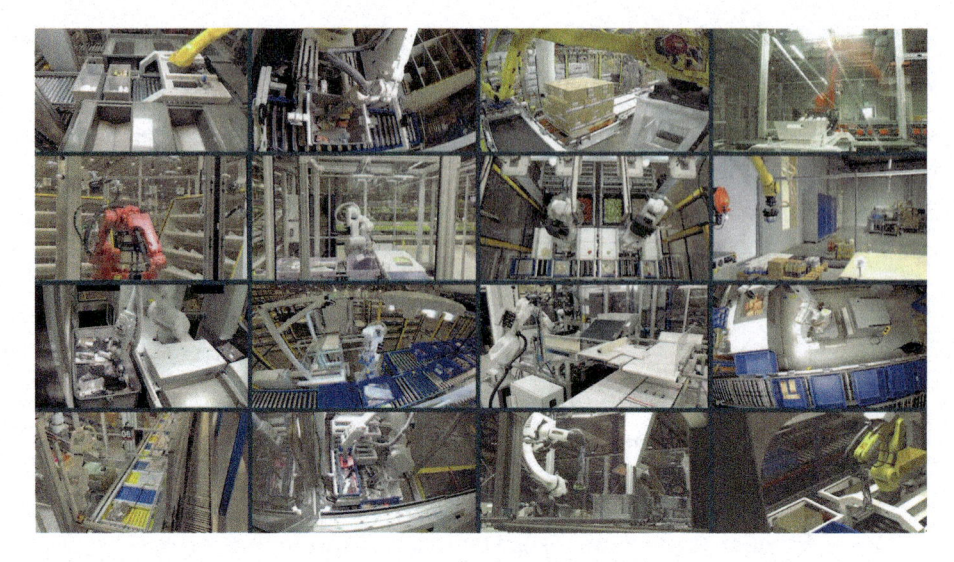

Figure 9-9.   RFM-1 system by Covariant AI performing various tasks. Used with permission, source: `https://covariant.ai/insights/introducing-rfm-1-giving-robots-human-like-reasoning-capabilities/[30]`

This method shows the power of large-scale data collection and benefits that multimodal data can bring to improving interactions with robots in pick-and-place environments. To learn more, we recommend reading this release[30] on RFM-1.

Performance Metrics

The success of deep learning-powered pick-and-place robots is primarily measured by a variety of performance, efficiency, and reliability metrics. These metrics help assess how well the robots do the tasks they were designed for. A few important metrics include:

- **Accuracy of object detection:** How accurately a robot can detect objects is crucial because it determines the robot's capacity to recognize and categorize things in its surroundings. High object detection accuracy improves the ability of the robot to work efficiently within diverse environments.

- **Grasping success rate:** The robot's capacity to pick and move items without dropping them or damaging them should be measured. Pick-and-place operations must be reliable and precise, which requires algorithms that have a high grasping success rate.

- **Cycle time:** The robot's time to execute a full pick-and-place task, including detecting the object, planning, grabbing, and placement, is measured in terms of cycle time. Reducing this cycle time is important for improving output and productivity in industrial settings.

- **Collision avoidance rate:** The robot's ability to avoid hitting objects or obstacles while navigating its surroundings is measured by its collision avoidance rate. For safety and effectiveness, reducing collisions and near-miss accidents is critical.

- **Task completion rate:** This measures the proportion of pick-and-place tasks that were successfully completed out of all tasks that were tried. It measures how dependable and efficient the robot is in completing the tasks that are assigned to it.

CHAPTER 9   INDUSTRIAL ROBOTICS

Safety Considerations

When employing deep learning for industrial robotics, safety must always come first in order to safeguard people and avoid accidents. Here are some safety points to remember:

- Use safety sensors to keep an eye on the robot's surroundings and spot any sudden movements or obstacles. When required, these sensors can initiate emergency stops.

- When using cobots, make sure they follow safety regulations by designing safety features like speed monitoring and force-limiting joints to lessen the possibility of accidents when interacting with people.

- Implement software capabilities that are linked to safety, such as path tracking and collision detection. These features can help the robot identify and react to unforeseen objects or changes in its surroundings.

Summary

In this chapter, you learned the following:

- Industrial robots perform tasks such as pick-and-place, peg-in-hole, welding, and warehouse operations. Deep learning allows robots to adapt to their environments, improving their ability to identify objects, plan paths, and execute tasks in unstructured environments.

- Industrial robots come in various forms depending on their design and application. Standalone robots include articulated robots with rotary joints, parallel robots for

CHAPTER 9 INDUSTRIAL ROBOTICS

high-speed assembly, and gantry robots. Collaborative robots (cobots) work alongside humans with safety features like force-limiting joints. Mobile robots, such as AGVs and AMRs, handle inventory movement, while humanoid robots mimic human movements.

- Robotics in manufacturing facilities is expanding due to decreasing costs, improved capabilities, and the need to automate complex tasks.

- The chapter explained a system design for pick-and-place robots that are useful for assembly, sorting, and packaging. The design includes a robotic arm equipped with various end effectors, like grippers and vacuum cups. Deep learning is used to improve grasping accuracy through computer vision and reinforcement learning, which allow the robot to adapt to diverse environments and objects. Multimodal systems use various sensory data, which can be useful for the robot to process visual and tactile information.

- Performance metrics like object detection accuracy, grasping success rate, cycle time, collision avoidance, and task completion rate are crucial for improving the efficiency and reliability of industrial robots in real time. Ensuring safety is equally important by requiring robots to have sensors for collision detection, emergency stops, and force-limiting joints to prevent accidents when working alongside humans.

The next chapter covers the hardware and software setup for humanoid robots, including approaches for manipulation, walking, teleoperation, and whole body manipulation.

CHAPTER 9 INDUSTRIAL ROBOTICS

References

[1] Quinlivan, Joseph. "How Amazon Deploys Robots in Its Operations Facilities." US Amazon, 26 June 2023, www.aboutamazon.com/news/operations/how-amazon-deploys-robots-in-its-operations-facilities.

[2] https://www.fanucamerica.com/solutions/industries/automotive-robots

[3] https://new.abb.com/products/robotics

[4] https://new.abb.com/products/robotics/robots/articulated-robots

[5] https://www.fanucamerica.com/products/robots/series/m-20/m-20ia-35m

[6] https://www.kuka.com/en-ca/products/robotics-systems/industrial-robots/kr-quantec

[7] Liu, Gene. "Tesla Factory Robots Named after X-Men Superheroes." TESLARATI, 18 Nov. 2014, www.teslarati.com/tesla-factory-upgrade-facts-figures/.

[8] Black, Thomas. "GM Hooking 30,000 Robots to Internet to Keep Factories Humming." *Bloomberg*, 4 Apr. 2017, www.bloomberg.com/news/articles/2017-04-04/gm-hooking-30-000-robots-to-internet-to-keep-factories-humming?embedded-checkout=true.

[9] https://www.robots.com/articles/how-toyota-uses-automation-to-improve-processes

[10] https://www.fanucamerica.com/products/robots/series/arc-mate

[11] https://new.abb.com/products/robotics/robots/articulated-robots/irb-6700

CHAPTER 9   INDUSTRIAL ROBOTICS

[12] "Industrial Robotics." McKinsey & Company, July 2019, www.mckinsey.com/~/media/mckinsey/industries/advanced%20electronics/our%20insights/growth%20dynamics%20in%20industrial%20robotics/industrial-robotics-insights-into-the-sectors-future-growth-dynamics.ashx.

[13] "Fine Foods Packing Plant Uses ABB Cobots to Keep Ahead of Competition." *ABB News*, ABB Group, 7 Apr. 2022, new.abb.com/news/detail/89351/prsrl-fine-foods-packing-plant-uses-abb-cobots-to-keep-ahead-of-competition

[14] Labs, Wayne. "Robotic Vision Systems and Roles for Cobotics." *Food Engineering RSS, Food Engineering*, 3 Aug. 2020, www.foodengineeringmag.com/articles/99045-a-further-look-at-robotic-vision-systems-and-roles-for-cobotics

[15] https://www.gudel.com/products/robots/gantry-robot

[16] https://www.universal-robots.com/products/

[17] https://www.waredock.com/magazine/what-is-amazon-robotic-fulfillment-center/

[18] https://www.zebra.com/us/en/products/autonomous-mobile-robots.html

[19] https://www.prnewswire.com/news-releases/figure-announces-commercial-agreement-with-bmw-manufacturing-to-bring-general-purpose-robots-into-automotive-production-302036263.html

[20] Carter, Tom. "Tesla Has Put 2 Optimus Robots to Work on Its Factory Floor." *Business Insider,* 12 June 2024, www.businessinsider.com/tesla-says-two-optimus-humanoid-robots-working-in-factory-autonomously-2024-6.

[21] Davies, Alan, et al. "Getting Warehouse Automation Right." *McKinsey & Company,* McKinsey & Company, 1 Dec. 2023, www.mckinsey.com/capabilities/operations/our-insights/getting-warehouse-automation-right

[22] https://market.us/report/industrial-robotics-market/

[23] Fleck, Anna. "Infographic: The Giants of Industrial Robotics." *Statista Daily Data,* 13 May 2024, www.statista.com/chart/32239/global-market-share-of-industrial-robotics-companies/.

[24] https://bostondynamics.com/products/spot/

[25] Moraes, Cassiano Ferro. "Pick and Place Robots: An in-Depth Guide to Their Functionality and Applications." *Wevolver,* 25 Mar. 2024, www.wevolver.com/article/pick-and-place-robots-an-in-depth-guide-to-their-functionality-and-applications

[26] "Everything You Need to Know about Pick and Place Robots." *Robotic Automation Systems,* 30 June 2023, www.roboticautomationsystems.com/blog/everything-you-need-to-know-about-pick-and-place-robots/

[27] Levine, Sergey, et al. "Learning hand-eye coordination for robotic grasping with deep learning and large-scale data collection." *The International Journal of Robotics Research* 37.4-5 (2018): 421-436.

[28] Herzog, Alexander, et al. "Deep RL at scale: Sorting waste in office buildings with a fleet of mobile manipulators." *arXiv preprint arXiv:*2305.03270 (2023).

[29] Lee, Kuang-Huei, et al. "PI-QT-Opt: Predictive information improves multi-task robotic reinforcement learning at scale." Conference on Robot Learning. PMLR, 2023.

[30] "Introducing RFM-1: Giving Robots Human-like Reasoning Capabilities." *Covariant*, 11 Mar. 2024, `covariant.ai/insights/introducing-rfm-1-giving-robots-human-like-reasoning-capabilities/`

[31] "Automation, Upgraded: Robotic Induction." Covariant.ai, 2022, `covariant.ai/insights/automation-upgraded-robotic-sorter-induction/`.

[32] Zhu, Zuyuan and Huosheng Hu. "Robot learning from demonstration in robotic assembly: A survey." *Robotics* 7.2 (2018): 17.

[33] Phan, Gia-Hoang. "Integrating long short-term memory for optimal control of 6-DOF welding robot arm." *Advances in Mechanical Engineering* 16.6 (2024): 16878132241260525.

[34] "Industrial Robotics Market." Market.us, `market.us/report/industrial-robotics-market/`

CHAPTER 10

Humanoid Robotics

Humanoid robotics concerns the study of robots that assume a human form factor. The term is loosely applied to robots that try to match a human form factor in various ways, and exact matching is not required. Usually, this means mobile robots (walking and sometimes on wheels), with human height (5 to 6 feet, although sometimes only as tall as children), and bimanual platforms, with grippers or multi-fingered dexterous hands for end effectors. Actuating a full humanoid—with legs, multi-fingered hands, and torso and head mobility—may be harder, for which reason roboticists may opt to simplify the platform in various ways (wheels instead of legs, grippers instead of hands, etc.).

The Case for Humanoids

There are several reasons why humanoid robotics is a popular strategy in robotics:

1. **The physical world is often designed for the human form factor.** Cups handles are made to be manipulated by fingers, coke cans fit conveniently in hands, kitchen counters are at heights accessible by humans, and so on. This makes a case for why having a human form factor may allow a robot to perform lots of different tasks in environments

designed for humans without modification, and theoretically could reach human proficiency in physical skills.

2. **Internet scale models transfer general intelligence to robotics.** However, most of the data powering Internet scale models is the Internet itself and the Internet is a collection of human experiences. Optimal control policies may differ based on the bodies and intelligence may transfer better from a human body to a humanoid form factor than to other form factors, because the embodiment gap is smaller.

3. **A lot of valuable physical labor in the world is performed by humans.** Humanoids point to an opportunity to supplement labor markets and provide economic value without the need to invent new markets.

Alternative Approaches

A humanoid approach to robotics is to a greater degree vertically integrated. This contrasts the extreme cross-embodied approach, which builds general physical intelligence that may transfer to any robot body. It is likely that solving robotics necessitates both approaches:

1. End-to-end ownership on deployment and data with vertical integration and optimization of a robot body.

2. Large cross-embodied foundation models that provide general intelligence and reduce data requirements to train a well actuated body.

Humanoids are also general bodies that can perform a wide variety of tasks with the same body. This contrasts the approach of building specialized robots for specialized, repeatable tasks. Most robots deployed in industries today are specialized robots:

1. Task specialization via general-purpose manipulator arms; for example, a generic ABB arm[1] assembling Printed Circuit Boards (PCBs).

2. Embodiment specialization for tasks; for example, snake robots for pipe inspections.

The humanoid strategy cuts across task and embodiment specialization toward generalization.

Humanoid Markets

The humanoid space has multiple players at the moment, as shown in Figure 10-1:

1. **Intelligence providers:** Companies that ship foundation models, like Google[2], OpenAI[3], Mistral[4], Meta[5], and so on. These companies are building foundation models used for humanoid development, have teams that are doing humanoid research, or they are funding/partnering with companies to conduct humanoid research.

2. **Hardware providers:** Companies that build humanoid hardware, such as Boston Dynamics[6], Fourier Intelligence[7], Unitree[8], Apptronik[9], and Agility[10].

CHAPTER 10   HUMANOID ROBOTICS

3. **Fully integrated providers:** Companies that solve intelligence and hardware: Tesla[11], Figure AI[12], 1x Technologies[13], and so on.

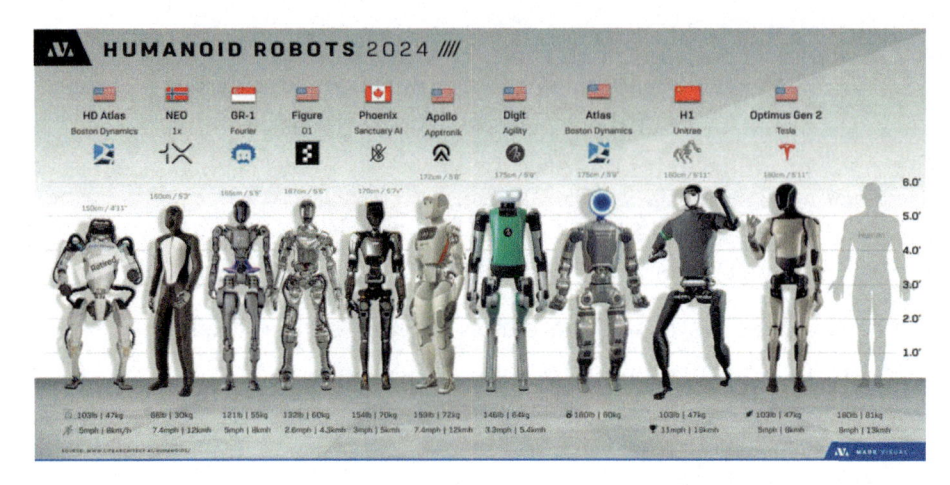

Figure 10-1.   The humanoid hardware market. Used with permission, source: https://lifearchitect.ai/humanoids/ by https://lifearchitect.ai/[32]

These companies share several key ingredients that have contributed to their success in humanoid development, including[14]:

1. **Capital:** Creating advanced robots can be expensive, so having funding to support research, development, and production of humanoid robots is essential. Given the impact and growth of this industry, investors are increasing funding toward humanoid companies[15]. For example, Figure AI, a California-based startup, raised $675 million in a Series B round including LG, Samsung, and Microsoft[15]. Similarly, Sanctuary AI, based in Vancouver, has raised significant funds, including recent investments from Accenture Ventures and Magna[15]. These funds will be used to further

CHAPTER 10   HUMANOID ROBOTICS

their work in general-purpose humanoid robots. The growing investment in humanoid technology highlights the increasing belief in their potential and the resources needed to bring these robots to market.

2. **Foundation Models:** Large-scale AI models that enable humanoids to understand and learn from vast amounts of data are critical. These foundation models demonstrate the cognitive abilities of humanoid robots, allowing them to perform tasks autonomously and adapt to new situations. Further, these large-scale generalized models form the basis to be fine-tuned for specialized tasks.

3. **Data:** Generally, the more data a robot has access to, the better it can learn to perform tasks, recognize objects, and understand human behavior. For humanoids, this data can include anything from images and videos to text and sensor data from real-world environments.

4. **Robots:** The physical construction of humanoid robots involves hardware and biomechanics such as sensors, and actuators to create physical machinery that can mimic human actions. Companies like Boston Dynamics and Tesla invest heavily in this field, developing robots that can lift heavy objects or perform other complex movements with agility.

5. **Compute:** As you scale model size and data, having compute infrastructure is critical to support the training and testing of models on a vast amount of data.

409

CHAPTER 10   HUMANOID ROBOTICS

The market for humanoid robots is projected to grow significantly, with an estimated value of $4.85 trillion by 2035[14]. This is mostly driven by the increasing adoption of humanoid robots in industrial and household settings. In industries like manufacturing, humanoid robots are expected to take over about 35 percent of current human tasks, which would contribute to a $1.75 trillion market[14]. For households, these robots could become common for tasks like cleaning and elderly care and grow to a $2.8 trillion market[14]. As more advancements happen in tech and prices drop, humanoid robots are likely to follow a similar adoption path as electric vehicles, allowing them to become a daily part of our life and improving industries and homes.

How to Build a Humanoid

This section contains a rough overview of the techniques used to build and control a humanoid.

Hardware

Hardware design choices [16][17]:

1. **Mobility:** There are two commonly accepted approaches to humanoid mobility:

    a. *Legged humanoids:* These humanoids are equipped with legs for walking. The advantage of this method is that the humanoid can traverse multiple terrains, environments with curbs or steps, and so on. A second advantage is that the footprint of a legged humanoid is smaller, and its dexterity toward whole body manipulation is higher. For example, it can traverse cluttered spaces and orient its legs in ways that distribute moment for lifting heavy objects, manipulating far away objects on tables, and so on, which could increase the workspace and payload of a humanoid. (When humans lift a heavy box, they often take a wide stance or put one leg in front to distribute the load.) Downsides include the fact that walking is a harder control problem and it introduces an instability into the system. This makes failures more catastrophic (if the humanoid falls, it could damage itself, surrounding objects, and cause injury). An interesting design choice for legs is the reverse knee from agility[18].

b. *Wheeled humanoids:* The lower torso of the humanoid is a wheeled platform, similar to Eve from 1X technologies[19]. The main advantage of this design choice is that stabilizing the robot is easier. Wheels also allow the robot to be heavier since distributing the weight doesn't require specific control. A downside is that the robot may be useless in harder terrains and uneven environments. Additionally, its ability to move through cluttered spaces and do whole body manipulation might be limited, therefore limiting its workspace and flexibility.

2. **End effectors:** Humanoids may be equipped with a variety of end effectors:

a. *Grippers:* Grippers are commonly used in robotics and are rather easy to control due to the lower degrees of freedom. This makes them a popular choice for humanoids.

b. *Dexterous hands:* Humanoids may also come equipped with multi-fingered hands. Three-fingered and five-fingered hands are popular choices. Each finger adds additional degrees of freedom, which then must be modeled and controlled, adding to the complexity of control.

   c. *End effectors with tactile feedback:* Force feedback on hands allows for proprioceptive control, in addition to visual, and is useful for manipulating deformable and fragile objects. There is a tradeoff with increased cost.

3. **Wrist and head camera vs head camera only:** It has been shown that incorporating wrist-mounted cameras can improve manipulation performance compared to using only head-mounted cameras[20]. Having wrist cameras provides additional visual feedback, improving spatial awareness and accuracy in teleoperation and human-robot interactions.

4. **Camera only vs camera and stereo/LiDAR depth:** Camera-only systems are simpler and cheaper but face challenges in depth perception and object recognition. Advanced ML techniques can help overcome some of these limitations. Most humanoid designs on the market do not use LiDAR to keep the bill of materials smaller. Therefore, methods using only a camera as a sensor are preferred and may require more algorithmic development.

5. **Linear/rotary actuators, hydraulic actuators:** Linear/rotary actuators are more accurate than hydraulic ones. It is also messier to develop with hydraulic actuators because of hydraulic systems maintenance and potential fluid leaks, and so on. As such, most humanoid companies, such as Boston Dynamics, are moving to fully electric motors and away from hydraulics[21].

A humanoid robot may have the following control parameters:

1. **Head/hip position:** This parameter controls the vertical and horizontal positioning of the head and hips, which allows the robot to orient itself and interact with its environment from various angles.

2. **Head rotation:** This parameter controls the rotational movement of the robot's head, which allows it to look in different directions. Head rotation is important for focusing on specific objects or areas.

3. **Shoulder extensions/rotations:** These parameters manage the extension and rotational movements of the robot's shoulders. They position the arms correctly for various tasks like reaching or lifting objects.

4. **Elbow flex:** This parameter controls the bending and straightening of the robot's elbows. Elbow flexion is useful for adjusting arm length and positioning during manipulation tasks.

5. **Wrist joint rotations:** This parameter monitors the rotational movements of the wrist joints. Wrist joint control is important for fine motor tasks.

6. **Finger joints/gripper open-close:** This parameter controls the movement of individual finger joints or the opening and closing of the robot's gripper. It allows the robot to grasp, hold, and release objects with different degrees of force and precision.

7. **Torso joint rotations/extensions:** This parameter monitors the rotational and extension movements of the robot's torso. It allows for adjustments in the robot's upper body posture, which is important for maintaining balance and reaching different areas.

8. **Hip joints:** These parameters control the movement of the robot's hips, like rotation and tilt. Hip joint control is important for walking, sitting, and adjusting the way the robot stands.

9. **Knee flexions:** This parameter controls the bending and straightening of the robot's knees. Knee flexion is used during walking, squatting, and maintaining stability when doing activities.

10. **Ankle joint rotations:** This parameter manages the rotational movements of the robot's ankles. It helps the robot adapt to uneven surfaces and maintain balance while standing or moving.

11. **Foot positions:** This parameter controls the placement and orientation of the robot's feet. Adjusting foot positions is important for walking stability, posture, and navigation over different terrains.

While most people expect humanoids to be modeled after the human form factor, humanoid design can happen in a way where they've fewer or greater control parameters. The argument for fewer degrees of freedom is that it is simplified and easier to control. The argument to make it more complex is that machines are not bound to the limitations of the human form factor: that they can have full rotations at various joints. A few companies experimenting with the latter are Booster Robotics[22] and Boston Dynamics.

Software

This section describes the software stack of a typical humanoid and approaches to learning or scripting humanoid control.

Approaches to Manipulation

Humanoid manipulation is a problem of bimanual dexterous manipulation and various control algorithms and techniques, discussed in Chapters 4 and 5. As such, large-scale deployment of humanoids beyond demos is still an unsolved problem, which points to unsolved research questions on generalizable ways to control a full humanoid. Imitation-learned approaches with scalable data generation are common in research, a recent example being Humanoid Shadowing Transformer[23], which takes images from humanoids' cameras and their proprioception (joint positions) as inputs and uses a decoder only transformer to learn control parameters from imitating humans. An overview of this architecture is shown in Figure 10-2.

CHAPTER 10   HUMANOID ROBOTICS

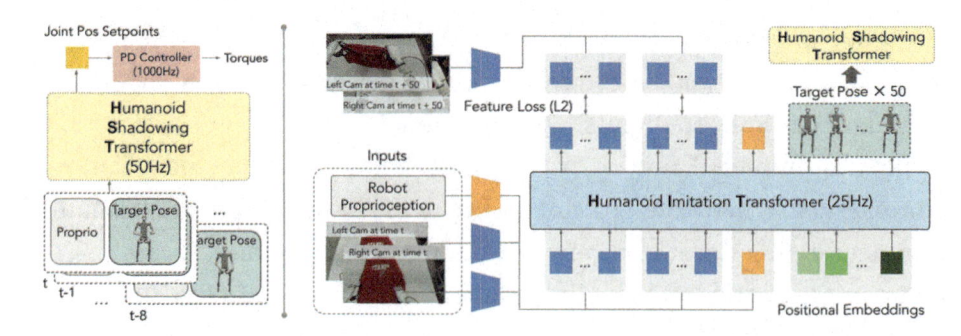

Figure 10-2. *Architecture of the humanoid shadowing transformer. Used with permission, source:* `https://arxiv.org/html/2406.10454v`. *HumanPlus: Humanoid Shadowing and Imitation from Humans by Zipeng Fu, Qingqing Zhao, Qi Wu, Gordon Wetzstein, and Chelsea Finn at Stanford University[23].*

Learning from large video data is still the Holy Grail in terms of unlocking general-purpose humanoid intelligence and the methods covered in Chapters 4 and 5. Using simulation data should point in that direction. Human manipulation data has the lowest embodiment gap to humanoid control, which makes transfer more likely and easier than for any other embodiment due to structural similarity of policies.

Approaches to Walking

Traditional approaches to solving walking included scripted approaches that did not utilize machine learning. Since this book focuses on machine learned approaches, this section covers some new and experimental approaches to walking. A notable recent approach is from Radosavovic, et al.[24], which treats humanoid walking as a next token prediction problem. They train a walking neural network controller on many different types of data, where some include action outputs and some don't. The four types of datasets they train on are:

CHAPTER 10   HUMANOID ROBOTICS

1. **Neural network policies:** Observation action pairs generated in sim by an RL policy trained in sim.

2. **Data from model based controllers:** Trajectories without actions, generated by a humanoid company's controller (in this case, Agility Robotics).

3. **Mocap (motion capture) data:** Data of humans with markers on their bodies, such as KIT[25]. This data was retargeted to the humanoid by using the humanoid's inverse kinematics model.

4. **YouTube data of human poses:** YouTube videos of humans doing things, with pose estimation applied to the videos and then retargeted onto humanoids with the inverse kinematics model.

In order to train on data with lots of different modalities, the authors use a mask token, initialized as a random vector to replace a missing modality (e.g., action). They also use casual masking, meaning that each token only attends to previous inputs. Figure 10-3 shows a schematic of this. Due to these methods and the data sources used, they are able to outperform reinforcement learned state-of-the-art methods and perform zero-shot walking in unseen scenarios. Training in this manner exhibits scaling with data, larger context lengths, and model size, indicating that very large models with large datasets and generic capabilities may be built by scaling strategies similar to ones deployed in large foundation model training.

417

CHAPTER 10   HUMANOID ROBOTICS

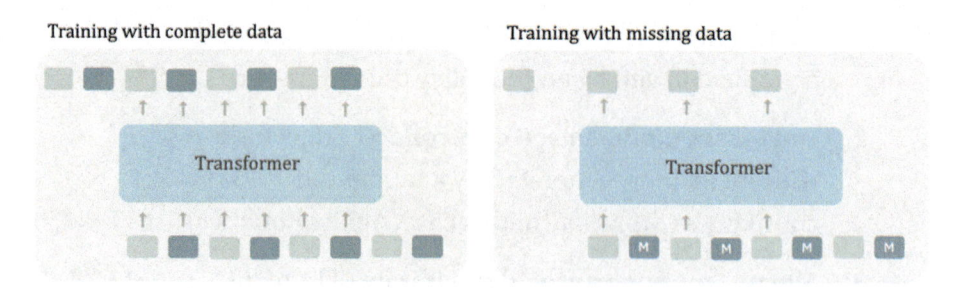

Figure 10-3. Training with missing data. Used with permission, source: https://arxiv.org/pdf/2402.19469[24]

Approaches to Teleoperation and Data

One way to control humanoids is through teleoperation, which provides many ways to improve the functionally of humanoids. Teleoperation techniques for humanoid bimanual operation include several design choices that may be relevant:

1. **VR headset teleoperation vs line-of-sight:** In a VR headset teleoperation mode[29], the operator who controls the humanoid is only subjected to the images from the humanoid's camera. This ensures that the observation for training learned policies have as much information to do the task as the operator does. Since the operator can succeed at the task, one can assume that a minimum set of features to finish the task is captured in the dataset. However, a robot's camera stream may have several downsides, including inaccurate depth perception compared to humans. This would lower the success of tasks and the throughput of data collection for imitation learning. One way to deal with this inaccurate depth perception is by doing line-of-sight

CHAPTER 10 HUMANOID ROBOTICS

teleoperation, such as in ALOHA[26] and RT_1[28]. While line-of-sight teleoperation is very natural for humans, it may not capture head movements (i.e., information about where to look/where to direct the camera). As such, line-of-sight teleoperation is more useful for stationary setups and operation on table tops than for mobile robots or whole body manipulation.

2. **Puppeteering vs VR:** In a puppeteering setup, there are two sets of robot arms—one is the leader and the other the follower (the puppet). In practice, data from puppeteering is cleaner because the leader and follower are often identical, as in ALOHA[26], and the leader is physically moved by a teleoperating human. The movement and similarity of the two arms ensures that the targets conceived by the leader are achievable by the follower. In a VR setup, targets set by a VR controller operated by a human are retargeted on to the robot and may include tracking errors or infeasible inverse kinematics (IK) since the human arm is not exactly identical to the robot arm. However, since puppeteering involves two sets of robots (the puppet and the puppeteer), it may become more expensive depending on the setup.

3. **Motion capture:** In motion capture teleoperation[30][31], a computer vision algorithm is run on video streams of humans doing things, such as detecting poses of body parts, and these targets are retargeted on to a robot to control the robot. Some of these algorithms are quite similar to the ones covered in Chapter 4.

CHAPTER 10   HUMANOID ROBOTICS

Approaches to Whole Body Manipulation

Whole body manipulation (walking and performing a task at the same time) may be taught entirely by learning/modeling all joints of a robot. But whole body *teleoperation* is much harder and requires full exoskeletons. This section introduces a hybrid approach to whole body manipulation—one that uses learning for the upper body/arms and scripting for the lower body. In a world where we expect robots to do a lot of manual labor, manipulation may be a more valuable skill than navigation, and a hybrid approach allows manipulation or the placement of hands/arms to drive the modeling of the rest of the robot. So, for example, accurate foot placement is driven by the goal of opening the door, and correct hand placement and movement for it, but here the role of the lower body is simply to allow for these arm/hand movements while keeping the body stable and moving.

One such approach is TRILL (Teleoperation and Imitation Learning for Loco-manipulation)[27] from UT Austin, where robot arm manipulation is learned via deep imitation of human arms/hands via VR teleoperation. The method is hierarchical and consists of a high-level learned policy that generates trajectories based on language goals and image observations, and a low-level whole body controller that converts these learned trajectories into joint torques. Figure 10-4 shows a schematic of TRILL. Modeling high-level behaviors via a learned policy allows it to model semantics (move hand) as opposed to modeling individual joints, and makes the robot efficient in learning longer horizon behaviors, even with a smaller modeling capacity.

CHAPTER 10 HUMANOID ROBOTICS

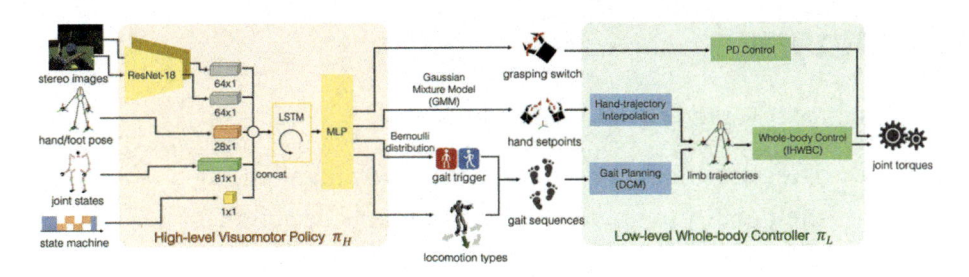

Figure 10-4. Overview of how TRILL works. Used with permission, source: https://arxiv.org/pdf/2309.01952 [27]

Conclusion

Widely deployed, generally intelligent, dexterous humanoids could transform the labor market. Recent advances in embodied AI provide an opportunity to unlock this futuristic technology. They include:

1. The potential market use cases for humanoids and cost-benefit tradeoffs

2. Hardware design choices for building effective humanoids

3. Learning and data approaches for scaling humanoid intelligence

The authors believe that humanoids will be an important robotics area in the future and we recommend digging deeper using the references and papers cited in this chapter.

CHAPTER 10   HUMANOID ROBOTICS

Summary

This chapter covered the following points:

- The usefulness of humanoid robots by highlighting their ability to operate in environments designed for humans. This includes tasks like handling objects designed for human hands or navigating spaces like kitchens and offices. The humanoid robot market is divided into three main categories: foundational model providers (e.g., Google, OpenAI); hardware providers (e.g., Boston Dynamics, Unitree) that build physical robots; and fully integrated companies (e.g., Tesla, Figure AI) that combine AI and hardware for complete humanoids.

- Key hardware design decisions include choosing between legged and wheeled mobility, simple grippers versus dexterous hands, and different camera setups (head-mounted, wrist-mounted, or additional sensors). Linear and rotary actuators are often preferred over hydraulic ones due to their precision and easier maintenance.

- Methods like imitation learning and advanced walking algorithms can improve humanoid performance. Teleoperation techniques include VR headsets, line-of-sight control, puppeteering, and motion capture, each of which has unique benefits and challenges.

- Hybrid systems like TRILL combine learning-based control for upper-body manipulation with scripted stability for the lower body and help balance flexibility and control.

The next chapter brings together all the concepts we've covered by exploring data infrastructure, training and deployment strategies, and large-scale robotic data collection, all of which are essential for real-world robotics.

References

[1] https://new.abb.com/products/robotics/robots

[2] https://research.google/research-areas/robotics/

[3] https://www.prnewswire.com/news-releases/figure-raises-675m-at-2-6b-valuation-and-signs-collaboration-agreement-with-openai-302074897.html

[4] https://mistral.ai/

[5] https://ai.meta.com/research/

[6] https://bostondynamics.com/

[7] https://fourierintelligence.com/

[8] https://www.unitree.com/

[9] https://apptronik.com/

[10] https://agilityrobotics.com/

[11] https://www.tesla.com/

[12] https://www.figure.ai/

[13] https://www.1x.tech/

[14] Caspi, Ido. "The Rise of Humanoids, Explained." *Global X ETFs,* 1 Mar. 2024, www.globalxetfs.com/the-rise-of-humanoids-explained/.

CHAPTER 10   HUMANOID ROBOTICS

[15] Andonov, Kaloyan. "Humanoid Robot Startups Attract Corporate Investment." *Global Corporate Venturing*, 26 June 2024, globalventuring.com/corporate/investment/humanoid-robot-startups-attract-corporate-investment/.

[16] Ficht, Grzegorz, and Sven Behnke. "Bipedal humanoid hardware design: A technology review." *Current Robotics Reports* 2 (2021): 201-210.

[17] Mesbahi, Mina. "Design Considerations for Humanoid Robots." *Wevolver*, 23 Oct. 2020, www.wevolver.com/article/design-considerations-for-humanoid-robots.

[18] Ackerman, Evan. "Agility Robotics Raises $8 Million for Commercial Bipedal Robots." IEEE Spectrum, 22 Mar. 2018, spectrum.ieee.org/agility-robotics-raises-8-million-for-commercial-bipedal-robots.

[19] https://www.1x.tech/androids/eve

[20] Wood, Chris. "Hand-Mounted Cameras Make Robots Better at Mapping Their Environments." *New Atlas*, 17 May 2016, newatlas.com/hand-mounted-camera-robots-manipulation/43375/.

[21] https://bostondynamics.com/blog/electric-new-era-for-atlas/

[22] https://www.boosterobotics.com/

[23] Fu, Zipeng, et al. "HumanPlus: Humanoid Shadowing and Imitation from Humans." *arXiv preprint arXiv*:2406.10454 (2024).

[24] Radosavovic, Ilija, et al. "Humanoid locomotion as next token prediction." *arXiv preprint arXiv*:2402.19469 (2024).

[25] Plappert, Matthias, Christian Mandery, and Tamim Asfour. "The kit motion-language dataset." *Big Data* 4.4 (2016): 236-252.

[26] Fu, Zipeng, Tony Z. Zhao, and Chelsea Finn. "Mobile aloha: Learning bimanual mobile manipulation with low-cost whole-body teleoperation." *arXiv preprint arXiv*:2401.02117 (2024).

[27] Seo, Mingyo, et al. "Deep imitation learning for humanoid loco-manipulation through human teleoperation." 2023 IEEE-RAS 22nd International Conference on Humanoid Robots (Humanoids). IEEE, 2023.

[28] Brohan, Anthony, Noah Brown, Justice Carbajal, Yevgen Chebotar, Joseph Dabis, Chelsea Finn, Keerthana Gopalakrishnan, et al. "Rt-1: Robotics transformer for real-world control at scale." *arXiv preprint arXiv:2212.06817* (2022).

[29] Allspaw, Jordan, Gregory LeMasurier, and Holly Yanco. "Implementing Virtual Reality for Teleoperation of a Humanoid Robot." *arXiv preprint arXiv:2104.11826* (2021).

[30] https://www.movella.com/resources/cases/humanoid-robots-learning-human-movement-using-xsens-motion-capture

[31] Kim, Seungsu, ChangHwan Kim, and Jong Hyeon Park. "Human-like arm motion generation for humanoid robots using motion capture database." In *2006 IEEE/RSJ International Conference on Intelligent Robots and Systems*, pp. 3486-3491. IEEE, 2006.

[32] "Humanoid Robots Ready for LLMs." Dr Alan D. Thompson. LifeArchitect.ai, Aug. 2023, lifearchitect.ai/humanoids/.

CHAPTER 11

Data-Driven Robotics in Practice

Data-driven robotics is sustained by far more activities than developing algorithms and programming robots. In fact, a large part of the bet is about scaling robot datasets[1] to accumulate skills that are as general as possible.

This chapter discusses the practical aspects of data-driven robotics. Data flywheels are the heart of data-driven robotics, but this is a relatively small body of published work. We invite our readers to refer to [10], [12], and [16] to learn about practical aspects of imitation learning flywheels and to [14] for RL flywheels.

Robot Operations

Robot operations encompass the day-to-day management and control of robots in their working environments. For data-driven robot learning, robot data is the lifeline of the operations. Robot operations involve managing the following:

1. **Deployment of robots:** Installing, configuring, and powering robot hardware in the target environment. For navigation research on legged robots, this also involves setting up gantries to catch unstable robots.

CHAPTER 11   DATA-DRIVEN ROBOTICS IN PRACTICE

2. **Monitoring and maintenance:** Involves continuously monitoring robot performance and system health. Also involves calibrating sensors and replacing parts/repair of robots.

3. **Organization:** Teleoperator scheduling to perform human aided data collection.

Most robotics teams have dedicated operations that are in charge of maintaining the hardware and software builds.

Safe operation of robots is an important component of operations. Common safety procedures used in robotics include[10]:

1. E-stops (emergency stops) to freeze robots in place via manual intervention during emergencies or routine operations

2. Hard e-stops to cut power to robot actuators during emergencies

3. Hardware modifications for low impact upon collisions[18]

4. Onboard software limitations on maximum speed and acceleration of actuators, and maximum allowed force on end effectors

5. Human-in-the-loop intervention in real time to handle tricky scenarios that an onboard model cannot handle[18]

6. Foundation model safety for contextual evaluation of safety scenarios[10]

Organizations that prioritize safety invest a lot of resources to ensure safe deployment of robots. This can have huge consequences for the trajectory of the program and the whole industry. Past incidents in self-driving safety have taken human lives[8], caused grave injuries[9], and crippled the companies and set back entire industries. Bringing a new technology to market and allowing it to win the trust of consumers requires acting responsibly.

Data Infrastructure

Data infrastructure plays a critical role in collecting, storing, processing, and analyzing data generated by robots, from sensor readings to performance metrics (see Figure 11-1). Key components of data infrastructure include:

1. **Data collection and logging:** Making sure that data coming from the robots is synchronized to represent an MDP/POMDP in (state, action, reward) tuples. Sometimes this requires that sensors that update at a faster rate are downsampled to match control frequency.

2. **Data storage:** Involves building frameworks to store large quantities of robot data. TFexamples is a commonly used storage format[2].

3. **Data labeling:** This involves building pipelines to get labeled data from sources like mechanical turk or Scale AI that utilize human labelers to examine and rate robot data. Human labeling is used in robot learning in a variety of ways:

    a. Success detection labeling for episodic data. Labelers can act as an independent arbitrator of success and define why an episode failed with visual grounding (which frames had the incident) [18][12][10].

b. Interleaved labeling for video segmentation. This can teach models about what happened during what part of policy execution, where policies went wrong during execution, how to reason about data quality, the smoothness of trajectories, and so on[18].

c. Visual question-answering datasets to learn reasoning over robotics data[18][19].

4. **Data processing and loading:**

   a. Loading teleoperator collected data for training, and allowing it to train on specific partitions of the dataset for ablation experiments

   b. Integrating hindsight relabeling back into datasets

   c. Allowing post processing of datasets for additional feature extractions, such as what's done in RTTrajectory[3] (where trajectories were added back into datasets for training) and RobotMOO (where object-centric labels were added back into training)

5. **Data analysis:** Employing machine learning and data analytics techniques to extract insights from robot data.

   a. Offline analysis of data include areas like determining spread of data over objects, actions, and 3D trajectories[20]

   b. Offline metrics provide proxy objectives to evaluate the capacity of a dataset to give rise to generalization[12][10]

CHAPTER 11   DATA-DRIVEN ROBOTICS IN PRACTICE

c. Additionally, offline metrics can be used to guide online data collection (more of which tasks to collect for, which object configurations are missing, etc.)[10]

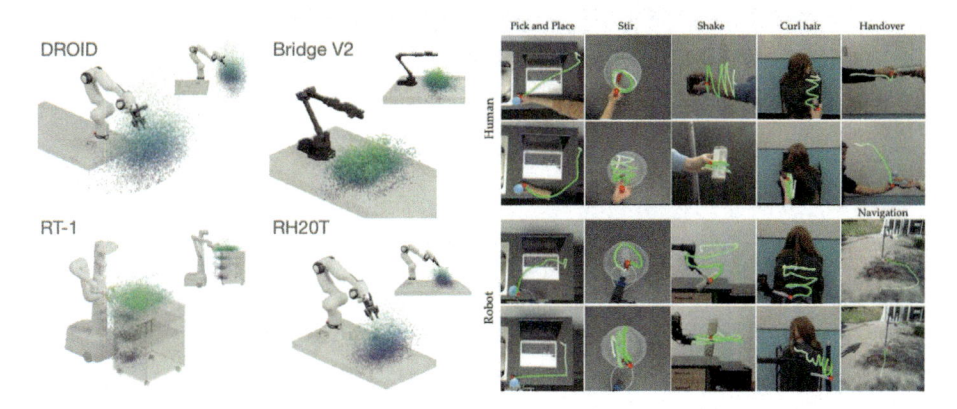

Figure 11-1. *Top: A 3D span of interactions in many datasets, source[20] Bottom: Trajectory visualizations in MotIF1k, source[13]. Figure used under CC 4.0*

A well-designed data infrastructure saves a lot of developer time by making it easier to gather valuable insights into robot performance, as well as to run experiments.

The Training and Deployment Infrastructure

Training robots to perform complex tasks requires a dedicated infrastructure that provides the necessary resources for simulation, learning, and evaluation. Training infrastructure encompasses:

1. **Training algorithms:** This involves developing and implementing machine learning algorithms that enable robots to learn from data and experience. Generally, organizations develop either imitation

learning based algorithms or reinforcement learning based algorithms, both of which require a similar infrastructure. Both types require breaking robotics data down into (state, action, reward) tuples, where action becomes the target to train upon for imitation and reward is largely used only in reinforcement learning setups.

Since RL requires improvement from experience, bootstrapping becomes a prerequisite. Gathering good data requires a good policy, and this policy is often first trained in simulation, before being deployed in the real world, such as in QtOpt[5]. Hence, simulation has a larger role in an RL framework over an imitation learning framework.

2. **Performance evaluation:** This involves assessing robot performance through metrics and benchmarks to measure progress and identify areas for improvement. Evaluation involves the following types:

   a. *Offline evaluation:* This is inferred by using the loss on validation sets or by running inference on test splits of data and comparing results against ground truth[10][12].

   b. *Evaluation in simulation:* If simulation closely mocks real-world performance, simulation can be used to scale up evaluation[6][12]. Generative AI can be used to build very realistic simulation environments to test model performance[7].

CHAPTER 11   DATA-DRIVEN ROBOTICS IN PRACTICE

3. **Distributed infrastructure:** For load allocation during training and for faster inference during deployment, efficient use of allocated compute is necessary for ensuring optimality of foundation models trained. This was explored in Chapter 4.

Robot Data Flywheels

The successful integration of robot operations, data infrastructure, and training/deployment infrastructure is crucial for maximizing the value of robots. Combined, this system is called the *robot data flywheel* (see Figure 11-2) and can result in models with increasing capabilities via acquiring new data. Data flywheels are the heart of any AI-driven robotics research lab.

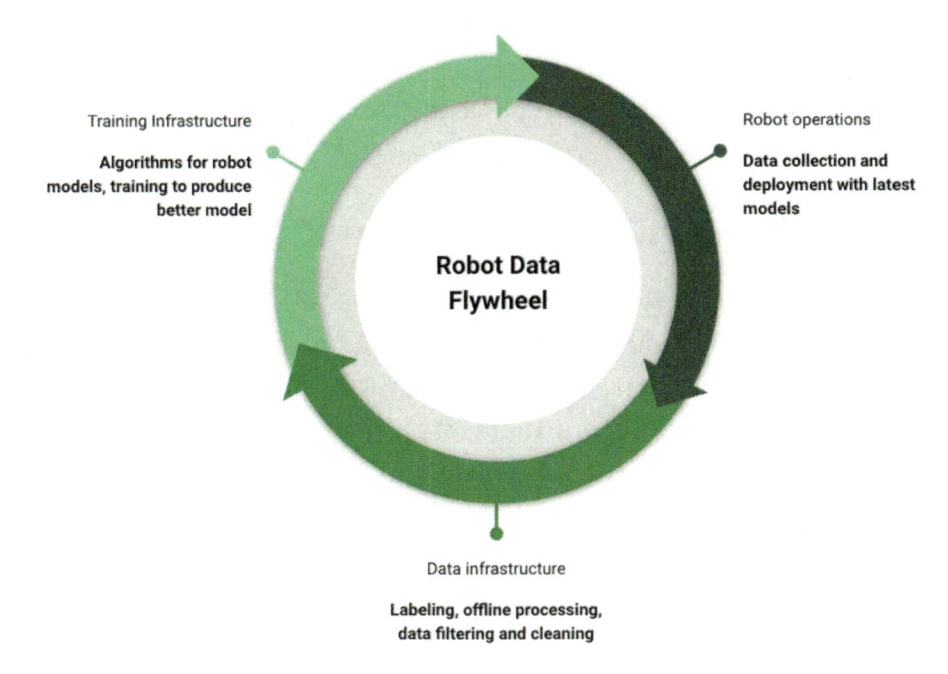

Figure 11-2. Schematic diagram of a robot data flywheel

CHAPTER 11 DATA-DRIVEN ROBOTICS IN PRACTICE

With a robot data flywheel, organizations can:

1. **Continuously improve robot performance:** Leverage data insights to identify areas for improvement and refine robot training processes.

2. **Adapt to changing environments:** Collect new environments to quickly adapt a foundation model to new situations and unseen tasks.

3. **Optimize resource utilization:** Analyze data to optimize robot deployment, scheduling, and maintenance strategies.

In conclusion, the integration of robot operations, data infrastructure, and training infrastructure forms the foundation for successful robot deployment and utilization. By effectively managing robots, harnessing data insights, and supporting ever-changing research processes, organizations can move fast in robotics.

The next section examines a couple of data-driven robotics projects.

Large-Scale Robotic Data Collection

AutoRT[10] is a recent system from Google DeepMind that uses LLMs and VLMs in the loop for large-scale robot orchestration and data collection for in-the-wild scenarios (see Figure 11-3). This method uses a variant of curiosity-driven exploration, where robots roam areas that may be inhabited by humans, and reason using VLMs and LLMs for feasible tasks to do, then call a remote human/onboard action model to perform inference based on affordance estimation of the task vis-à-vis capabilities of autonomous models. After collection of that episode, the episode gets a diversity rating based on its first and last images.

CHAPTER 11   DATA-DRIVEN ROBOTICS IN PRACTICE

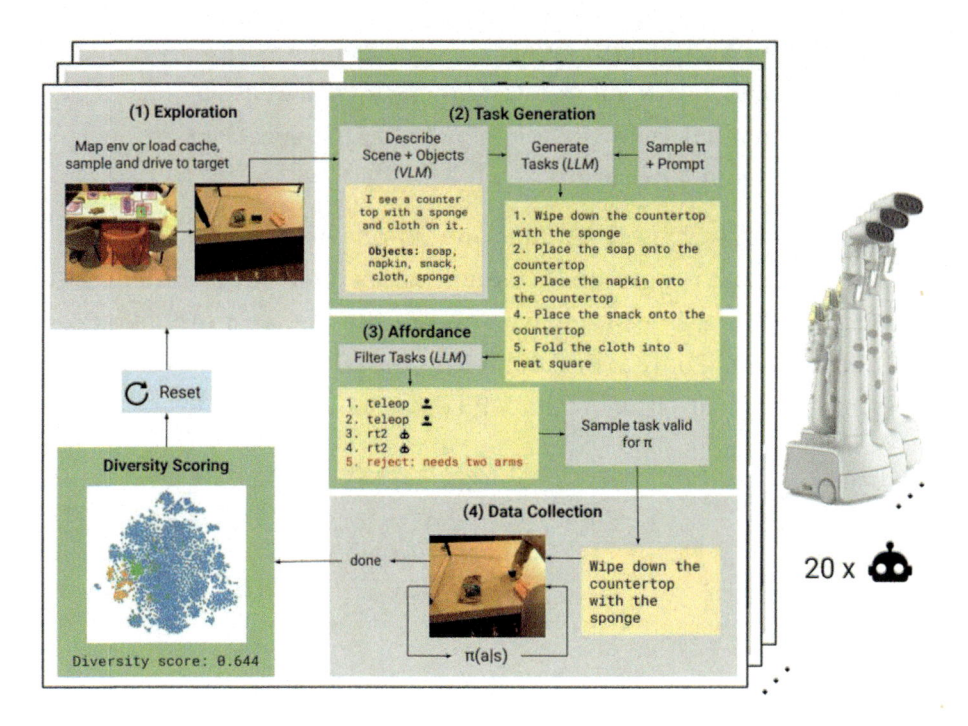

Figure 11-3.   Schematic diagram of AutoRT, source: `https://arxiv.org/pdf/2401.12963`*[10]. Figure used under CC 4.0*

To decide which tasks are feasible and useful to do, AutoRT introduces a robot constitution guide behavior. The constitution has three sections:

1. **Fundamental rules:** Derived from Asimov's laws, these rules talk about how robots should not do anything that would harm a human.

2. **Safety rules:** Describes tasks that are unsafe to do based on current capabilities in deployment, such as not to deal with sharp objects, electrical appliances, or living beings.

435

CHAPTER 11 DATA-DRIVEN ROBOTICS IN PRACTICE

3. **Embodied specific rules:** These rules inform the robot of its own affordance; for example, that it is a unimanual robot, meaning that it has to reject tasks that require two arms.

Scaling robotic datasets requires one-on-one human supervision, which means that the number of human teleoperators are a bottleneck to scaling. AutoRT was able to achieve a one-to-five human-to-robot ratio, with humans acting only to intervene. This introduced a way to scale robot deployment and learn from in-the-wild data. See Figure 11-4.

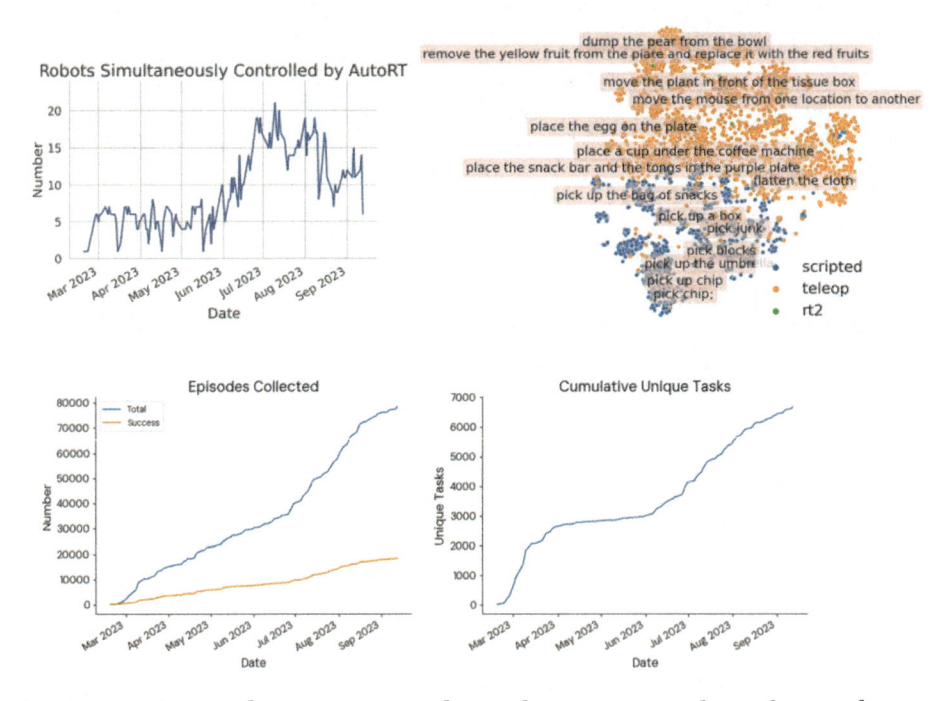

Figure 11-4. *Scaling unique tasks in datasets, number of episodes, number of robots deployed simultaneously. AutoRT controlled 55 robots over its six month lifetime with a peak load of 20, source: https://arxiv.org/pdf/2401.12963 [10]. Figure used under CC 4.0*

CHAPTER 11   DATA-DRIVEN ROBOTICS IN PRACTICE

While AutoRT was successful in collecting highly diverse data, the method informs future data collection efforts in the following ways:

1. The best dataset is not the largest dataset or the most diverse dataset, it is the dataset that leads to the best policy improvement. While diversity and size are proxy objectives, it is important to verify that you are collecting the right data by doing frequent trainings and evaluation of the collected data.

2. While autonomously collected data may be a promising way to harvest data without linearly scaling humans involvement, learning from that data efficiently is less than solved. Most foundation models sit atop of scaled imitation learning algorithms, but *imitation learning shows worse performance when lower quality data is added*, and imitation learning loses RL's ability to trajectory stitch from suboptimal trajectories. Combining imitation and reinforcement in a complementary manner in robotics may become a winning recipe.

3. Highly diverse data may be too wide of a distribution for current models to learn. The construction of a dataset should depend on the sample efficiency of the underlying policies. Training from a smaller, but carefully curated high-quality dataset may yield more improvement than training from large quantities of weak data. That means that tracking quality while scaling data becomes important.

CHAPTER 11   DATA-DRIVEN ROBOTICS IN PRACTICE

Recipes for the Future

At the precipice of where robotics stands today, there are enough compelling pieces of evidence that the key to solving robotics and building general purpose, generally intelligent robots lies in building very good robot foundation models. As such, finding a way to make robotics look more and more like a vision-language problem will be necessary to bring the two worlds together coherently.

- Motion generalization may be the last remaining fundamental research problem in robotics. As argued in [13] and [11], VLMs at the moment cannot sufficiently understand trajectories, and their ability to come up with new motions is limited to their datasets, which are costly to acquire. Improving reasoning about motions via motion-centric visual representations (such as in [3] and [13]) or via learning from Internet videos[15] may be necessary to unlock and scale-motion generalization, and therefore endow VLMs with the ability to reason about motions as easily as they do about images and language.

- Understanding how VLAs scale and what the empirical scaling laws may be that govern them is critical to understanding and projecting how the field evolves.

- The next era will also see robots and robot foundation models increasingly deployed in the real world for dexterous manipulation. There's no scaling without scale itself, and we need to bring robots into the real world to capture diversity and scale.

- Safety and alignment in general-purpose robots will be a necessary condition to large-scale deployment, and making breakthroughs in this area will be crucial to push for real-world usage.

- Autonomy through hill-climbing with scaled, semi-autonomous systems, with the performance gap bridged via intervention with humans-in-the loop, may be another key trend we will see in the future.

Acting intelligently in the physical space is an emergent property of a large audio-visual language agent, and intelligence at the most fundamental level is the same, whether its expression is digital or physical. The authors are excited to see how this field evolves, and how you, the reader, will go forth and shape it!

References

[1] Banfield, Richard. "Physical Intelligence Raises $70M to Build AI-Powered Robots for Any Application." *Maginative*, 12 Mar. 2024, www.maginative.com/article/physical-intelligence-raises-70m-to-build-ai-powered-robots-for-any-application/.

[2] https://www.tensorflow.org/api_docs/python/tf/train/Example

[3] Gu, Jiayuan, Sean Kirmani, Paul Wohlhart, Yao Lu, Montserrat Gonzalez Arenas, Kanishka Rao, Wenhao Yu, et al. "Rt-trajectory: Robotic task generalization via hindsight trajectory sketches." *arXiv preprint arXiv:2311.01977* (2023).

[4] Stone, Austin, Ted Xiao, Yao Lu, Keerthana Gopalakrishnan, Kuang-Huei Lee, Quan Vuong, Paul Wohlhart, et al. "Open-world object manipulation using pre-trained vision-language models." *arXiv preprint arXiv:2303.00905* (2023).

[5] Kalashnikov, Dmitry, Alex Irpan, Peter Pastor, Julian Ibarz, Alexander Herzog, Eric Jang, Deirdre Quillen, et al. "Scalable deep reinforcement learning for vision-based robotic manipulation." In *Conference on Robot Learning*, pp. 651-673. PMLR, 2018.

[6] Li, Xuanlin, Kyle Hsu, Jiayuan Gu, Karl Pertsch, Oier Mees, Homer Rich Walke, Chuyuan Fu, et al. "Evaluating Real-World Robot Manipulation Policies in Simulation." *arXiv preprint arXiv:2405.05941* (2024).

[7] https://www.1x.tech/discover/1x-world-model

[8] https://www.wired.com/story/ubers-fatal-self-driving-car-crash-saga-over-operator-avoids-prison/

[9] https://fortune.com/2024/05/16/inside-gm-cruise-self-driving-car-accident-san-francisco-what-really-happened/

[10] Ahn, Michael, Debidatta Dwibedi, Chelsea Finn, Montse Gonzalez Arenas, Keerthana Gopalakrishnan, Karol Hausman, Brian Ichter, et al. "AutoRT: Embodied foundation models for large scale orchestration of robotic agents." *arXiv preprint arXiv:2401.12963* (2024).

[11] Brohan, Anthony, Noah Brown, Justice Carbajal, Yevgen Chebotar, Xi Chen, Krzysztof Choromanski, Tianli Ding, et al. "Rt-2: Vision-language-action models transfer web knowledge to robotic control." *arXiv preprint arXiv:2307.15818* (2023).

[12] Brohan, Anthony, Noah Brown, Justice Carbajal, Yevgen Chebotar, Joseph Dabis, Chelsea Finn, Keerthana Gopalakrishnan, et al. "Rt-1: Robotics transformer for real-world control at scale." *arXiv preprint arXiv:2212.06817* (2022).

[13] Hwang, Minyoung, Joey Hejna, Dorsa Sadigh, and Yonatan Bisk. "MotIF: Motion Instruction Fine-tuning." *arXiv preprint arXiv:2409.10683* (2024).

[14] Herzog, Alexander, Kanishka Rao, Karol Hausman, Yao Lu, Paul Wohlhart, Mengyuan Yan, Jessica Lin, et al. "Deep RL at scale: Sorting waste in office buildings with a fleet of mobile manipulators." *arXiv preprint arXiv:2305.03270* (2023).

[15] Hou, Shuaiying, Hongyu Tao, Junheng Fang, Changqing Zou, Hujun Bao, and Weiwei Xu. "Learning Human Motion from Monocular Videos via Cross-Modal Manifold Alignment." *arXiv preprint arXiv:2404.09499* (2024).

[16] Jang, Eric, Alex Irpan, Mohi Khansari, Daniel Kappler, Frederik Ebert, Corey Lynch, Sergey Levine, and Chelsea Finn. "Bc-z: Zero-shot task generalization with robotic imitation learning." In *Conference on Robot Learning*, pp. 991-1002. PMLR, 2022.

[17] https://evjang.com/2024/08/31/motors.html

[18] Sermanet, P., Ding, T., Zhao, J., Xia, F., Dwibedi, D., Gopalakrishnan, K., Chan, C., Dulac-Arnold, G., Maddineni, S., Joshi, N.J. and Florence, P., 2024, May. Robovqa: Multimodal long-horizon reasoning for robotics. In *2024 IEEE International Conference on Robotics and Automation (ICRA)* (pp. 645-652). IEEE.

[19] Du, Yilun, Mengjiao Yang, Pete Florence, Fei Xia, Ayzaan Wahid, Brian Ichter, Pierre Sermanet, et al. "Video language planning." *arXiv preprint arXiv:2310.10625* (2023).

[20] Khazatsky, Alexander, Karl Pertsch, Suraj Nair, Ashwin Balakrishna, Sudeep Dasari, Siddharth Karamcheti, Soroush Nasiriany, et al. "Droid: A large-scale in-the-wild robot manipulation dataset." *arXiv preprint arXiv:2403.12945* (2024).

Index

A

Accelerometers, 51
Action chunking, 171
Active Domain Randomization (ADR), 227
Additive attention, 82
Advanced driver-assistance system (ADAS), 40
AI-driven robots, 199
Artificial general intelligence (AGI), 19, 23
Artificial Intelligence (AI), 19
Artificial Super Intelligence (ASI), 19
Automated guided vehicles (AGVs), 379, 382
Autonomous Mobile Robots (AMRs), 382
AutoRT, 436, 437

B

BEVFusion, 132
Bird's-eye view (BEV), 134
Boston Dynamics
    spot robot, 312
Bullet Physics simulator, 237

C

Carnegie Mellon University (CMU), 353
Charge-Coupled Device (CCD), 37
ChatGPT, 3
Classifier-free guidance, 187
Classifier-guided diffusion, 186
Collaborative Robots (Cobots), 381
Complementary filters, 52
Complementary Metal-Oxide Semiconductor (CMOS), 37
Compound Annual Growth Rate (CAGR), 4
ConceptFusion, 279
Consumer Price Index (CPI), 1
Contrastive Video Representation Learning (CVRL), 193
Convolutional neural networks (CNNs), 390
    EfficientNet, 76
    faster R-CNN, 67–69
    FC layers, 61
    layers, 62–66, 97
    mask R-CNN, 71
    ResNet, 72–75
Cross entropy method, 236
Cross-modal fusion strategy, 128

INDEX

Cyberattacks, 368
CycleGAN, 231

D

Data-driven robotics
    algorithms and programming robots, 427
    data infrastructure, 429–431
    LLMs/VLMs, 434, 435, 437
    robot data flywheel, 433, 434
    robot operations, 427–429
    training/deployment infrastructure, 431, 432
    vision-language problem, 438, 439
DeepFusion, 128, 129
Deep learning (DP), 19
    hybrid strategy, 267–269
    labeled datasets, 303
    methods, 267
    traditional methods, 266
    typical mobile robot setup, 269
Deep learning (DL), 19
Deep Planning Network (PlaNet), 330
Degrees of freedom (DoF), 12
Denial-of-Service (DoS), 194
Denoising Diffusion Implicit Models (DDIM), 183
Denoising Diffusion Probabilistic Models (DDPMs), 180
Detection transformers (DETR), 98

Direct preference optimization (DPO), 143, 198, 347
Disengagement, 354
Domain randomization, 224
Dynamic Graph CNN (DGCNN), 133

E

EfficientNet, 76
Embodied PaLM, 157
End-to-end robot control
    action diffusion, 188–190
    autoregressive transformers, 169–176
    conditional generation, 186, 187
    DDIM, 183–185
    DDPMs, 180–183
    diffusion models, 176–180
    generalization, 168
End-to-End Self-Driving (E2E), 357
Event-based cameras, 40

F

Failsafe mechanisms, 195
Fast R-CNN, 67
Foundation models
    AI safety, 194–197
    components, 141
    compositional approach, 140, 141
    evaluating language models, 148–150

INDEX

human annotators, 142, 143
large multimodal model, 140
LLM post-training strategy, 144
pre-trained language
    model, 143
scaling laws, language
    models, 144–148

G

Gazebo, 221, 254
Generative Adversarial Network
    (GAN), 231
Goal-conditioned reinforcement
    learning (GCRL), 23
Google's MT-Opt, 346
Gradient-based optimization, 324
Graphics Processing Units
    (GPUs), 16
Grippers, 411
Guided Domain Randomization
    (GDR), 227
Gyroscopes, 51

H

Humanoid robotics
    approaches, 406
    categories, 422
    dexterous, 421
    hardware, 410–414
    human form factor, 405
    humanoid case, 405
    hybrid systems, 422
    markets, 407–409
    methods, 422
    software
        manipulation approaches,
            415, 416
        teleoperation/data, 418, 419
        walking approaches, 416–418
        whole body
            manipulation, 420
Humanoid robots, 384

I, J

Image classification, 54
Image encoder, 140
Image segmentation
    convolutional encoder-decoder
        architecture, 55
    object detection, 60
    pixels, 54
    types, 55
        instance, 57–59
        semantic, 56, 57
Industrial robots
    Cobots, 381, 382
    deep learning techniques, 375
    humanoid, 384
    industrial and manufacturing
        settings, 375
    market opportunity, 384–386
    mobile, 382, 383
    performance metrics, 399
    safety, 398
    standalone, 380

445

INDEX

Industrial robots (*cont.*)
  tasks, 398
    peg-in-hole, 377, 378
    pick-and-place, 376, 377
    warehouse, 379
    welding, 378
Inertial measurement units
      (IMUs), 5, 36, 37, 45
Instance segmentation, 57
Intersection over Union (IoU), 57

K

Kalman filters, 52, 266

L

Language
  human knowledge, 151
  mapping, 160, 162
  planning
    approaches, 152
    challenges, 158, 159
    closed loop, 155, 156
    LLM, 152
    multimodal, 156–158
    open loop SayCan, 153, 154
  reward, 162, 163
  robot code, 164, 165, 167
Language Embedded Radiance
      Fields (LeRFs), 278
Language models, 3
Large language
      models (LLMs), 134

Learnable Alignment
      (LearnableAlign), 130
Learning from video
      demonstrations (LfV)
  challenges, 191
  generalization, 190
  world model, 191–193
LiDAR-camera fusion methods, 133
Light Detection and Ranging
      (LiDAR), 45
Llama 3, 140
Localization, 265
  2D-to-2D, 281, 282, 284
  2D-to-3D, 284, 286
  3D-to-3D, 286, 287

M

Machine learning (ML), 19
Machine learning perception
      systems, 35
Magnetometers, 52
Mapping, 265
  geometric
    definition, 270
    depth
      representation, 270–272
    NeRFs, 274–277
    voxel, 272–274
  scene map, 270
  semantic
    ConceptFusion, 279, 280
    definition, 277
    LeRF, 278

Markov Decision Processes
(MDPs), 223, 296, 320, 346
Mean average overlap (mAP), 58
MineCLIP, 242
MineDojo, 254
Model-based reinforcement
learning, 327, 328
Model-free reinforcement
learning, 318
Moore's law, 2
Mordor Intelligence, 4
Multi-Joint Dynamics with Contact
(MuJoCo), 221, 254
Multimodal Instruction Navigation
with Demonstration Tours
(MINT), 288
Multimodal perception/
sensor fusion
fusion outputs, 121–123
LiDAR-camera fusion
point-level fusion
techniques, 126–133
proposal level fusion
methods, 123–125
LiDARs, 116
raw data, 118–120
strategies, deep learning, 116
uses, 117
MuZero method, 329

N

Navigation, 265
exploration, 291–293

legged robots, locomotion
hardware, 294, 295
learning-based methods,
296, 297, 300–303
MDP formulation, 296
simulation, 295
VLA, 288, 289, 304
VLM, 288–291
Neural Radiance Fields (NeRFs),
274, 304
NLMap-SayCan, 160

O

One-stage detectors
model comparison, 79, 80
SSD, 78, 79
YOLO, 77
Open Motion Planning Library
(OMPL), 220

P, Q

Pick-and-place robots, 399
components, 387
hardware, 387–389
software, 390, 391, 393
multimodal grasping
system, 394–396
performance
metrics, 396, 397
RL
components, 393
PI-QT-Opt, 394

INDEX

PlaNet, 319
Point-level fusion techniques, 126
PointNet, 110, 124
PointPainting method, 128
Pretrained language model, 143
Proximal Policy Optimization (PPO), 297, 319, 326
PyBullet, 219, 254–257

R

Rectified Linear Unit (ReLU), 64
Recurrent Neural Networks (RNNs), 286
Recurrent State Space Model (RSSM), 329, 330
RedNose, 266
Red teaming, 196
Region of Interest (RoI), 120
Region Proposal Network (RPN), 69
Reinforcement learning, 19, 313
    application/challenges, 333
    deep Q Learning, 321
    emerging trends, 344, 345
    intelligent systems, 311
    MDP, 314, 316, 317
    model-based RL, 327–330
    model-free *vs.* model-based RL, 318–321
    offline, 331, 332
    play soccer, robots, 339
    policy gradient methods, 322, 323
    Q Learning, 321
    RLAIF, 342
    RLHF, 340
    scaling up, 334–338
    trust region policy optimization, 324–327
Reinforcement Learning from AI Feedback (RLAIF), 342, 347
Reinforcement Learning from Human Feedback (RLHF), 143, 340, 347
Reinforcement learning (RL), 222, 393
RelocNet, 284
Residual Neural Networks (ResNet), 55, 73
RFM-1, 394
RL-CycleGAN, 233, 254
Robot data flywheel, 433
Robotics Transformer 1 (RT-1), 169
Robotic vision systems, 84
Robot Operating System (ROS), 218, 221
Robots
    AGI, 23
    capabilities, 3
    components, 6
    deep learning, 15
        benefits, 16
        frameworks, 19–21
        infrastructure paradigms, 17, 18
    design
        DoF, 12

INDEX

end effector/workspaces, 13
kinematics, 13, 14
robotic manipulators, 11
design principles, 27
frameworks, 22, 23, 27
language models, 25
large-scale manufacturing, 1
machine learning, 4, 27
real-world applications, 26
science fiction, 1
sensors, 5
tasks, 2
types, 7, 9, 10

S

Scene coordinate regression, 284
Segment Anything Model (SAM), 95
Self-driving vehicles
    economic opportunity, 353–355
    perception, 358–361
    planning, 363, 364, 366, 370
    prediction, 362, 363
    safety
        AI systems, 366, 367
        algorithmic robustness, 368
        cybersecurity, 368
        data privacy, 369
        ethical decisions, 368
        fail-safe mechanisms, 369
        human machine interaction, 369

sensor reliability, 367
software, 370
system design, 355–357
Semantic segmentation, 56
Sensor fusion localization (SLAM), 266
Sensors
    depth, 42–44
    IMUs, 51–53
    LiDARs, 36, 37
    range
        LiDAR, 46–49
        ultrasonic sensors, 49–51
    vision
        camera, 38
        CCD, 37
        CMOS, 37
        components, camera, 39
        event-based cameras, 40, 42
        light, 37
Sim2Real
    domain adaptation, 223, 224
    domain randomization, 224, 226
    GDR, 227, 229
    RL, 230–235
Simulation
    benefits, 212, 213
    boostrapping RL, 236, 237
    components
        features, 216–218
        Gazebo, 221, 222

INDEX

Simulation (*cont.*)
    MuJoCo, 221
    PyBullet, 219, 220
    real-world object
        behavior, 218
    rigid body, 214, 215
  foundation agents, 238–243
  imitation learning, 249–253
  limitations, 214
  reward design, 244–246
  robotics training pipelines, 211
  world modelling, 246–249
Simulation Description Format
    (SDFormat), 219
Simultaneous Localization and
    Mapping (SLAM), 280
Single-shot detector (SSD), 78
Skill library, 239
Stein Variational Policy Gradient
    (SVPG), 228
Structure from motion (SfM), 266
Supervised fine-tuning (SFT), 142,
    198, 341
Supervised learning, 18, 311, 313

T

Tensor Processing Units (TPUs), 16
3D data processing, 107
  data representation, 108, 109
  point clouds,
    processing, 110–115
  research opportunities, 115
3D sensor data
    industrial and consumer
        robots, 107
Time-Contrastive Networks
    (TCN), 193
Training actor-critic RL models, 143
Transformer
  attention mechanism, 82–84
  learning joint image-language
    features, 92, 93
  multi-head attention, 85, 86
  open vocabulary object
    detection, 93, 94
  promptable open vocabulary
    segmentation, 95–97
  scaling vision, 89, 90, 92
  ViT, image classification, 87, 89
Trust Region Policy Optimization
    (TRPO), 297

U

Ultrasonic sensors, 37
U-Net, 185
Unified Predictive Decision
    Process (UPDP), 193
Unified Robotics Description
    Format (URDF), 219
Unsupervised learning, 313

V, W, X

Variational Autoencoder (VAE), 184
Video-conditioned policy
    learning, 193

Vision language models (VLMs), 287, 289
Vision sensors, 97
Vision transformers (ViT), 87, 98

Y, Z

You Only Look Once (YOLO), 77

Printed in Great Britain
by Amazon